Oxidative Stress and Antioxidant Defense in Crop Plants

Oxidative Stress and Antioxidant Defense in Crop Plants

Guest Editors

María C. Romero-Puertas
Marta Jańczak-Pieniążek

MDPI
Basel • Beijing • Wuhan • Barcelona • Belgrade • Novi Sad • Cluj • Manchester

Guest Editors
María C. Romero-Puertas
Department of Stress, Development and Signaling in Plants
Estación Experimental del Zaidín-Spanish Council for Scientific Research Agency (EEZ-CSIC)
Granada
Spain

Marta Jańczak-Pieniążek
Department of Crop Production
University of Rzeszów
Rzeszów
Poland

Editorial Office
MDPI AG
Grosspeteranlage 5
4052 Basel, Switzerland

This is a reprint of the Special Issue, published open access by the journal *Antioxidants* (ISSN 2076-3921), freely accessible at: https://www.mdpi.com/journal/antioxidants/special_issues/954044JQ7U.

For citation purposes, cite each article independently as indicated on the article page online and as indicated below:

Lastname, A.A.; Lastname, B.B. Article Title. *Journal Name* **Year**, *Volume Number*, Page Range.

ISBN 978-3-7258-5983-2 (Hbk)
ISBN 978-3-7258-5984-9 (PDF)
https://doi.org/10.3390/books978-3-7258-5984-9

Contents

About the Editors

María C. Romero-Puertas

María C. Romero-Puertas is a senior researcher at the Estación Experimental del Zaidín-CSIC in Granada, Spain. She specializes in analyzing the response of plants to abiotic and biotic interactions and how this is regulated by ROS/RNS, with a particular focus on peroxisome-dependent signaling. Her results have shown that peroxisomes have an extraordinary ability to adapt. They are highly dynamic and plastic in terms of their morphology, number, mobility, and metabolic pathways, and are key organelles in the early perception of stress and the subsequent regulation of the stress response. Since joining the CSIC, she has participated in around 30 projects. The knowledge she has acquired during her career, and the results she has obtained, have led her to author around 80 articles published in highly regarded scientific journals, as well as 1 book and 13 book chapters. She has also participated in outreach activities and the "Biología Fundamental y de Sistemas" PhD Program at the University of Granada, and has been a member of the Academic Committee of this Program's Master's course since 2012. She is currently Head of the Stress, Development and Signalling in Plants Department and President-Elect of the European Plant Oxygen Group (POG) and serves on the editorial board of various journals. She is also among the world's top 2% of scientists (Ranking Univ. of Stanford).

Marta Jańczak-Pieniążek

Marta Jańczak-Pieniążek is an Associate Professor at the Department of Crop Production, Faculty of Technology and Life Sciences, University of Rzeszów, Poland. Her research focuses on crop physiology, antioxidant metabolism in plants, and the biochemical and physiological responses of crops to abiotic stress. She has participated in numerous national and international projects on sustainable crop production and stress physiology and has authored and co-authored many papers published in peer-reviewed journals. She teaches undergraduate and graduate courses in agriculture, environmental protection, and food technology. Her work contributes to a better understanding of plant adaptation mechanisms to changing environmental conditions and supports the development of strategies to improve stress tolerance and productivity in agricultural systems.

Preface

This Reprint seeks to bring together recent advances in understanding oxidative stress responses in crop plants and to emphasize their importance for sustainable food production. The collected works demonstrate how plants activate antioxidant defenses at multiple levels—molecular, biochemical, and physiological—to maintain redox homeostasis under abiotic stress. By combining studies on different plant species and stress conditions, this volume reflects the diversity of adaptive mechanisms operating in nature and under cultivation. We believe that this Reprint will support future innovations in crop management and breeding for stress tolerance.

María C. Romero-Puertas and Marta Jańczak-Pieniążek
Guest Editors

antioxidants

MDPI

Article

Arginine and Spermine Ameliorate Water Deficit Stress in Fenugreek (*Trigonella foenum-graecum* L.) by Enhancing Growth and Physio-Biochemical Processes

Ali A. Badawy [1,*], Wadha Kh. Alshammari [2], Noura F. G. Salem [3], Woroud S. Alshammari [4] and Hebat-Allah A. Hussein [4]

1 Botany and Microbiology Department, Faculty of Science, Al-Azhar University, Nasr City, Cairo 11884, Egypt
2 Biology Department, College of Science, University of Hafr Al Batin, Hafr Al-Batin 31991, Saudi Arabia
3 Botany and Microbiology Department, Faculty of Science (Girls Branch), Al-Azhar University, Cairo 11754, Egypt
4 Department of Science and Technology, University College at Nairiyah, University of Hafr Al Batin, Nairiyah 31981, Saudi Arabia
* Correspondence: ali.abdalhalim@azhar.edu.eg

Abstract: Plants face various stresses, particularly water deficit, which negatively impacts photosynthesis, growth, and development, thereby limiting agricultural production. Utilizing growth regulators, such as amino acids and polyamines, to enhance osmotic stress tolerance is a crucial area of research in sustainable agriculture. This study investigates the impact of arginine and spermine treatments on various growth attributes, enzymatic and non-enzymatic antioxidants, photosynthetic pigments, protein and lipid peroxidation, and yield traits of fenugreek plants under both normal and drought conditions. The results indicate that drought conditions significantly reduce morphological characteristics, leaf pigments, and yield traits. However, the application of arginine and spermine enhances these parameters, with spermine showing a more pronounced effect. Additionally, treatments boost antioxidant enzymes activities and improve the levels of non-enzymatic antioxidants and osmolytes, contributing to better stress tolerance and growth performance. Principal component analysis confirms that drought significantly alters plant physiology, increasing proline and malondialdehyde levels, while arginine and spermine alleviate drought stress by enhancing antioxidant activity and osmolyte accumulation. The current investigation aims to evaluate the effectiveness of spermine and arginine treatments on various growth attributes and stress tolerance of fenugreek plants under normal and drought conditions, focusing on their comparative efficacy.

Keywords: growth regulators; fenugreek; antioxidants; osmolytes; yield traits; drought stress

1. Introduction

Fenugreek (*Trigonella foenum-graecum* L.) is a versatile annual herb from the Fabaceae family, extensively grown in India and across the Mediterranean region, Northern Africa, China, parts of Europe, Australia, and recently North America [1]. Traditionally used as a food and medicine, fenugreek seeds are incorporated into wheat and maize flour in Egypt and serve as a spice, a vegetable, forage for cattle, and for medicinal purposes [2]. The plant is rich in biochemical constituents such as steroids, saponins, polysaccharides, alkaloids, volatile oils, fixed oil, proteins, sugars, mucilage, and flavonoids, which contribute to its medicinal and pharmaceutical significance [3]. Its known for its immunological,

https://doi.org/10.3390/antiox14030329

anticarcinogenic, antioxidant, antidiabetic, and hypocholesterolemic activities, making it valuable in treating ailments like diabetes and hyperglycemia [4]. Additionally, it is a rich source of protein, lysine, essential nutrients, dietary fiber, and steroid saponins, which are commercially useful for steroid hormone synthesis [5]. Fenugreek's diverse applications and health benefits underscore its importance as a medicinal and economical plant.

Plant growth and productivity are significantly impacted by various abiotic and biotic stress factors, including low temperature, salt, drought, flooding, heat, oxidative stress, heavy metal toxicity, and pathogens [6–12]. Drought stress, in particular, poses a major challenge, affecting approximately one-third of the potentially viable land globally due to inadequate water supply. This stress leads to various physiological and biochemical effects on plants, disrupting growth, metabolism, development, productivity, and molecular expressions [13–15]. It induces disturbances in the photosynthetic process and carbon metabolism and causes partial stomatal closure, reducing carbon dioxide availability and causing imbalances in nitrogen and carbon metabolism [16]. The upcoming water shortage in certain regions around the world presents a significant challenge to agricultural development and crop production. Under drought conditions, plants store osmolytes like sugars and amino acids to regulate water uptake [17]. Sugars are more efficient than proline in replacing water, forming a hydration shell around biomolecules. Proteins are essential for all physiological activities in plant cells, and their levels rise under drought stress [18]. Drought also increases phenols and flavonoids in plants, enhancing their stress tolerance [19]. Stress-induced reactive oxygen species (ROS) deactivate enzyme functionality and cause oxidative disruption to lipids, proteins, and nucleic acids [20], leading to disruptions in water relationships and membrane stability [21]. Also, it has been shown that fenugreek plants experienced significant reductions in growth, photosynthetic pigments, and proteins under water stress [5].

Plants possess various methods to face environmental stressors and enhance their physiological activity. One effective strategy is the application of chemical compounds, such as plant growth regulators like amino acids and polyamines. These regulators are not only cost-effective but also play a crucial role in boosting plant stress tolerance.

L-arginine is a highly versatile amino acid in living cells, serving as a constituent of proteins and a precursor for the biosynthesis of polyamines, proline, agmatine, and cell signaling molecules like nitric oxide and glutamine [22]. It plays a fundamental role in stress tolerance due to its involvement in numerous physiological processes, including protein synthesis, osmotic potential, stomatal activity, and vegetative growth [23,24]. Investigations have confirmed that L-arginine is a vital modulator in various processes within higher plants and in their response to stress factors like salinity, water deficiency, and disease [25–27]. Arginine is particularly important in nitrogen metabolism in germinating seeds and developing seedlings [24]. It is widely used to increase growth, chemical constituents, and yields of crops. Arginine induces enzymes responsible for antioxidation, such as ascorbate peroxidase, superoxide dismutase, and glutathione reductase, thereby alleviating several stressors effects [28].

Polyamines, such as spermine, spermidine, and putrescine, are low-molecular-weight aliphatic amines found in plant cells [29]. They are essential for various fundamental processes, including cell division, differentiation, root elongation, floral development, leaf senescence, fruit ripening, protein translation, transcript expression, and chromatin organization [30]. Several studies have clarified the essential role of polyamines in enhancing plant tolerance to abiotic stresses. They protect against oxidative damage by preventing lipid peroxidation and neutralizing free radicals [31,32]. Spermine, in particular, modulates several physio-biochemical processes to reduce oxidative damage and enhance plant protection against multiple abiotic and biotic stresses [33]. It acts as a secondary messenger in signaling pathways, regulating plant development and boosting tolerance mechanisms [34]. Spermine applications on various plants have been found to positively impact drought responses by boosting osmolyte accumulation, elevating free polyamine levels, and regulating polyamine biosynthetic genes [35].

The aim of this study is to evaluate the effects of arginine and spermine treatments on the growth attributes, enzymatic and non-enzymatic antioxidants, pigments, osmolytes, and yield traits of fenugreek plants under both normal and drought conditions. This study seeks to determine how these treatments can mitigate the adverse effects of drought stress and enhance plant performance, with a particular focus on the comparative efficacy of arginine and spermine.

2. Materials and Methods

2.1. Layout of the Experiment

The experiment was carried out at the Botanical Research Station, Botany and Microbiology Department, Faculty of Science, Al-Azhar University, Cairo, Egypt. Seeds of the fenugreek, Giza 3 variety, obtained from the Institute of Crops Research, Agricultural Research Centre, Giza, Egypt, were utilized. Thirty fenugreek seeds were sown in 40 cm diameter earthenware pots containing 7 kg of soil. These pots were divided into two groups: control (unstressed) and drought-stressed (induced by PEG-stimulated drought stress). Each group was further divided into three sub-groups: control (untreated), foliar-treated with 1 mM arginine, and foliar-treated with 1 mM spermine. Seven plants remained in each pot after thinning and were irrigated as needed. Arginine and spermine were applied as a foliar spray twice, on the 25th and 35th days after sowing. Randomized plant samples were harvested on the 45th day after sowing for morphological, biochemical, and physiological analysis.

2.2. Plant Lengths and Biomass

Five fenugreek plants from each group were randomly collected to measure growth characteristics, including shoot length (cm), root length (cm), number of leaves, fresh and dry weights of shoots (g), and fresh and dry weights of roots (g).

2.3. Enzymatic Antioxidants

Fenugreek buds were used to extract superoxide dismutase (SOD), peroxidase (POD), polyphenol oxidase (PPO), and catalase (CAT), as described by [36]. SOD activity was determined using the method of Marklund and Marklund [37], which involves measuring pyrogallol reduction at 325 nm. POD activity was measured using the method outlined by Bergmeyer [38], focusing on the increase in pyrogallol absorbance at 470 nm. PPO activity was assessed using the method of Matta [39], which measures changes in catechol absorbance at 495 nm. CAT activity was estimated by measuring the cleavage of hydrogen peroxide following the technique of Aebi [40], with spectrophotometric readings at 240 nm.

2.4. Non-Enzymatic Antioxidants and Osmolytes

Free proline was determined using the method of Bates et al. [41], with proline levels measured at 520 nm. Total soluble sugars were determined following the described method in [42], with absorbance measured at 620 nm. Phenolic compounds were determined using the procedure in [43], with absorbance measured at 725 nm. Free amino acids were extracted and estimated using a series of standard solutions and a ninhydrin reagent, with absorbance measured at 570 nm [44].

2.5. Leaf Pigments

Chlorophylls and carotenoids in fenugreek leaves were quantified according to the described method by Vernon and Seely [45], with absorbance readings taken at 470 nm, 649 nm, and 665 nm.

2.6. Proteins and Malondialdehyde

Total soluble proteins were assayed using the technique in [46], with absorbance measured at 750 nm. Malondialdehyde (MDA) was estimated using the method in [47], with absorbance read at 532 nm, 600 nm, and 450 nm.

2.7. Yield

At the yield stage, five fenugreek plants were randomly selected to assess yield attributes, including weight of pods/plant, number of pods/plant, weight of seeds/plant, number of seeds/plant, and 100-seed weight.

2.8. Statistical Analysis

A two-way ANOVA followed by Tukey's test was conducted to assess the significance among treatments at a significance level of $\alpha = 0.05$, ensuring data normality and homogeneity of variances using Shapiro–Wilk's and Levene's Median tests, respectively [48]. When assumptions were violated, data transformations were applied. Principal component analysis (PCA) was carried out to illustrate the relationship between treatments and physiological parameters using PC-ORD version 5.

3. Results

3.1. Morphological Attributes

The provided results in Table 1 illustrate the response of various morphological traits of fenugreek plants, including root length, root fresh weight, root dry weight, shoot length, shoot fresh weight, shoot dry weight, and number of leaves, to arginine and spermine treatments under both normal and drought conditions.

Subjecting fenugreek plants to drought conditions caused notable reductions in various morphological characteristics. Specifically, there were decreases of approximately 12.5% in root length, 33.05% in root fresh weight, 50% in root dry weight, 13.71% in shoot length, 31.32% in shoot fresh weight, 36.15% in shoot dry weight, and 29.33% in the number of leaves, compared to plants under normal conditions.

The treatment with arginine and spermine resulted in noticeable improvements in the measured morphological parameters of fenugreek plants under normal conditions. These improvements included increases of approximately 0.8% and 2.7% in root length, 8.15% and 9.01% in root fresh weight, 25% and 41.67% in root dry weight, 13.31% and 17.74% in shoot length, 16.39% and 60.27% in shoot fresh weight, 12.21% and 41.78% in shoot dry weight, and 14.67% and 34.66% in the number of leaves in response to arginine and spermine, respectively.

Table 1. Effect of arginine and spermine treatments on the morphological growth characteristics of fenugreek plants under normal and drought conditions. Values are expressed as mean ± standard error. Distinct letters denote significant variations among the means.

Conditions \ Treatments	Control	Arginine	Spermine
	Root length (cm)		
Normal	12.8 ± 0.44 ab	12.9 ± 0.43 ab	13.14 ± 0.48 a
Drought	11.2 ± 0.60 b	11.76 ± 0.57 ab	12.3 ± 0.48 ab
	Root fresh weight (g)		
Normal	0.47 ± 0.014 a	0.504 ± 0.016 a	0.508 ± 0.021 a
Drought	0.31 ± 0.021 c	0.356 ± 0.013 bc	0.4 ± 0.02 b
	Root dry weight (g)		
Normal	0.07 ± 0.008 bc	0.09 ± 0.007 ab	0.10 ± 0.006 a
Drought	0.04 ± 0.007 d	0.05 ± 0.007 cd	0.07 ± 0.006 bc
	Shoot length (cm)		
Normal	24.8 ± 0.48 a	28.1 ± 1.01 a	29.2 ± 1.03 a
Drought	21.4 ± 1.05 a	23.4 ± 1.13 a	23.9 ± 0.88 a
	Shoot fresh weight (g)		
Normal	5.64 ± 0.42 bc	6.56 ± 0.39 b	9.04 ± 0.45 a
Drought	3.87 ± 0.23 d	5.25 ± 0.29 c	5.55 ± 0.36 bc
	Shoot dry weight (g)		
Normal	0.85 ± 0.065 b	0.96 ± 0.062 b	1.21 ± 0.055 a
Drought	0.54 ± 0.067 c	0.76 ± 0.058 b	0.84 ± 0.049 b
	Number of leaves		
Normal	30 ± 1.68 c	34.4 ± 1.20 b	40.4 ± 1.33 a
Drought	21.2 ± 0.95 d	28 ± 1.08 c	28.4 ± 1.20 c

Both under normal and adverse (drought) conditions, treating fenugreek plants with arginine and spermine showed significant responses in the morphological growth parameters. These included improvements of approximately 5% and 9.8% in root length, 14.1% and 28.21% in root fresh weight, 50% and 94.44% in root dry weight, 9.35% and 11.68% in shoot length, 35.64% and 43.32% in shoot fresh weight, 38.97% and 54.78% in shoot dry weight, and 32.08% and 33.96% in the number of leaves, respectively, when compared with untreated plants under stress.

3.2. Enzymatic Antioxidants

The data presented in Figure 1 demonstrate how different antioxidant enzymes of fenugreek plants, such as superoxide dismutase, peroxidase, polyphenol oxidase, and catalase, respond to treatments with arginine and spermine under both normal and drought conditions.

Administering arginine and spermine to fenugreek plants led to substantial inductions in the activities of various enzymatic antioxidants under normal conditions. Specifically, superoxide dismutase increased by approximately 14.89% and 42.55%, peroxidase increased by 10.68% and 12.62%, polyphenol oxidase increased by 57.69% and 92.31%, and catalase increased by 26.42% and 32.08%, respectively, in response to arginine and spermine.

The activities of enzymes of fenugreek plants, such as superoxide dismutase, peroxidase, polyphenol oxidase, and catalase, boosted significantly by approximately 89.36%, 33.98%, 115.38%, 39.62%, respectively, when subjected to drought conditions, as opposed to normal conditions.

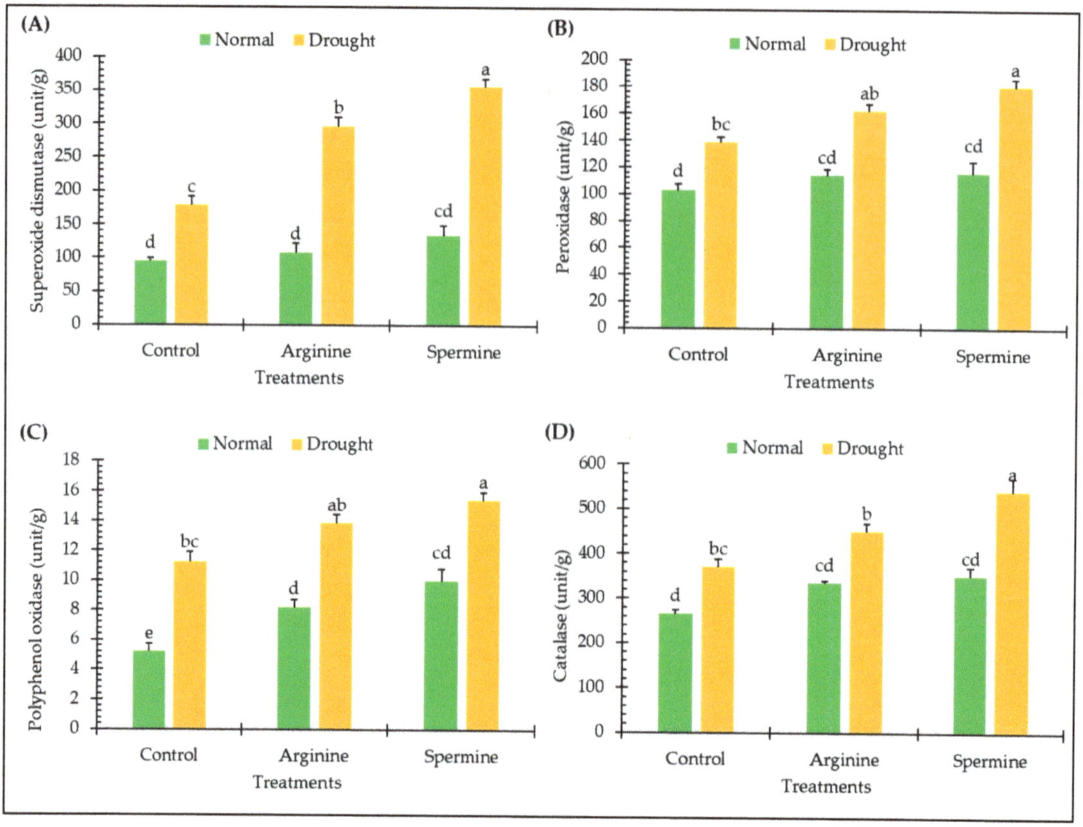

Figure 1. Effect of arginine and spermine treatments on the activities of superoxide dismutase (**A**), peroxidase (**B**), polyphenol oxidase (**C**), and catalase (**D**) in fenugreek plants under normal and drought conditions. Bars are expressed as the mean ± the standard error. Distinct letters denote significant variations among the means.

Moreover, under drought conditions, fenugreek plants treated with arginine and spermine exhibited notable improvements in the activities of various antioxidant enzymes. These included enhancements of approximately 66.29% and 100% in superoxide dismutase, 17.39% and 30.43% in peroxidase, 23.21% and 37.5% in polyphenol oxidase, and 21.62% and 45.95% in catalase, respectively, in comparison with untreated plants under stress.

3.3. Non-Enzymatic Antioxidants and Osmolytes

The findings in Figure 2 illustrate the impact of arginine and spermine treatments on various non-enzymatic antioxidants and osmolytes of fenugreek plants, such as sugars, proline, amino acids, and phenolics, under both normal and drought conditions.

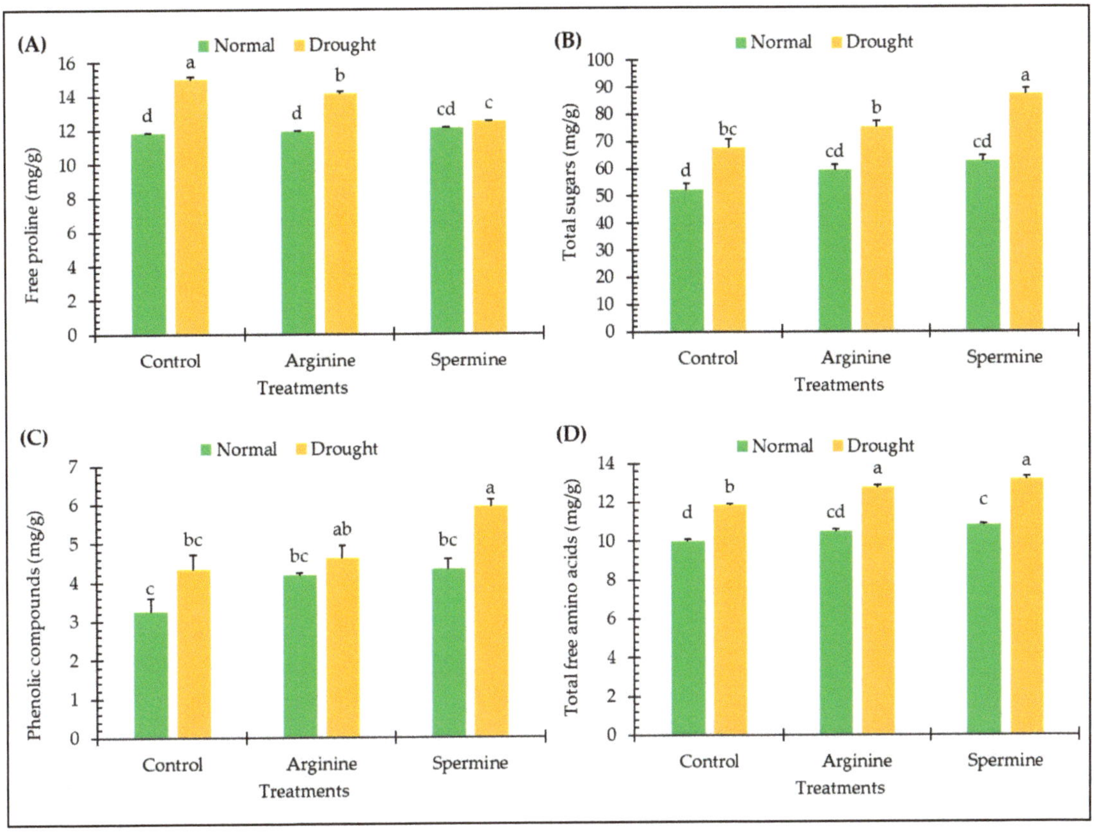

Figure 2. Effect of arginine and spermine treatments on the levels of free proline (**A**), total sugars (**B**), phenolic compounds (**C**), and total free amino acids (**D**) in fenugreek plants under normal and drought conditions. Bars are expressed as the mean ± the standard error. Distinct letters denote significant variations among the means.

Under drought conditions, fenugreek plants exhibited significant promotions in the levels of free proline, total sugars, phenolic compounds, and total free amino acids by approximately 26.84%, 29.16%, 33.57%, and 18.38%, respectively, in comparison to plants grown under normal conditions.

Application of arginine and spermine resulted in noticeable improvements in the levels of free proline, total sugars, phenolic compounds, and total free amino acids in fenugreek plants under normal conditions. These enhancements included increases of approximately 0.92% and 2.43% in proline, 13.33% and 19.63% in sugars, 29.18% and 33.38% in phenolics, and 5% and 8.23% in amino acids, respectively, in response to arginine and spermine.

Treating fenugreek plants with arginine and spermine showed significant variations in the abovementioned attributes not only under normal conditions but also under unfavorable (drought) conditions. These included decreases of approximately 5.55% and 16.79% in proline levels and increases of approximately 11.06% and 28.81% in sugars, 6.77% and 37.06% in phenolics, and 7.96% and 11.21% in amino acids, respectively, compared to untreated plants under stress.

3.4. Photosynthetic Pigments

Figure 3 presents the effects of arginine and spermine treatments on certain leaf pigments of fenugreek plants, including chlorophyll *a*, chlorophyll *b*, total chlorophylls, and carotenoids, in normal and drought conditions.

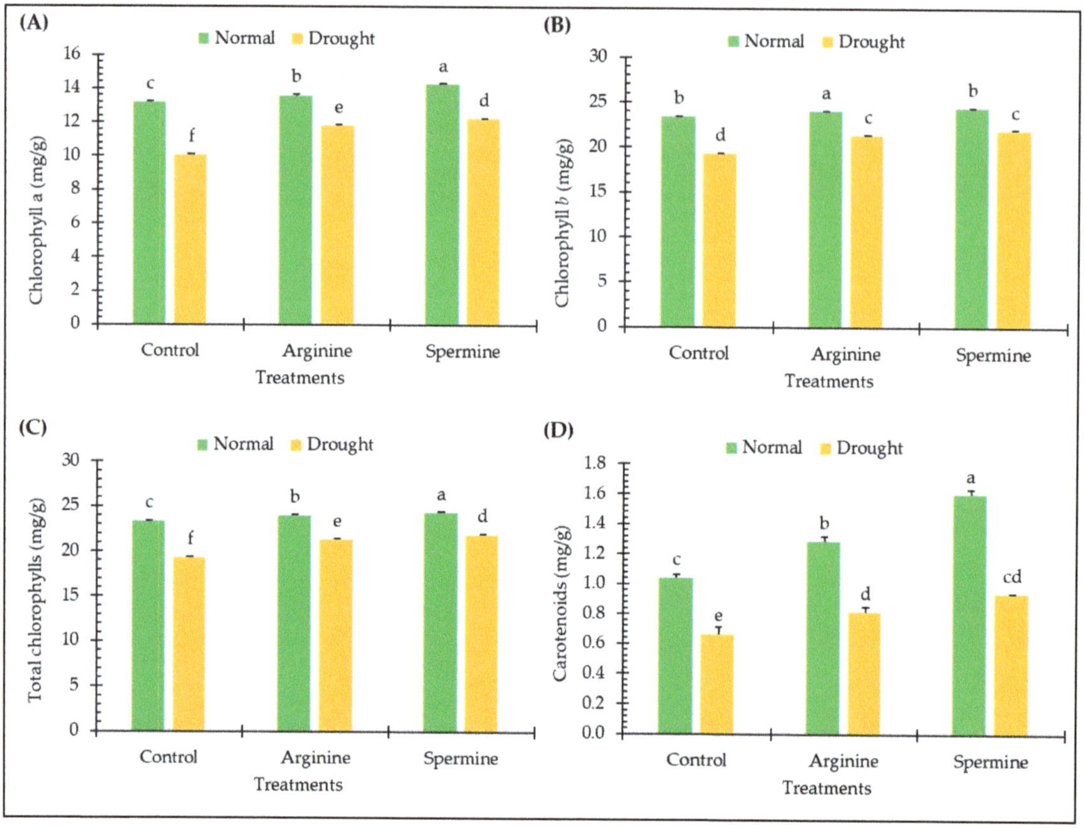

Figure 3. Effect of arginine and spermine treatments on the contents of chlorophyll *a* (**A**), chlorophyll *b* (**B**), total chlorophylls (**C**), and carotenoids (**D**) in fenugreek plants under normal and drought conditions. Bars are expressed as the mean ± the standard error. Distinct letters denote significant variations among the means.

Drought exposure resulted in substantial decreases in the levels of photosynthetic pigments of fenugreek leaves, including chlorophyll *a* (23.60%), chlorophyll *b* (9.69%), total chlorophylls (17.52%), and carotenoids (36.06%), in comparison to plants under normal conditions.

The treatment of fenugreek plants with arginine and spermine under normal conditions led to slight enhancements in leaf pigments especially in carotenoids which were promoted by approximately 23.69% and 53.7% in carotenoids. Furthermore, treating the tested plants with arginine and spermine resulted in significant increases of approximately 17.37% and 21.45% in chlorophyll *a*, 3.18% and 4.92% in chlorophyll *b*, 10.59% and 13.55% in total chlorophylls, and 22.67% and 41.14% in carotenoids, respectively, compared to untreated plants under stress.

3.5. Proteins and Malondialdehyde

The results displayed in Figure 4 highlight the impact of arginine and spermine treatments on the amounts of total proteins and malondialdehyde in fenugreek plants in both normal and drought conditions.

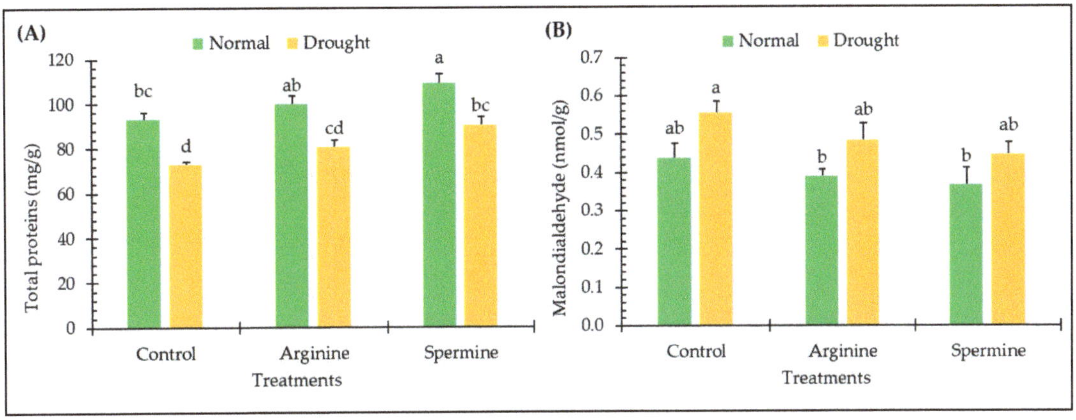

Figure 4. Effect of arginine and spermine treatments on the amounts of total proteins (**A**) and malondialdehyde (**B**) in fenugreek plants under normal and drought conditions. Bars are expressed as the mean ± the standard error. Distinct letters denote significant variations among the means.

The drought stress led to notable variations, with reductions of about 21.68% in total proteins contents, while increments of about 27.06% were observed for the malondialdehyde contents of fenugreek plants, relative to unstressed plants.

Remarkable enhancements of 6.98% and 17.19% for protein contents and noticeable inhibitions of 10.87% and 16.02% for malondialdehyde levels in fenugreek plants were observed under normal conditions following treatment with arginine and spermine. Furthermore, under drought conditions, fenugreek plants treated with arginine and spermine showed obvious accumulations in protein amounts by about 10.55% and 23.91%, while observed suppressions in malondialdehyde contents (as an indicator for the suppression of lipid peroxidation) reached about 12.84% and 19.55%, respectively, compared to untreated plants under stress.

3.6. Yield Attributes

The data in Table 2 show how fenugreek plants' yield indices, such as pod weight, pod number, seed weight, seed number, and 100-seed weight, are affected by arginine and spermine treatments under both normal and drought conditions.

When exposed to drought conditions, fenugreek plants showed significant declines in yield traits, such as weight of pods/plant, number of pods/plant, weight of seeds/plant, number of seeds/plant, and 100-seed weight, by approximately 34.72%, 27.08%, 30.51%, 24.55%, and 18.18%, respectively, compared to those under normal conditions.

The application of arginine and spermine led to noticeable improvements in the yield characteristics of fenugreek plants under normal conditions. These enhancements included increases of approximately 29.27% and 36.63% in the weight of pods, 22.91% and 75% in the number of pods, 41.2% and 64.75% in the weight of seeds, 30.9% and 41.14% in the number of seeds, and 11.9% and 24.69% in the 100-seed weight, respectively, in response to arginine and spermine.

Table 2. Effect of arginine and spermine treatments on the yield characteristics of fenugreek plants under normal and drought conditions. Values are expressed as the mean ± the standard error. Distinct letters denote significant variations among the means.

Conditions \ Treatments	Control	Arginine	Spermine
	Weight of pods/plant (g)		
Normal	6.09 ± 0.27 ab	7.87 ± 0.31 a	8.32 ± 0.28 a
Drought	3.97 ± 0.20 c	4.47 ± 0.41 c	5.15 ± 0.38 ab
	Number of pods/plant		
Normal	19.2 ± 1.11 c	23.6 ± 1.20 b	33.6 ± 0.66 a
Drought	14 ± 1.08 c	15.6 ± 0.88 de	17.8 ± 1.25 cd
	Weight of seeds/plant (g)		
Normal	3.97 ± 0.22 b	5.6 ± 0.27 a	5.82 ± 0.30 a
Drought	2.76 ± 0.33 c	3.17 ± 0.29 bc	3.53 ± 0.23 bc
	Number of seeds/plant		
Normal	211 ± 8.76 b	276.2 ± 11.54 a	297.8 ± 13.11 a
Drought	159.2 ± 8.83 c	182.4 ± 13.30 bc	197.4 ± 5.94 b
	100-seeds weight (g)		
Normal	1.78 ± 0.08 bc	1.99 ± 0.06 ab	2.22 ± 0.07 a
Drought	1.46 ± 0.05 d	1.65 ± 0.11 cd	1.78 ± 0.09 bc

Both under normal and adverse (drought) conditions, treating fenugreek plants with arginine and spermine showed significant responses in yield parameters. These included increases of approximately 12.48% and 29.64% in the weight of pods, 11.43% and 27.14% in the number of pods, 14.95% and 28.16% in the weight of seeds, 14.57% and 23.99% in the number of seeds, 13.17% and 21.81% in the 100-seed weight, respectively, compared to untreated plants under stress.

3.7. Principal Component Analysis

The provided image (Figure 5) from the PCA plot shows distinct clustering of treatments, with drought-stressed treatments (T4, T5, and T6) clearly separated from non-drought treatments (T1, T2, and T3), indicating different physiological responses. Drought alone (T4) induces unique physiological responses, while arginine (T5) and spermine (T6) treatments mitigate drought effects, leading to similar but distinct profiles compared to untreated plants. Non-drought treatments cluster closely, suggesting minimal changes in plant physiology under normal conditions. Physiological parameters such as proline and malondialdehyde (MDA) are strongly associated with drought-induced stress (T4), while antioxidant enzymes (POD, SOD, and CAT), sugars, phenols, and amino acids are linked to T5 and T6, highlighting their role in mitigating drought stress. Growth-related parameters are associated with non-drought treatments, reflecting better growth performance under normal irrigation.

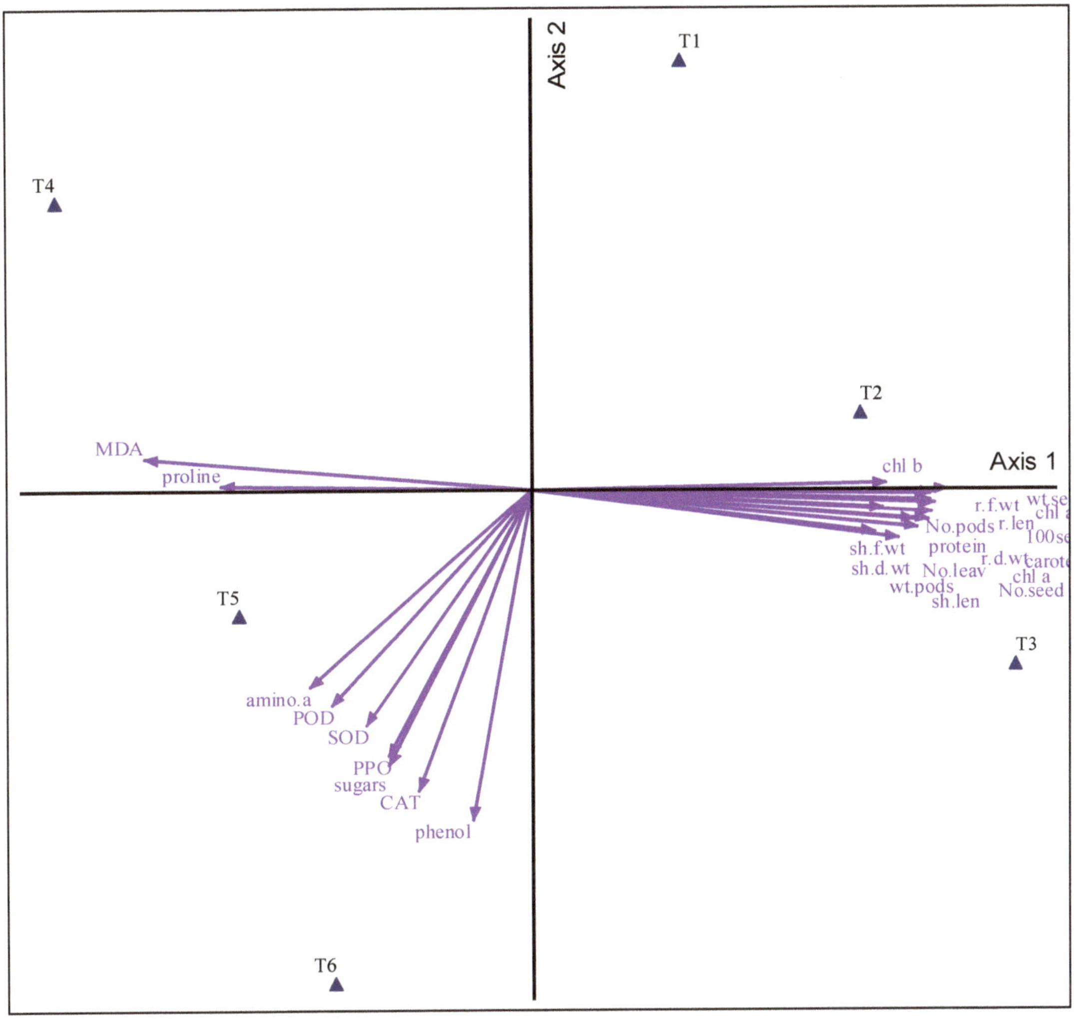

Figure 5. Principal component analysis visualizes the relationships of fenugreek growth attributes with treatments. Different abbreviations used in the chart are as follows: no treatment under normal irrigation conditions (control) (T1), arginine without drought (T2), spermine without drought (T3), drought (T4), arginine with drought (T5), spermine with drought (T6), r.len (root length), r.f.wt (root fresh weight), r.d.wt (root dry weight), sh.len (shoot length), sh.f.wt (shoot fresh weight), sh.d.wt (shoot dry weight), No.leav (number of leaves), wt.pods (weight of pods), No.pods (number of pods), wt.seed (weight of seeds), No.seed (number of seeds), 100 seed (100-seeds weight), chl a (chlorophyll *a*), chl b (chlorophyll *b*), chl a + b (total chlorophyll), caroten (carotenoids), protein, amino.a (amino acids), proline, phenol (phenolic compounds), sugars, MDA (malondialdehyde), SOD (speroxide dismutase), POD (peroxidase), PPO (polyphenol oxidase), and CAT (catalase).

4. Discussion

Enhancing plant growth, development, productivity, and resistance to climatic stress are crucial areas in agriculture and plant-based biotechnologies [49]. Throughout their life cycle, plants face numerous biotic and abiotic stresses, with drought being one of the most severe, leading to significant reductions in agricultural productivity and posing a threat to global food security [50,51]. Drought stress affects plants on both the morphological and molecular levels, decreasing growth and productivity [13]. Addressing plant drought tolerance is a major challenge in modern agriculture, where biostimulants, like amino acids and polyamines, play a vital role [52]. These substances help mitigate the harmful effects of stress and offer essential protection against oxidative damage, ultimately enhancing plant development and productivity [53].

In the current study, drought stress significantly reduced the number of fenugreek morphological parameters, including plant height, root depth, fresh and dry weights of shoots and roots, and the number of leaves. This aligns with previous studies on various crops, showing decreased germination and growth under drought stress [54–57]. The growth inhibition is attributed to reduced cell turgor, suppressing cell elongation and development, and tissue water loss, hindering cell division and elongation [58,59]. Arginine is a crucial amino acid that significantly contributes to plant growth. Research indicates that its application results in notable enhancements in the morphological growth attributes of various crops [28,60,61]. Our results indicate that foliar application of arginine mitigates drought effects on fenugreek plant growth. Arginine may enhance plant responses to drought stress through its conversion into proline and nitric oxide, which are essential for drought adaptation. Furthermore, arginine's ability to counteract abiotic stresses could be linked to the production of polyamines, which play significant roles in various biological processes such as growth, metabolism, and stress responses [26,27]. Similarly, the application of spermine in this study resulted in marked improvements in the growth indices of fenugreek under normal or drought conditions, consistent with previous findings on other crops [62–64]. Spermine plays a significant role in cell division, elongation, and protein synthesis [65]. It is particularly involved in shoot and root development, floral induction, fruit set, leaf senescence, DNA synthesis, osmolyte balance, chlorophyll protection, gene transcription, and protein translation [62,66]. Spermine plays a vital role in enabling plants to effectively respond to environmental stresses, including drought [35,67].

The activation of enzymatic antioxidants is essential for mitigating stress [68,69]. Enzymatic antioxidant activities increase when mung bean plants face water deficit conditions [70]. Additionally, previous investigations reported that drought stress induces antioxidant enzyme activities in various plants [71–73], aligning with our findings. Enhanced antioxidant enzyme activities under water stress are attributed to elevated hydrogen peroxide and singlet oxygen levels. In response, plants typically elevate their antioxidant activities to scavenge reactive oxygen species and mitigate stress [74]. Arginine pretreatment increased enzyme activity in different plants under normal and stress conditions [61,75]. Arginine, an amino acid, plays a significant role in alleviating drought stress [76]. Its application promotes enzyme activity, aiding in converting free radicals into water and oxygen, protecting the cell [26]. Similarly, spermine treatment ameliorated drought-induced osmotic stress by increasing catalase, superoxide dismutase, peroxidase, and polyphenol oxidase activities in several crops [14,77]. Enhanced antioxidant activities resulting from polyamine treatments are associated with improved molecular signaling, which supports adaptive plant responses to water stress [70]. Earlier studies indicate that polyamines help stabilize membranes and neutralize free radicals by boosting antioxidant activities [31].

The current study demonstrated that water scarcity in fenugreek plants led to the accumulation of certain osmolytes and non-enzymatic antioxidants, including proline, sugars, phenolics, and amino acids. These findings align with several studies reporting high levels of osmo-protectants in different crops under drought stress [78–81]. Additionally, the results showed that arginine and spermine applications increased the content of amino acids, sugars, and phenolics while reducing proline levels, as previously documented in various investigations [32,61,76,82]. The accumulation of these biomolecules serves as a tolerance strategy to reduce the oxidative damage caused by stress. Increases in soluble sugars, proline, and free amino acids in stressed plants help the cells adapt to drought conditions [83,84]. These osmolytes can scavenge free radicals, inhibit cellular redox potential, adjust osmotic pressure, stabilize membranes and proteins, and maintain the relative water content necessary for plant growth and metabolism [6,85,86]. It is noteworthy that proline levels decreased when drought-stressed plants were treated with arginine and spermine, indicating reduced plant sensitivity to drought. This indicates that polyamines and their precursor arginine play crucial roles as modulators in higher plants, influencing growth, physiological processes, development, and responses to stress factors [24,87].

Consistent with our findings, drought stress led to a reduction in chlorophyll and carotenoid levels in cotton [88], wheat [89], peanut [79], barley [28], and rice [90] and other important crops. Drought conditions have been reported to damage the photosynthetic system, reduce gas exchange, and decrease growth parameters and productivity [91]. The decline in the net photosynthetic rate under drought stress is due to biochemical disruptions caused by lipid oxidation and protein denaturation, which are crucial for pigment and chloroplast structures [92]. Conversely, the application of arginine to both normal and stressed fenugreek plants was found to significantly enhance leaf pigment content. Our findings align with previous studies on different plants [60,61,78]. The role of arginine in boosting pigment content can be attributed to its function as an amino acid that serves as a nitrogen source for chlorophyll formation [23]. The ability of arginine to alleviate stress and enhance growth characteristics is likely due to the production of polyamines, which participate in various biological processes such as growth, development, and responses to abiotic stresses [26]. Moreover, spermine greatly enhances the biosynthesis of chlorophyll pigments and PSII function through stomatal regulation, modulation of electron transfer chains to PSI receptors, and improvement in CO_2 assimilation rates, plant growth, and biomass yield under stress conditions [93]. This enhancement is linked to the increased stability of thylakoid membranes, plastid biogenesis, and the prevention of chlorophyll degradation [94,95]. Additionally, spermine promotes chlorophyll synthesis by increasing the uptake of magnesium ions, which are essential components of chlorophyll [96].

In the current investigation, drought conditions led to a reduction in protein content and an increase in malondialdehyde (a product of membrane lipid peroxidation). These findings are consistent with studies on maize [97], barley [55], wheat [62], lettuce [67], and soybean [98], regarding lipids, as well as fenugreek [5], soybean [33,98], and cowpea [99], regarding proteins, under various stress conditions. Under stress, plants may enhance proteolytic enzymes, leading to protein degradation, and accumulate excessive reactive oxygen species (ROS), which destabilize cell membranes and cause damage to DNA, pigments, proteins, and lipids [64,100]. The inhibitory effects of drought may be linked to reduced photosynthesis, as carbohydrates, the primary photosynthetic product, are essential for forming vital biomolecules [101]. Supplementing fenugreek plants, whether under natural or stress conditions, with arginine or spermine enhances protein content and reduces malondialdehyde accumulation from lipid oxidation, as documented in previous studies [27,75,77,97]. L-arginine plays a crucial role in physiological processes by modulating polypeptides involved in oxidative stress. It contributes to polyamine synthesis, membrane stability, osmotic balance, signal transduction, and electron transport [25,102]. Arginine's role in counteracting abiotic stresses may involve polyamine production, which supports growth, metabolism, and stress responses [26]. Spermine also acts a substantial role in post-transcriptional protein modifications, stabilizing protein conformation and function [103]. Exogenous spermine application alleviates stress effects by reducing lipid peroxidation and increasing total polyphenols, catalase, and superoxide dismutase activities [64,104]. Polyamines are essential for protein homeostasis, ROS detoxification, and antioxidative machinery activation under stress conditions [66]. They maintain membrane stability and permeability, enhance catalase activity, and reduce H_2O_2 content, ROS markers, and lipid peroxidation, thereby providing broad-spectrum tolerance against various stresses [62,105].

A water deficit significantly reduces yield attributes in crops such as tomatoes [106], wheat [107,108], and cotton [80] by negatively impacting growth and productivity. The reduction in growth and yield is associated with the excessive production of reactive oxygen species, which cause damage to cell membranes and components [97,109]. However, the application of arginine significantly increases plant yield under drought stress by promoting protein, proline, and polyamine biosynthesis, enhancing stomatal activity, osmotic potential, and overall growth [23,110]. Amino acids also provide essential substances for protein formation and function as osmo-regulators, increasing cellular osmotic components [31,111]. In this respect, polyamines play a crucial role in physiological processes such as reproductive organ development, tuberization, floral initiation, fruit development, and ripening [66], in addition to their role in maintaining turgor pressure [70,80].

PCA analysis confirms that drought significantly alters plant physiology, increasing proline and MDA levels. Arginine and spermine effectively alleviate drought stress by enhancing antioxidant activity and osmolyte accumulation, supporting their role as protective agents. These substances have a vital impact, highlighting their potential as targeted treatments for enhancing crop resilience to water deficit stress.

5. Conclusions

Applying arginine and spermine has been shown to effectively alleviate the adverse effects of drought stress on fenugreek plants. These treatments enhance both morphological growth characteristics and yield traits, while also improving the activities of antioxidant enzymes, as well as the levels of non-enzymatic antioxidants and osmolytes. Notably, spermine exhibits greater efficacy in promoting growth and stress tolerance. This study underscores the potential of utilizing arginine and spermine as eco-friendly and cost-effective solutions for enhancing plant performance under both normal and drought conditions.

Author Contributions: Conceptualization, A.A.B., N.F.G.S. and H.-A.A.H.; methodology, A.A.B. and N.F.G.S.; software, A.A.B. and W.K.A.; validation, A.A.B. and W.S.A.; formal analysis, A.A.B. and N.F.G.S.; investigation, A.A.B. and W.S.A.; resources, A.A.B. and W.K.A.; data curation, A.A.B. and H.-A.A.H.; writing—original draft preparation, A.A.B., N.F.G.S. and H.-A.A.H.; writing—review and editing, A.A.B., W.K.A., N.F.G.S., W.S.A. and H.-A.A.H.; visualization, A.A.B., W.K.A. and W.S.A.; supervision, A.A.B. and N.F.G.S.; project administration, A.A.B. and H.-A.A.H.; funding acquisition, A.A.B., W.K.A., N.F.G.S., W.S.A. and H.-A.A.H. All authors have read and agreed to the published version of the manuscript.

Funding: This research received no external funding.

Institutional Review Board Statement: Not applicable.

Informed Consent Statement: Not applicable.

Data Availability Statement: The data presented in this study are available upon request from the corresponding author.

Acknowledgments: The authors would like to thank the faculty members who helped with advice, materials, or methods in producing this research in its final form.

Conflicts of Interest: The authors declare no conflicts of interest.

References

1. Acharya, K.; Chakraborty, N.; Chatterjee, S. Fungal Diseases of Fenugreek. *Am. J. Soc. Issues Humaities* **2014**, 56–59.

2. Pandey, H.; Awasthi, P. Effect of Processing Techniques on Nutritional Composition and Antioxidant Activity of Fenugreek (*Trigonella foenum-graecum*) Seed Flour. *J. Food Sci. Technol.* **2015**, *52*, 1054–1060. [CrossRef] [PubMed]
3. Hanafy, R.S.; Akladious, S.A. Physiological and Molecular Studies on the Effect of Gamma Radiation in Fenugreek (*Trigonella foenum-graecum* L.) Plants. *J. Genet. Eng. Biotechnol.* **2018**, *16*, 683–692. [CrossRef]
4. Wani, S.A.; Kumar, P. Fenugreek: A Review on Its Nutraceutical Properties and Utilization in Various Food Products. *J. Saudi Soc. Agric. Sci.* **2018**, *17*, 97–106. [CrossRef]
5. Ahmed, M.A.; Shalaby, M.A.F.; El-housini, E.A.; Khater, M.A. Alleviation of Drought Stress on Fenugreek (*Trigonella foenum-graecum* L.) Plants by Foliar Application of Polyamines Compounds. *Middle East J. Appl. Sci.* **2018**, *8*, 883–894.
6. Badawy, A.A.; Alamri, A.A.; Hussein, H.-A.A.; Mashlawi, A.M.; Kenawy, S.K.M.; El-Shabasy, A. Glycine Betaine Mitigates Heavy Metal Toxicity in *Beta vulgaris* (L.): An Antioxidant-Driven Approach. *Agronomy* **2024**, *14*, 797. [CrossRef]
7. Attia, M.S.; Hashem, A.H.; Badawy, A.A.; Abdelaziz, A.M. Biocontrol of Early Blight Disease of Eggplant Using Endophytic *Aspergillus terreus*: Improving Plant Immunological, Physiological and Antifungal Activities. *Bot. Stud.* **2022**, *63*, 26. [CrossRef]
8. Osman, M.S.; Badawy, A.A.; Osman, A.I.; Abdel Latef, A.A.H. Ameliorative Impact of an Extract of the Halophyte Arthrocnemum Macrostachyum on Growth and Biochemical Parameters of Soybean under Salinity Stress. *J. Plant Growth Regul.* **2021**, *40*, 1245–1256. [CrossRef]
9. Ewais, E.A.; Ismail, M.A.; Amin, M.A.; Badawy, A.A. Efficiency of Salicylic Acid and Glycine on Sugar Beet Plants Grown under Heavy Metals Pollution. *Egypt. J. Biotechnol.* **2015**, *48*, 112–126.
10. Sayed, A.I.; Hanafy, M.S. Impact of Sea Salt Stress on Growth and Some Physiological Attributes of Two Soybean (*Glycine max* L.) Cultivars. *Al-Azhar J. Agric. Res.* **2021**, *46*, 88–100.
11. Hussein, H.-A.A.; Darwesh, O.M.; Mekki, B.B. Environmentally Friendly Nano-Selenium to Improve Antioxidant System and Growth of Groundnut Cultivars under Sandy Soil Conditions. *Biocatal. Agric. Biotechnol.* **2019**, *18*, 101080. [CrossRef]
12. Omer, A.M.; Osman, M.S.; Badawy, A.A. Inoculation with *Azospirillum brasilense* and/or *Pseudomonas geniculata* Reinforces Flax (*Linum usitatissimum*) Growth by Improving Physiological Activities under Saline Soil Conditions. *Bot. Stud.* **2022**, *63*, 15. [CrossRef] [PubMed]
13. Timmusk, S.; Abd El-Daim, I.A.; Copolovici, L.; Tanilas, T.; Kännaste, A.; Behers, L.; Nevo, E.; Seisenbaeva, G.; Stenström, E.; Niinemets, Ü.; et al. Drought-Tolerance of Wheat Improved by Rhizosphere Bacteria from Harsh Environments: Enhanced Biomass Production and Reduced Emissions of Stress Volatiles. *PLoS ONE* **2014**, *9*, e96086. [CrossRef] [PubMed]
14. Farooq, M.; Wahid, A.; Lee, D.-J. Exogenously Applied Polyamines Increase Drought Tolerance of Rice by Improving Leaf Water Status, Photosynthesis and Membrane Properties. *Acta Physiol. Plant.* **2009**, *31*, 937–945. [CrossRef]
15. Eid, A.M.; Salim, S.S.; Hassan, S.E.D.; Ismail, M.A.; Fouda, A. Role of Endophytes in Plant Health and Abiotic Stress Management. In *Microbiome in Plant Health and Disease*; Springer: Singapore, 2019. [CrossRef]
16. Kasim, W.A.; Hamada, E.A.M.; El-Din, N.G.S.; Eskander, S. Influence of Seaweed Extracts on the Growth, Some Metabolic Activities and Yield of Wheat Grown under Drought Stress. *Int. J. Agron. Agric. Res.* **2015**, *7*, 173–189.
17. Hassanein, R.A.; Amin, A.A.E.-S.; Rashad, E.-S.M.; Hussein, H.-A.H. Effect of Thiourea and Salicylic Acid on Antioxidant Defense of Wheat Plants under Drought Stress. *Int. J. ChemTech Res.* **2015**, *7*, 346–354.
18. Badawy, A.A.; Husen, A.; Salem, S.S. Proteomic Study on the Effects of Silver Nanoparticles Under Abiotic Stress. In *Plant Response to Silver Nanoparticles: Plant Growth, Development, Production, and Protection*; Husen, A., Ed.; Springer Nature: Singapore, 2025; pp. 79–91, ISBN 978-981-97-7352-7.
19. Sallam, A.; Alqudah, A.M.; Dawood, M.F.A.A.; Baenziger, P.S.; Börner, A. Drought Stress Tolerance in Wheat and Barley: Advances in Physiology, Breeding and Genetics Research. *Int. J. Mol. Sci.* **2019**, *20*, 3137. [CrossRef]
20. Gill, S.S.; Tuteja, N. Reactive Oxygen Species and Antioxidant Machinery in Abiotic Stress Tolerance in Crop Plants. *Plant Physiol. Biochem.* **2010**, *48*, 909–930. [CrossRef]
21. Abid, M.; Ali, S.; Qi, L.K.; Zahoor, R.; Tian, Z.; Jiang, D.; Snider, J.L.; Dai, T. Physiological and Biochemical Changes during Drought and Recovery Periods at Tillering and Jointing Stages in Wheat (*Triticum aestivum* L.). *Sci. Rep.* **2018**, *8*, 4615. [CrossRef]
22. Liu, J.-H.; Nada, K.; Honda, C.; Kitashiba, H.; Wen, X.-P.; Pang, X.-M.; Moriguchi, T. Polyamine Biosynthesis of Apple Callus under Salt Stress: Importance of the Arginine Decarboxylase Pathway in Stress Response. *J. Exp. Bot.* **2006**, *57*, 2589–2599. [CrossRef]
23. Hamid, Z.H.; Amer, M.; Wahab, A. Effect of Arginine on Growth and Yield of Tomato Plant (*Lycopersicon esculentum*) under Drought Stress. *Plant Arch.* **2019**, *19*, 4441–4444.
24. Winter, G.; Todd, C.D.; Trovato, M.; Forlani, G.; Funck, D. Physiological Implications of Arginine Metabolism in Plants. *Front. Plant Sci.* **2015**, *6*, 534. [CrossRef] [PubMed]
25. Matysiak, K.; Kierzek, R.; Siatkowski, I.; Kowalska, J.; Krawczyk, R.; Miziniak, W. Effect of Exogenous Application of Amino Acids L-Arginine and Glycine on Maize under Temperature Stress. *Agronomy* **2020**, *10*, 769. [CrossRef]
26. Nasibi, F.; Yaghoobi, M.M.; Kalantari, K.M. Effect of Exogenous Arginine on Alleviation of Oxidative Damage in Tomato Plant Underwater Stress. *J. Plant Interact.* **2011**, *6*, 291–296. [CrossRef]

27. Hussein, H.-A.A.; Alshammari, S.O.; Kenawy, S.K.M.; Elkady, F.M.; Badawy, A.A. Grain-Priming with L-Arginine Improves the Growth Performance of Wheat (*Triticum aestivum* L.) Plants under Drought Stress. *Plants* **2022**, *11*, 1219. [CrossRef]
28. Shalaby, M.A.F.; Ahmed, M.A.; Khater, M.A. Physiological Responses of Some Barley Cultivars to Foliar Treatments with Arginine under Water Stress Conditions. *Middle East J. Agric. Res.* **2018**, *7*, 1102–1123.
29. Alcázar, R.; Bueno, M.; Tiburcio, A.F. Polyamines: Small Amines with Large Effects on Plant Abiotic Stress Tolerance. *Cells* **2020**, *9*, 2373. [CrossRef]
30. Pascual, L.S.; López-Climent, M.F.; Segarra-Medina, C.; Gómez-Cadenas, A.; Zandalinas, S.I. Exogenous Spermine Alleviates the Negative Effects of Combined Salinity and Paraquat in Tomato Plants by Decreasing Stress-Induced Oxidative Damage. *Front. Plant Sci.* **2023**, *14*, 1193207. [CrossRef]
31. Velikova, V.; Yordanov, I.; Edreva, A. Oxidative Stress and Some Antioxidant Systems in Acid Rain-Treated Bean Plants: Protective Role of Exogenous Polyamines. *Plant Sci.* **2000**, *151*, 59–66. [CrossRef]
32. Talaat, N.B.; Shawky, B.T. Dual Application of 24-Epibrassinolide and Spermine Confers Drought Stress Tolerance in Maize (*Zea mays* L.) by Modulating Polyamine and Protein Metabolism. *J. Plant Growth Regul.* **2016**, *35*, 518–533. [CrossRef]
33. Basit, F.; Bhat, J.A.; Guan, Y.; Jan, B.L.; Tyagi, A.; Ahmad, P. Nitric Oxide and Spermine Revealed Positive Defense Interplay for the Regulation of the Chromium Toxicity in Soybean (*Glycine max* L.). *Environ. Pollut.* **2022**, *308*, 119602. [CrossRef]
34. Diao, Q.; Song, Y.; Qi, H. Exogenous Spermidine Enhances Chilling Tolerance of Tomato (*Solanum lycopersicum* L.) Seedlings via Involvement in Polyamines Metabolism and Physiological Parameter Levels. *Acta Physiol. Plant.* **2015**, *37*, 230. [CrossRef]
35. Ebeed, H.T.; Hassan, N.M.; Aljarani, A.M. Exogenous Applications of Polyamines Modulate Drought Responses in Wheat through Osmolytes Accumulation, Increasing Free Polyamine Levels and Regulation of Polyamine Biosynthetic Genes. *Plant Physiol. Biochem.* **2017**, *118*, 438–448. [CrossRef] [PubMed]
36. Mukherjee, S.P.; Choudhuri, M.A. Implications of Water Stress-Induced Changes in the Levels of Endogenous Ascorbic Acid and Hydrogen Peroxide in Vigna Seedlings. *Physiol. Plant.* **1983**, *58*, 166–170. [CrossRef]
37. Marklund, S.; Marklund, G. Involvement of the Superoxide Anion Radical in the Autoxidation of Pyrogallol and a Convenient Assay for Superoxide Dismutase. *Eur. J. Biochem.* **1974**, *47*, 469–474. [CrossRef] [PubMed]
38. Bergmeyer, H.U. *Methods of Enzymatic Analysis*, 2nd ed.; Academic Press Inc.: New York, NY, USA, 1974; Volume 1, p. 974.
39. Matta, A. Accumulation of Phenols in Tomato Plants Infected by Different Forms of *Fusarium oxysporum*. *Phytopathology* **1969**, *59*, 512–513.
40. Aebi, H. Catalase in Vitro. *Methods Enzimol.* **1984**, *105*, 121–126.
41. Bates, L.S.; Waldren, R.P.; Teare, I.D. Rapid Determination of Free Proline for Water-Stress Studies. *Plant Soil* **1973**, *39*, 205–207. [CrossRef]
42. Umbreit, W.W.; Burris, R.H.; Stauffer, J.F. *Manometric Techniques: A Manual Describing Methods Applicable to the Study of Tissue Metabolism*; Burgess Publishing Company: Minneapolis, MN, USA, 1964.
43. Dai, G.H.; Andary, C.; Cosson-Mondolot, L.; Boubals, D. Polyphenols and Resistance of Grapevines to Downy Mildew. *Int. Symp. Nat. Phenols Plant Resist.* **1993**, *381*, 763–766. [CrossRef]
44. Sadasivam, S.; Manickam, A. *Biochemical Methods*, 3rd ed.; New Age International: New Delhi, India, 2008.
45. Vernon, L.P.; Seely, G.R. (Eds.) *The Chlorophylls*; Academic Press: New York, NY, USA, 1966.
46. Lowry, O.H.; Rosebrough, N.J.; Farr, A.L.; Randall, R.J. Protein Measurement with the Folin Phenol Reagent. *J. Biol. Chem.* **1951**, *193*, 265–275. [CrossRef]
47. Heath, R.L.; Packer, L. Photoperoxidation in Isolated Chloroplasts: I. Kinetics and Stoichiometry of Fatty Acid Peroxidation. *Arch. Biochem. Biophys.* **1968**, *125*, 189–198. [CrossRef]
48. Snedecor, G.W.; Cochran, W.G. *Statistical Methods*, 8th ed.; The Iowa State University Press: Ames, IA, USA, 1989.
49. Badawy, A.A.; Husen, A.; Salem, S.S. Use of Nanobiotechnology in Augmenting Soil–Plant System Interaction for Higher Plant Growth and Production. In *Essential Minerals in Plant-Soil Systems*; Husen, A., Ed.; Plant Biology, Sustainability and Climate Change; Elsevier: Amsterdam, The Netherlands, 2024; pp. 423–443. ISBN 978-0-443-16082-0.
50. Ma, Y.; Dias, M.C.; Freitas, H. Drought and Salinity Stress Responses and Microbe-Induced Tolerance in Plants. *Front. Plant Sci.* **2020**, *11*, 591911. [CrossRef] [PubMed]
51. Helalia, A.A.; Hassuba, M.M. Efficacy of Several Chemical Fungicides and Biofungicides for Controlling Damping-off and Root Rot Diseases in Common Bean under Field Conditions. *Al-Azhar J. Agric. Res.* **2021**, *46*, 154–167.
52. Gupta, S.; Agarwal, V.P.; Gupta, N.K. Efficacy of Putrescine and Benzyladenine on Photosynthesis and Productivity in Relation to Drought Tolerance in Wheat (*Triticum aestivum* L.). *Physiol. Mol. Biol. Plants* **2012**, *18*, 331–336. [CrossRef]
53. Calvo, P.; Nelson, L.; Kloepper, J.W. Agricultural Uses of Plant Biostimulants. *Plant Soil* **2014**, *383*, 3–41. [CrossRef]
54. Fathi, A.; Tari, D.B. Effect of Drought Stress and Its Mechanism in Plants. *Int. J. Life Sci.* **2016**, *10*, 1–6. [CrossRef]
55. Ahmad, M.A.; Saleem, A.; Tahir, M.; Khilji, S.A.; Sajid, Z.A.; Landry, K.B.; El-Sheikh, M.A.; Ahmad, P. Modulation of the Polyamines, Osmolytes and Antioxidant Defense System to Ameliorate Drought Stress Tolerance in *Hordeum vulgare* L. Using Ascorbic Acid. *S. Afr. J. Bot.* **2024**, *171*, 726–736. [CrossRef]

56. Saikia, J.; Sarma, R.K.; Dhandia, R.; Yadav, A.; Bharali, R.; Gupta, V.K.; Saikia, R. Alleviation of Drought Stress in Pulse Crops with ACC Deaminase Producing Rhizobacteria Isolated from Acidic Soil of Northeast India. *Sci. Rep.* **2018**, *8*, 3560. [CrossRef]
57. Faraji, J.; Sepehri, A. Ameliorative Effects of TiO_2 Nanoparticles and Sodium Nitroprusside on Seed Germination and Seedling Growth of Wheat under PEG-Stimulated Drought Stress. *J. Seed Sci.* **2019**, *41*, 309–317. [CrossRef]
58. Bogati, K.; Walczak, M. The Impact of Drought Stress on Soil Microbial Community, Enzyme Activities and Plants. *Agronomy* **2022**, *12*, 189. [CrossRef]
59. Hellal, F.; Abdel-Hady, M.; Khatab, I.; El-Sayed, S.; Abdelly, C. Yield Characterization of Mediterranean Barley under Drought Stress Condition. *AIMS Agric. Food* **2019**, *4*, 518–533. [CrossRef]
60. Nejadalimoradi, H.; Nasibi, F.; Kalantari, K.M.; Zanganeh, R. Effect of Seed Priming with L-Arginine and Sodium Nitroprusside on Some Physiological Parameters and Antioxidant Enzymes of Sunflower Plants Exposed to Salt Stress. *Agric. Commun.* **2014**, *2*, 23–30.
61. Ramadan, A.A.; Abd Elhamid, E.M.; Sadak, M.S.; Elhamid, E.M.A.; Sadak, M.S. Comparative Study for the Effect of Arginine and Sodium Nitroprusside on Sunflower Plants Grown under Salinity Stress Conditions. *Bull. Natl. Res. Cent.* **2019**, *43*, 118. [CrossRef]
62. Hassan, N.; Ebeed, H.; Aljaarany, A. Exogenous Application of Spermine and Putrescine Mitigate Adversities of Drought Stress in Wheat by Protecting Membranes and Chloroplast Ultra-Structure. *Physiol. Mol. Biol. Plants* **2020**, *26*, 233–245. [CrossRef]
63. Ekıncı, M.; Yildirim, E.; Dursun, A.; Mohamedsrajaden, N.S. Putrescine, Spermine and Spermidine Mitigated the Salt Stress Damage on Pepper (*Capsicum annum* L.) Seedling. *Yuz. Yıl Univ. J. Agric. Sci.* **2019**, *29*, 290–299. [CrossRef]
64. Radhakrishnan, R.; Lee, I.J. Ameliorative Effects of Spermine against Osmotic Stress through Antioxidants and Abscisic Acid Changes in Soybean Pods and Seeds. *Acta Physiol. Plant.* **2013**, *35*, 263–269. [CrossRef]
65. Gull, A.; Lone, A.A.; Wani, N.U.I. *Biotic and Abiotic Stresses in Plants*; IntechOpen: London, UK, 2019; Volume 7, pp. 1–9. [CrossRef]
66. Tiburcio, A.F.; Altabella, T.; Bitrián, M.; Alcázar, R. The Roles of Polyamines during the Lifespan of Plants: From Development to Stress. *Planta* **2014**, *240*, 1–18. [CrossRef] [PubMed]
67. Liu, C.J.; Wang, H.R.; Wang, L.; Han, Y.Y.; Hao, J.H.; Fan, S.X. Effects of Different Types of Polyamine on Growth, Physiological and Biochemical Nature of Lettuce under Drought Stress. *IOP Conf. Ser. Earth Environ. Sci.* **2018**, *185*, 012010. [CrossRef]
68. Amin, M.A.; Ismail, M.A.; Badawy, A.A.; Awad, M.A.; Hamza, M.F.; Awad, M.F.; Fouda, A. The Potency of Fungal-Fabricated Selenium Nanoparticles to Improve the Growth Performance of *Helianthus annuus* L. and Control of Cutworm Agrotis Ipsilon. *Catalysts* **2021**, *11*, 1551. [CrossRef]
69. Osman, M.S.; Badawy, A.A. Influence of Arbuscular Mycorrhizal Fungi (*Glomus mosseae*) on Enhancing Salinity Tolerance of Two Common Bean Cultivars. *Egypt. J. Biotechnol.* **2020**, *60*, 145–162.
70. Ghassemi, S.; Farhangi-Abriz, S.; Faegi-Analou, R.; Ghorbanpour, M.; Lajayer, B.A. Monitoring Cell Energy, Physiological Functions and Grain Yield in Field-Grown Mung Bean Exposed to Exogenously Applied Polyamines under Drought Stress. *J. Soil Sci. Plant Nutr.* **2018**, *18*, 1108–1125. [CrossRef]
71. Mahdavian, M.; Sarikhani, H.; Hadadinejad, M.; Dehestani, A. Exogenous Application of Putrescine Positively Enhances the Drought Stress Response in Two Citrus Rootstocks by Increasing Expression of Stress-Related Genes. *J. Soil Sci. Plant Nutr.* **2021**, *21*, 1934–1948. [CrossRef]
72. Cui, G.; Zhao, Y.; Zhang, J.; Chao, M.; Xie, K.; Zhang, C.; Sun, F.; Liu, S.; Xi, Y. Proteomic Analysis of the Similarities and Differences of Soil Drought and Polyethylene Glycol Stress Responses in Wheat (*Triticum aestivum* L.). *Plant Mol. Biol.* **2019**, *100*, 391–410. [CrossRef] [PubMed]
73. Li, L.; Gu, W.; Li, C.C.; Li, W.; Li, C.C.; Li, J.; Wei, S. Exogenous Spermidine Improves Drought Tolerance in Maize by Enhancing the Antioxidant Defence System and Regulating Endogenous Polyamine Metabolism. *Crop Pasture Sci.* **2018**, *69*, 1076–1091. [CrossRef]
74. Mosa, K.A.; El-Naggar, M.; Ramamoorthy, K.; Alawadhi, H.; Elnaggar, A.; Wartanian, S.; Ibrahim, E.; Hani, H. Copper Nanoparticles Induced Genotoxicty, Oxidative Stress, and Changes in Superoxide Dismutase (SOD) Gene Expression in Cucumber (*Cucumis sativus*) Plants. *Front. Plant Sci.* **2018**, *9*, 872. [CrossRef] [PubMed]
75. Nasibi, F.; Barand, A.; Kalantari, K.M. The Effect of Arginine Pretreatment on Germination, Growth and Physiological Parameters in the Increase of Low Temperature Tolerance in *Pistacia vera* L. in Vitro Culture. *Int. J. Agric. Crop Sci.* **2013**, *5*, 1918–1925.
76. Feiz, F.S.; Hakimi, L.; Mousavi, A.; Jahromi, M.G. The Effects of Glycine Betaine and L-Arginine on Biochemical Properties of Pot Marigold (*Calendula officinalis* L.) under Water Stress. *Iran. J. Plant Physiol.* **2019**, *9*, 2795–2805.
77. Lee, R.R.I. Spermine Promotes Acclimation to Osmotic Stress by Modifying Antioxidant, Abscisic Acid, and Jasmonic Acid Signals in Soybean. *J. Plant Growth Regul.* **2013**, *32*, 22–30. [CrossRef]
78. Hasanuzzaman, M.; Nahar, K.; Rahman, A.; Inafuku, M.; Oku, H.; Fujita, M. Exogenous Nitric Oxide Donor and Arginine Provide Protection against Short-Term Drought Stress in Wheat Seedlings. *Physiol. Mol. Biol. Plants* **2018**, *24*, 993–1004. [CrossRef]
79. El-sayed, S.; Ramadan, A.A.E.; Hellal, F. Drought Stress Mitigation by Application of Algae Extract on Peanut Grown under Sandy Soil Conditions. *Asian J. Plant Sci.* **2020**, *19*, 230–239. [CrossRef]

80. Ahmed, A.H.H.; Darwish, E.; Alobaidy, M.G. Impact of Putrescine and 24-Epibrassinolide on Growth, Yield and Chemical Constituents of Cotton (*Gossypium barbadense* L.) Plant Grown under Drought Stress Conditions. *Asian J. Plant Sci.* **2017**, *16*, 9–23. [CrossRef]
81. Elewa, T.A.; Sadak, M.S.; Saad, A.M. Proline Treatment Improves Physiological Responses in Quinoa Plants under Drought Stress. *Biosci. Res.* **2017**, *14*, 21–33.
82. Abass, M.; Qin, C.; Maodong, Q.; Xue, X.; Ahmad, P. Spermine Application Alleviates Salinity Induced Growth and Photosynthetic Inhibition in *Solanum lycopersicum* by Modulating Osmolyte and Secondary Metabolite Accumulation and Di Ff Erentially Regulating Antioxidant Metabo. *Plant Physiol. Biochem.* **2019**, *144*, 1–13. [CrossRef]
83. Tan, Y.; Liang, Z.; Shao, H.; Du, F. Effect of Water Deficits on the Activity of Anti-Oxidative Enzymes and Osmoregulation among Three Different Genotypes of Radix Astragali at Seeding Stage. *Colloids Surf. B Biointerfaces* **2006**, *49*, 60–65. [CrossRef]
84. Badawy, A.A.; Alotaibi, M.O.; Abdelaziz, A.M.; Osman, M.S.; Khalil, A.M.A.; Saleh, A.M.; Mohammed, A.E.; Hashem, A.H. Enhancement of Seawater Stress Tolerance in Barley by the Endophytic Fungus *Aspergillus ochraceus*. *Metabolites* **2021**, *11*, 428. [CrossRef] [PubMed]
85. Iqbal, S. *Physiology of Wheat (Triticum aestivum L.) Accessions and the Role of Phytohormones under Water Stress*; Quaid-I-Azam University: Islamabad, Pakistan, 2009.
86. Abdel Latef, A.A.H.; Omer, A.M.; Badawy, A.A.; Osman, M.S.; Ragaey, M.M. Strategy of Salt Tolerance and Interactive Impact of Azotobacter Chroococcum and/or Alcaligenes Faecalis Inoculation on Canola (*Brassica napus* L.) Plants Grown in Saline Soil. *Plants* **2021**, *10*, 110. [CrossRef] [PubMed]
87. Liu, J.-H.; Wang, W.; Wu, H.; Gong, X.; Moriguchi, T. Polyamines Function in Stress Tolerance: From Synthesis to Regulation. *Front. Plant Sci.* **2015**, *6*, 827. [CrossRef] [PubMed]
88. Shallan, M.A.; Hassan, H.M.M.; Namich, A.A.M.; Ibrahim, A.A. Effect of Sodium Niroprusside, Putrescine and Glycine Betaine on Alleviation of Drought Stress in Cotton Plant. *Am. J. Agric. Environ. Sci.* **2012**, *12*, 1252–1265.
89. Hammad, S.A.R.; Ali, O.A.M. Physiological and Biochemical Studies on Drought Tolerance of Wheat Plants by Application of Amino Acids and Yeast Extract. *Ann. Agric. Sci.* **2014**, *59*, 133–145. [CrossRef]
90. Eissa, A.A.; Mourad, S.S.B.; Ibrahim, S.D. Evaluation of Some Rice (*Oryza sativa* L.) Genotypes under Drought Stress Conditions by Using Morphological, Physiological and Molecular Characteristics. *Al-Azhar J. Agric. Res.* **2023**, *48*, 179–193.
91. Anjum, S.A.; Wang, L.; Farooq, M.; Xue, L.; Ali, S. Fulvic Acid Application Improves the Maize Performance under Well-Watered and Drought Conditions. *J. Agron. Crop Sci.* **2011**, *197*, 409–417. [CrossRef]
92. Marcińska, I.; Czyczyło-Mysza, I.; Skrzypek, E.; Filek, M.; Grzesiak, S.; Grzesiak, M.T.; Janowiak, F.; Hura, T.; Dziurka, M.; Dziurka, K.; et al. Impact of Osmotic Stress on Physiological and Biochemical Characteristics in Drought-Susceptible and Drought-Resistant Wheat Genotypes. *Acta Physiol. Plant.* **2013**, *35*, 451–461. [CrossRef]
93. Li, Y.; Han, D.; Hu, G.; Sommerfeld, M.; Hu, Q. Inhibition of Starch Synthesis Results in Overproduction of Lipids in *Chlamydomonas reinhardtii*. *Biotechnol. Bioeng.* **2010**, *107*, 258–268. [CrossRef]
94. Borrell, A.; Carbonell, L.; Farras, R.; Puig-Parellada, P.; Tiburcio, A.F. Polyamines Inhibit Lipid Peroxidation in Senescing Oat Leaves. *Physiol. Plant.* **1997**, *99*, 385–390. [CrossRef]
95. Pandey, S.; Ranade, S.A.; Nagar, P.K.; Kumar, N. Role of Polyamines and Ethylene as Modulators of Plant Senescence. *J. Biosci.* **2000**, *25*, 291–299. [CrossRef]
96. Puyang, X.; An, M.; Xu, L.; Han, L.; Zhang, X. Protective Effect of Exogenous Spermidine on Ion and Polyamine Metabolism in Kentucky Bluegrass under Salinity Stress. *Hortic. Environ. Biotechnol.* **2016**, *57*, 11–19. [CrossRef]
97. Talaat, N.B.; Shawky, B.T.; Ibrahim, A.S. Alleviation of Drought-Induced Oxidative Stress in Maize (*Zea mays* L.) Plants by Dual Application of 24-Epibrassinolide and Spermine. *Environ. Exp. Bot.* **2015**, *113*, 47–58. [CrossRef]
98. Demirtas, Ç.; Yazgan, S.; Candogan, B.N.; Sincik, M.; Büyükcangaz, H.; Göksoy, A.T. Quality and Yield Response of Soybean (*Glycine max* L. Merrill) to Drought Stress in Sub--Humid Environment. *Afr. J. Biotechnol.* **2010**, *9*, 6873–6881.
99. Alsokari, S.S. Synergistic Effect of Kinetin and Spermine on Some Physiological Aspects of Seawater Stressed *Vigna sinensis* Plants. *Saudi J. Biol. Sci.* **2011**, *18*, 37–44. [CrossRef]
100. Miller, G.A.D.; Suzuki, N.; Ciftci-Yilmaz, S.; Mittler, R.O.N. Reactive Oxygen Species Homeostasis and Signalling during Drought and Salinity Stresses. *Plant Cell Environ.* **2010**, *33*, 453–467. [CrossRef]
101. Kaur, G.; Asthir, B. Molecular Responses to Drought Stress in Plants. *Biol. Plant.* **2017**, *61*, 201–209. [CrossRef]
102. Wang, T.; Liu, Q.; Wang, N.; Dai, J.; Lu, Q.; Jia, X.; Lin, L.; Yu, F.; Zuo, Y. Foliar Arginine Application Improves Tomato Plant Growth, Yield, and Fruit Quality via Nitrogen Accumulation. *Plant Growth Regul.* **2021**, *95*, 421–428. [CrossRef]
103. Huang, X.; Zhou, G.; Yang, W.; Wang, A.; Hu, Z.; Lin, C.; Chen, X. Drought-Inhibited Ribulose-1, 5-Bisphosphate Carboxylase Activity Is Mediated through Increased Release of Ethylene and Changes in the Ratio of Polyamines in Pakchoi. *J. Plant Physiol.* **2014**, *171*, 1392–1400. [CrossRef]
104. Nayyar, H.; Chander, S. Protective Effects of Polyamines against Oxidative Stress Induced by Water and Cold Stress in Chickpea. *J. Agron. Crop Sci.* **2004**, *190*, 355–365. [CrossRef]

105. Takahashi, T.; Kakehi, J.-I. Polyamines: Ubiquitous Polycations with Unique Roles in Growth and Stress Responses. *Ann. Bot.* **2010**, *105*, 1–6. [CrossRef] [PubMed]
106. Abdeldaym, E.A.; Sabra, M.A. The Root Endophytic Fungus *Piriformospora indica* Improves Growth Performance, Physiological Parameters and Yield of Tomato under Water Stress Condition. *Middle East J. Agric. Res.* **2018**, *7*, 1090–1101.
107. Raza, M.A.S.S.; Saleem, M.F.; Shah, G.M.; Khan, I.H.; Raza, A. Exogenous Application of Glycinebetaine and Potassium for Improving Water Relations and Grain Yield of Wheat under Drought. *J. Soil Sci. Plant Nutr.* **2014**, *14*, 348–364. [CrossRef]
108. Hussein, H.-A.A.; Alshammari, S.O.; Abd El-sadek, M.E.; Kenawy, S.K.M.; Badawy, A.A. The Promotive Effect of Putrescine on Growth, Biochemical Constituents, and Yield of Wheat (*Triticum aestivum* L.) Plants under Water Stress. *Agriculture* **2023**, *13*, 587. [CrossRef]
109. Shemi, R.; Wang, R.; Gheith, E.-S.M.S.; Hussain, H.A.; Hussain, S.; Irfan, M.; Cholidah, L.; Zhang, K.; Zhang, S.; Wang, L. Effects of Salicylic Acid, Zinc and Glycine Betaine on Morpho-physiological Growth and Yield of Maize under Drought Stress. *Sci. Rep.* **2021**, *11*, 3195. [CrossRef]
110. Sadak, M.S.; Abdelhamid, M.T.; Schmidhalter, U. Effect of Foliar Application of Amino Acids on Plant Yield and Some Physiological Parameters in Bean Plants Irrigated with Seawater. *Acta Biol. Colomb.* **2015**, *20*, 141–152. [CrossRef]
111. El-Bassiouny, H.M.; Mostafa, H.A.; El-Khawas, S.A.; Hassanein, R.A.; Khalil, S.I.; Abd El-Monem, A.A. Physiological Responses of Wheat Plant to Foliar Treatments with Arginine or Putrescine. *Aust. J. Basic Appl. Sci.* **2008**, *2*, 1390–1403.

antioxidants

MDPI

Article

Genome-Wide Identification and Drought-Responsive Functional Analysis of the *GST* Gene Family in Potato (*Solanum tuberosum* L.)

Ningfan Shi, Youfang Fan, Wei Zhang, Zhijia Zhang, Zhuanfang Pu, Zhongrun Li, Lijun Hu, Zhenzhen Bi, Panfeng Yao, Yuhui Liu, Zhen Liu, Jiangping Bai and Chao Sun *

College of Agronomy/State Key Laboratory of Aridland Crop Science, Gansu Agricultural University, Lanzhou 730070, China; shinf@st.gsau.edu.cn (N.S.); fanyf@st.gsau.edu.cn (Y.F.); 1073323120531@st.gsau.edu.cn (W.Z.); 1073323020278@st.gsau.edu.cn (Z.Z.); puzf@st.gsau.edu.cn (Z.P.); 1073324020399@st.gsau.edu.cn (Z.L.); 1073324020290@st.gsau.edu.cn (L.H.); bizz@gsau.edu.cn (Z.B.); yaopf@gsau.edu.cn (P.Y.); lyhui@gsau.edu.cn (Y.L.); liuzhen@gsau.edu.cn (Z.L.); baijp@gsau.edu.cn (J.B.)

* Correspondence: sunc@gsau.edu.cn

Abstract: Glutathione S-transferases (*GSTs*) play crucial roles in crop stress tolerance through protection against oxidative damage. In this study, we conducted genome-wide identification and expression analysis of the *GST* gene family in the autotetraploid potato cultivar Cooperative-88 (C88) using bioinformatic approaches. We identified 366 *GST* genes in the potato genome, which were classified into 10 subfamilies. Chromosomal mapping revealed that *StGSTs* were distributed across all 12 chromosomes, with 13 tandem duplication events observed in three subfamilies. Analysis of protein sequences identified 10 conserved motifs, with motif 1 potentially representing the *GST* domain. Analysis of cis-acting elements in the *StGSTs* promoter regions suggested their involvement in stress response pathways. RNA-seq analysis revealed that most *StGSTs* responded to both drought stress and DNA demethylation treatments. Quantitative PCR validation of 16 selected *StGSTs* identified four members that showed strong responses to both treatments, with distinct expression patterns between drought-tolerant (QS9) and drought-sensitive (ATL) varieties. Transient expression assays in tobacco demonstrated that these four *StGSTs* enhanced drought tolerance and may be regulated through DNA methylation pathways, though the precise mechanisms require further investigation. These findings provide a theoretical foundation for understanding the response and epigenetic regulation of potato *GST* genes under drought stress.

Keywords: potato; glutathione S-transferase; drought stress; DNA demethylation

1. Introduction

Plants exposed to abiotic stresses, including drought, salinity, and temperature fluctuations, accumulate reactive oxygen species (ROS), leading to oxidative damage of cellular membranes, biological macromolecules, and ultimately, cell death [1]. To counter these stresses, plants employ various enzymatic and non-enzymatic defense systems [2,3]. Glutathione S-transferase (GST) represents a crucial superfamily of enzymes that mitigate oxidative damage by eliminating excess ROS, thereby protecting cellular membrane integrity and protein function [4].

Glutathione S-transferase (GST, EC 2.5.1.18), an ancient enzyme superfamily present across prokaryotes and eukaryotes, primarily catalyzes the conjugation of reduced glutathione (γ-Glu-Cys-Gly, GSH) to electrophilic substrates, forming more soluble products [5–7]. Crystallographic studies reveal that *GSTs* typically exist as dimers composed of homologous or heterologous subunits, with each subunit comprising two domains: an N-terminal domain containing GSH-binding sites and a C-terminal domain housing electrophilic substrate binding sites [1,8,9]. *GSTs* exhibit remarkable functional diversity, participating in detoxification, metabolism, transport, and chelation of both endogenous and xenobiotic compounds. Additionally, *GSTs* function in ROS detoxification, neutralizing superoxide radicals, hydroxyl radicals, alkoxy radicals, hydrogen peroxide, and singlet oxygen generated in chloroplasts and mitochondria [10].

Plant *GSTs* comprise 14 distinct classes, as revealed by phylogenetic analyses of sequences from *Arabidopsis thaliana*, *Hordeum vulgare* L., *Oryza sativa* L., *Pinus tabuliformis Carr.*, *Populus trichocarpa*, and *Solanum lycopersicum*. These classes include phi (F-type), tau (U-type), theta (T-type), zeta (Z-type), lambda (L-type), hemerythrin, iota (I), ure2p, GHR (glutathione hydroquinone reductase), EF1B γ (eukaryotic translation elongation factor gamma-subunit), DHAR (dehydroascorbate reductase), TCHQD (tetrachlorohydroquinone dehalogenase), metaxin, and mPGES-2 (microsomal prostaglandin E synthase type 2) [1]. While zeta and theta classes are ubiquitous across organisms, lambda, tau, and DHAR classes are plant specific. Although phi classes were previously considered plant specific, homologous sequences have been identified in fungi, bacteria, and protists. The predominance of tau and phi classes in plants suggests their primary role in stress response.

Plant *GSTs* gained prominence due to their protective role against oxidative damage and stress conditions, facilitated by their broad substrate specificity [11,12]. Tau and phi *GSTs* exhibit GSH-dependent peroxidase activity (GPX, EC 1.11.1.9), catalyzing the conversion of hydrogen peroxide to water and organic hydroperoxides to alcohols, with concurrent formation of oxidized glutathione (GSSG) [9,11,13]. Theta class *GSTs* also demonstrate significant GSH-dependent peroxidase activity and are implicated in reducing oxidative stress metabolites, including secondary metabolites and fatty acid hydroperoxides [14–16]. DHAR class *GSTs* participate in the ascorbic acid/GSH cycle, where ascorbate peroxidase utilizes ascorbic acid to reduce hydrogen peroxide to monodehydroascorbic acid, which DHAR *GSTs* subsequently reduce to ascorbic acid using GSH as a substrate [17,18]. Beyond their catalytic functions, plant *GSTs* exhibit non-enzymatic activities through interactions with secondary metabolites, facilitating the transport of lipoproteins, anthocyanins, phenolics, and phytohormones. These interactions contribute to the degradation of harmful compounds and stress-signal transmission [10,18]. Additionally, zeta *GSTs* catalyze GSH-dependent isomerization reactions and participate in tyrosine catabolism [19].

GSTs in both plants and animals are associated with diverse cellular processes, including signal transduction, oxidative stress response, redox homeostasis regulation, apoptosis control, and secondary metabolite biosynthesis and transport [20–23]. In plants, *GSTs* are particularly crucial for drought-stress resistance, suggesting that modulation of *GST* expression and their regulatory pathways could enhance stress tolerance. Recent advances in plant epigenetics, particularly DNA methylation studies, have provided insights into drought-stress signal transduction mechanisms. While *GST* research in crops has primarily focused on gene cloning and functional validation, studies of epigenetic regulation in potato under drought stress remain limited. Previous studies documented the genome-wide characterization of the *GST* family of genes Doubled Monoploid (DM) potato [24]. The genome assembly of autotetraploid potato varieties (Collaboration 88, C88) was recently completed [25]; however, comprehensive analysis of the *GST* gene family in autotetraploid potatoes remains to be explored. This study presents a systematic analysis of autotetraploid *StGSTs*, encompassing subfamily classification, nomenclature, structural analysis, chromosomal distribution, sequence relationships, phylogenetic analysis, and expression patterns under drought stress and DNA demethylation, providing a foundation for understanding *StGST* function and developing stress-resistant potato varieties.

2. Materials and Methods

2.1. Genome-Wide Identification of Gst Genes in Potato C88

The potato C88 genome data were obtained from the Spud DB database (http://spuddb.uga.edu/, accessed on 5 August 2024). *GST* gene identification was performed using Hidden Markov Models (HMMs) for GST_C (PF00043) and GST_N (PF02798) domains downloaded from the Pfam database [26]. The C88 proteome was screened using HMMER v3.4 [27] with an E-value threshold of $\leq 1 \times 10^{-5}$. Additionally, BLASTp analysis was performed against *Arabidopsis GST* (*AtGST*) protein sequences using the same E-value cutoff [28]. Results from both approaches were combined, and redundant sequences were removed. Domain structures were verified using Pfam, yielding 366 *GST* genes. Physicochemical properties were predicted using ExPASy (http://web.expasy.org/protparam/, accessed on 5 August 2024) [29], and subcellular localization was determined using WoLF PSORT II (https://www.genscript.com/wolf-psort.html?src=leftbar, accessed on 5 August 2024) [30].

2.2. Phylogenetic Analysis and Classification

Protein sequences of 61 *AtGST* genes were retrieved from UniProt (https://www.uniprot.org/, accessed on 7 August 2024). Multiple sequence alignment of C88 and *Arabidopsis GST* proteins was performed using MUSCLE in MEGA X v10.2.6. Phylogenetic trees were constructed using the Neighbor-Joining method with 1000 bootstrap replicates, employing the Poisson correction model and pairwise deletion. Tree visualization and subfamily classification were performed using Evolview (https://evolgenius.info/helpsite/qst1.html, accessed on 7 August 2024).

2.3. Motif and Structural Analysis

Protein motif analysis of C88 *GST* family members was conducted using MEME (http://meme-suite.org/tools/meme, accessed on 10 August 2024) with a maximum motif number of 10 and default parameters [31]. Domain analysis was performed using NCBI CDD (https://www.ncbi.nlm.nih.gov/cdd/, accessed on 10 August 2024) [32]. Exon–intron structures were determined using GFF (General Feature Format) data. TBtools v2.056 were used to integrate and visualize phylogenetic relationships, motif patterns, domain structures, and gene architectures.

2.4. Chromosomal Distribution and Evolutionary Analysis

StGST genes were mapped to C88 chromosomes using TBtools. Gene density was calculated using 200-kb windows and visualized using a blue (low) to red (high) gradient, with blank regions indicating unmapped genomic segments. Comparative synteny analysis was performed between autotetraploid potato C88 and five species (*Arabidopsis thaliana* L., *Oryza sativa* L., *Solanum tuberosum DM* L., *Solanum lycopersicum* L., and *Capsicum annuum* L.).

2.5. Promoter Analysis

Promoter sequences (2000 bp upstream) of *StGST* genes were extracted using TBtools and analyzed for cis-regulatory elements using PlantCARE (http://bioinformatics.psb.ugent.be/webtools/plantcare/html/, accessed on 15 August 2024) [33]. Element distribution was visualized using TBtools.

2.6. Expression Analysis Under Drought Stress

Potato cultivars Qingshu 9 (QS9, drought resistant) and Atlantic (ATL, drought sensitive) were obtained from Gansu Agricultural University. Single-node stem segments were cultured in MS liquid medium under controlled conditions (22 ± 2 °C, 25.0–37.5 μmol $m^{-2}s^{-1}$, 16 h photoperiod). After 24 days, plantlets were treated with 200 mM mannitol and 60 μM 5-aza-deoxycytidine (5-Aza-dC, Sigma-Aldrich, St. Louis, Missouri, USA). Stem and leaf samples were collected at 0, 2, 6, 12, and 24 h post-treatment, flash-frozen in liquid nitrogen, and stored at −80 °C. Three biological replicates were maintained per treatment.

Total RNA was extracted using the Tiangen RNA extraction kit (TIANGEN Biotech (Beijing) Co., Ltd., Beijing, China), quality-assessed by electrophoresis, and quantified spectrophotometrically. cDNA was synthesized using the TOYOBO reverse transcription kit (TOYOBO CO., LTD. Life Sciences Department, Osaka, Japan). Gene-specific primers were designed with potato actin as internal reference. Primer sequence is shown in Supplementary Table S1. qPCR reactions (20 μL) contained 1 μL cDNA, 10.4 μL SYBR Premix Ex Taq (2×), 1.6 μL primers, and 7 μL ddH_2O. Amplification conditions were: 95 °C for 3 min; 35 cycles of 95 °C for 30 s, 58 °C for 30 s, and 72 °C for 30 s. Relative expression was calculated using the $2^{-\Delta\Delta Ct}$ method [34].

2.7. Protein Interaction Network and Functional Enrichment

StGST protein sequences were analyzed using the STRING database (https://string-db.org/, accessed on 17 August 2024) to predict protein–protein interactions based on *Arabidopsis* ortholog data. Interacting proteins were functionally characterized through GO and KEGG pathway enrichment analyses.

2.8. Transient Expression in Nicotiana Benthamiana

Nicotiana benthamiana plants were grown under controlled greenhouse conditions (22 ± 2 °C, 12 h photoperiod, 60% relative humidity). After 45 days, uniformly developed plants were selected and acclimated in Hoagland's solution for 2 days prior to transformation. Agrobacterium tumefaciens-mediated transformation was performed using leaf infiltration, with empty *pC2300s* vector serving as the negative control. Two days post-transformation, plants were exposed to 200 mM mannitol and 60 μM 5-Aza-dC in Hoagland's solution. Samples were collected after 6 h of treatment. Protein localization was visualized using confocal laser scanning microscopy(LSCM800 from Carl Zeiss AG located in Oberkochen, Germany).

2.9. Physiological Parameters Measurements

Physiological parameters (proline content (Pro), superoxide dismutase (SOD), peroxidase (POD), and catalase (CAT) activities) were measured with three biological replicates per assay. Each sample consisted of 0.2 g of frozen leaves, which were thoroughly homogenized in 2 mL of ice-cold 50 mM potassium phosphate buffer (pH 7.8) supplemented with 5 mM EDTA, 2 mM ascorbic acid, and 2% (*w*/*v*) polyvinyl pyrrolidone. The homogenate was centrifuged at 12,000 rpm for 10 min at 4 °C, and the supernatant was used as the crude extract for subsequent antioxidant enzyme activity assays.

The SOD activity was determined using the Nitrotetrazolium Blue chloride (NBT) method [35]. The assay medium was composed of 50 mM potassium phosphate buffer (pH 7.8), 1 mM EDTA, 130 mM methionine, 750 μM NBT, and 20 μM riboflavin. The reaction mixture was exposed to 4000 lux light for 20 min, and the absorbance was recorded at 560 nm.

The POD activity was determined using the Guaiacol method [36]. The reaction mixture comprised 200 mM potassium phosphate buffer (pH 7.0), 5.485 mM H_2O_2, 3.46 mM Guaiacol, and the crude enzyme extract. Absorbance changes were monitored at 470 nm.

The CAT activity was determined by UV absorption method [37]. The reaction mixture consisted of 50 mM potassium phosphate buffer (pH 7.0), 67 mM H_2O_2, and the crude enzyme extract. The activity of CAT was quantified by monitoring the decrease in the absorbance of hydrogen peroxide at 240 nm over a 2 min period.

Pro-content was determined using the acid ninhydrin method [38]: Cut 0.2 g of frozen leaves into small pieces and boil them in 5 mL of 3% (*w*/*v*) sulfosalicylic acid in a boiling-water bath for 10 min. Transfer 2 mL of the supernatant to a test tube and add 2 mL of 2.5% acidic ninhydrin reagent and 2 mL of glacial acetic acid. Boil the mixture in a boiling-water bath for 30 min and then cool to terminate the reaction. Extract the formed chromophore with 4 mL of toluene. Aspirate the toluene layer and centrifuge it at 3000 rpm for 5 min. Finally, measure the absorbance of the organic layer at a wavelength of 520 nm.

3. Results

3.1. Characterization of GST Gene Family in Potato C88 Genome

We identified 366 *GST* genes (*StGSTs*) in the potato C88 genome using data from the Spud DB database (http://spuddb.uga.edu/, accessed on 5 August 2024). Analysis of physicochemical properties and subcellular localization revealed substantial diversity among family members (Table S2). The predicted proteins ranged from 100 to 902 amino acids (09H1G010720–09H3G070630) with a mean length of 246 aa. Molecular weights varied from 11.304 to 103.295 kDa (05H2G041460–09H3G070630), averaging 28.36 kDa. Theoretical isoelectric points (pI) ranged from 4.12 to 9.71 (05H2G041460–08H1G022650), with most proteins exhibiting acidic characteristics (pI < 7). Subcellular localization analysis predicted that the majority of *StGSTs* localize to the cytoplasm (193/366) and chloroplasts (127/366), with smaller numbers in the nucleus (35/366), peroxisomes (6/366), mitochondrial matrix (3/366), and endoplasmic reticulum (2/366).

3.2. Phylogenetic Analysis and Classification of StGSTs

Phylogenetic analysis of 366 *StGSTs* and 61 *Arabidopsis GSTs* (*AtGSTs*) using TBtools (1000 bootstrap replicates) revealed ten distinct subfamilies: Phi, TCHQD, Theta, OMEGA, Zeta, MAPEG, DHAR, EF1Bγ, Lambda, and Tau (Figure 1). The Tau subfamily was the largest, comprising 281 members, followed by Lambda (22), Phi (15), Zeta (10), EF1Bγ, DHAR, and OMEGA (8 each), Theta (6), and MAPEG and TCHQD (4 each). Close clustering of multiple *StGSTs* with *AtGSTs* suggests high sequence homology, functional similarity, and evolutionary conservation.

3.3. Gene Structure and Conserved Motif Analysis of StGST Genes

To investigate the evolutionary diversity of the *StGST* gene family, the conserved motifs of 366 *StGST* proteins were analyzed using MEME, identifying 10 distinct conserved motifs (labeled motif 1–10) (Supplementary Figure S1A). Visualization with TBtools showed that genes within the same phylogenetic branch shared similar motifs, indicating potential functional similarities. The Tau subfamily displayed the highest number of conserved motifs, ranging from 2 to 8, with nearly all members containing motifs 1–5 and 7–8. Notably, the gene 09H3G062150 had 8 conserved motifs. In contrast, the Lambda subfamily contained 2–4 motifs, while the EF1By subfamily exhibited a unique motif (motif 10). The Omega subfamily contained only motif 6, while MAPEG (e.g., 05H2G041460) and Zeta (e.g., 01H3G123360) subfamilies contained motif 5 and motif 6, respectively. The Theta subfamily (e.g., 12H2G036350) did not exhibit any conserved motifs. Motif annotation revealed that motifs 1 and 4 were associated with GST_N domains, whereas motifs 2, 8, 9, and 10 were associated with GST_C domains. Domain analysis indicated that most Tau, Phi, and EF1By subfamily members contained both GST_N and GST_C domains, while other subfamilies had only one type of domain (Figure S1B).

Gene structure analysis revealed that members of the Tau, EF1By, Omega, TCHQD, and Phi subfamilies typically had 2–3 exons. In contrast, MAPEG and DHAR subfamily members had 5–8 exons, while Lambda and Theta subfamilies contained 8–10 exons. Most *StGST* genes contained 2 introns; however, the Theta subfamily gene 08H1G022650 contained 10 introns (Figure S1C). The distribution of exons and introns was consistent within subfamilies but varied significantly across subfamilies, suggesting structural divergence during evolution.

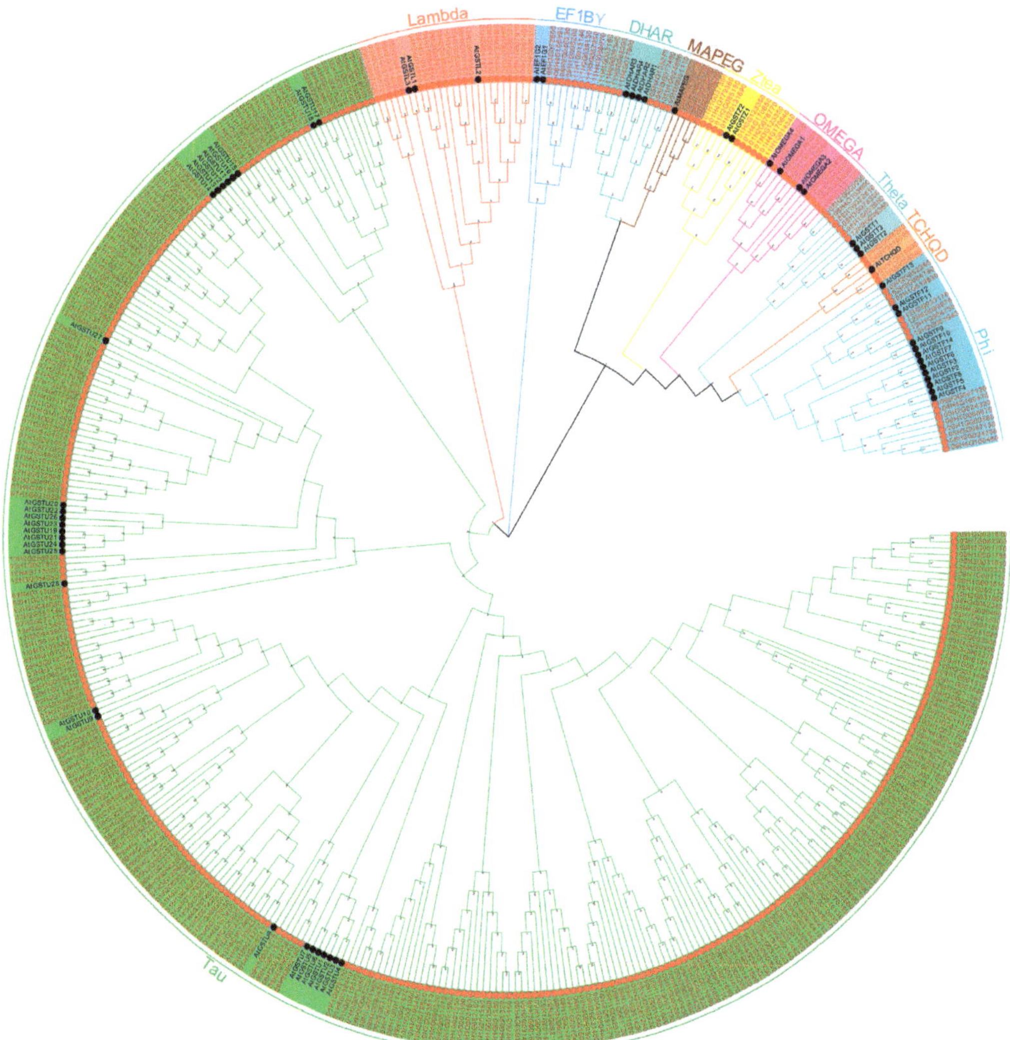

Figure 1. A phylogenetic tree of the relationship between *StGSTs* protein and *AtGSTs*, with Phi, TCHQD, Theta, OMEGA, Zeta, MAPEG, DHAR, EF1By, Lambda, Tau representing different subfamilies. Red fonts denote potato *GSTs*, and black fonts represent the *Arabidopsis GSTs*. Different colors in the background signify distinct subfamilies.

3.4. Analysis of Cis-Regulatory Elements in StGST Promoter Regions

Promoter sequences (2000 bp upstream) of *StGST* genes in potato C88 were analyzed using PlantCARE (http://bioinformatics.psb.ugent.be/webtools/plantcare/html/, accessed on 15 August 2024), and element distribution was visualized using TBtools (Figure S2). *StGST* promoters contained four major categories of cis-elements: hormone-responsive elements (ABRE, TGACG-motif, CGTGA-motif, TCA-element, and GARE-motif), stress-response elements (MBS, LTR, TC-rich repeats, and WUN-motif), metabolism-related elements (ARE, MBSI, and O2-site), and tissue-specific elements. The meristem-associated CAT-box was most abundant (15,272), followed by ABA-responsive elements (ABRE, 1126), antioxidant-responsive elements (ARE, 619), MeJA and salicylic acid-responsive elements (CGTGA-motif and TCA-element, 444), wound-responsive elements (WUN-motif, 387), and drought-responsive elements (MBS, 202) (Table S3).

3.5. Chromosomal Distribution and Duplication Analysis of StGST Genes

The 366 *StGST* genes showed non-uniform distribution across potato chromosomes, with the highest density on chromosome 9 (Figure S3). Genes within subfamilies exhibited random chromosomal distribution, predominantly clustering at chromosome termini. Using MCScanX v2022 and Circos v0.69-9 tools, we analyzed segmental and tandem duplications, identifying 13 tandem duplication events (defined as $\geq$2 genes within 200 kb): 01H3G123350/01H3G123360, 01H4G170640/01H4G170650, 03H1G027940/03H1G027950, 05H1G006990/05H1G007000, 05H1G007000/05H1G007010, 06H1G000570/06H1G000580, 07H2G048030/07H2G048040, 07H3G072090/07H3G072100, 07H4G101560/07H4G101570, 07H4G101580/07H4G101590, 09H1G001750/09H1G001760, 09H1G017760/09H1G017770, and 09H4G105940/09H4G105950. These duplications involved 25 genes across chromosomes 1, 3, 5, 6, 7, and 9, comprising nine pairs from the Tau subfamily, two from Zeta, and one from Phi. These findings suggest that tandem duplications significantly contributed to *StGST* family expansion during evolution.

3.6. Comparative Genomic Analysis of GST Proteins Across Species

To investigate evolutionary relationships of *GST* proteins, we conducted synteny analysis between autotetraploid potato C88 and five species: *Arabidopsis thaliana* (dicot), *Oryza sativa* (monocot), and three Solanaceae species (*Solanum tuberosum DM*, *Solanum lycopersicum*, and *Capsicum annuum*). The analysis revealed conserved syntenic relationships of 366 *StGSTs* across these species, with varying numbers of orthologous pairs: 23 in *Arabidopsis*, 8 in rice, 187 in potato DM, 134 in tomato, and 110 in pepper (Table S4, Figure S4). The limited synteny between autotetraploid potato C88 and rice suggests that most *GST* genes emerged after monocot–dicot divergence (Figure 2). In contrast, the extensive synteny (>100 gene pairs) between autotetraploid potato C88 and other Solanaceae species indicates high conservation within this family.

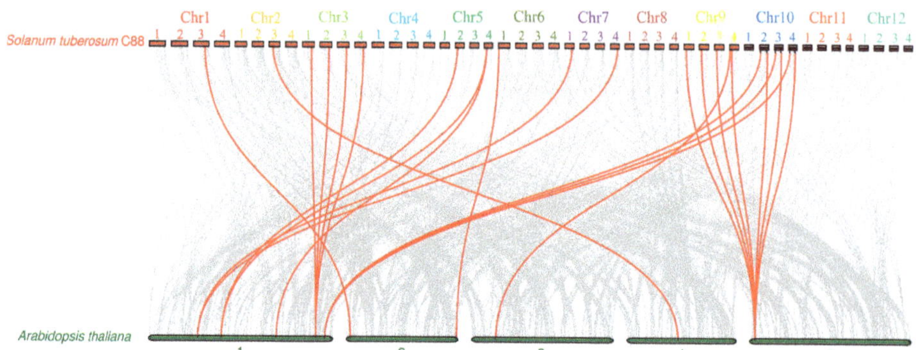

Figure 2. *Cont.*

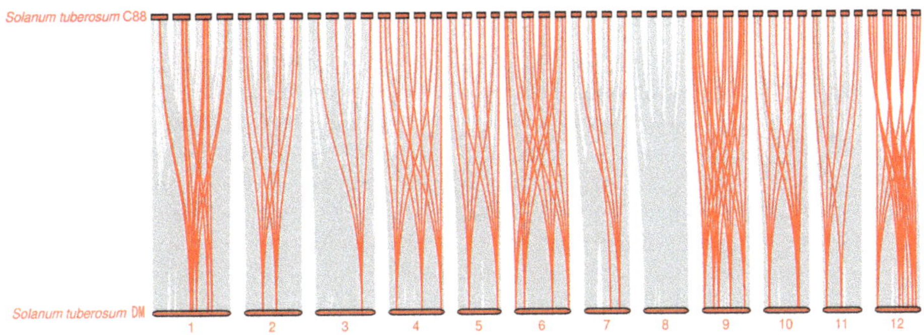

Figure 2. Syntenic relationships of *GST* genes between potato C88 and two species. Red lines highlight orthologous *GST* gene pairs between *Solanum tuberosum C88* and *Arabidopsis thaliana* and *Solanum tuberosum DM*.

3.7. Protein Interaction Network and Functional Enrichment Analysis

Protein interaction analysis of *StGST* family members using STRING database revealed a network comprising 32 nodes and 207 edges (average node degree: 12.9; clustering coefficient: 0.588; PPI p-value $< 1.0 \times 10^{-16}$). GO enrichment analysis showed significant enrichment in biological processes (BP) and molecular functions (MF), with 177 genes involved in cellular and metabolic processes. KEGG pathway analysis revealed enrichment in glutathione metabolism (265 genes) and metabolic pathways (270 genes) (Table S5). Interacting proteins showed significant GO enrichment in stimulus response, cellular metabolism, cellular processes, and toxin response. Major enriched molecular functions included glutathione transferase and antioxidant activities. Cellular components were enriched in cytosol, cytoplasm, and cellular anatomical entities. KEGG analysis identified significant enrichment in glutathione metabolism, general metabolic pathways, ascorbate/aldarate metabolism, and tyrosine metabolism (Figure 3).

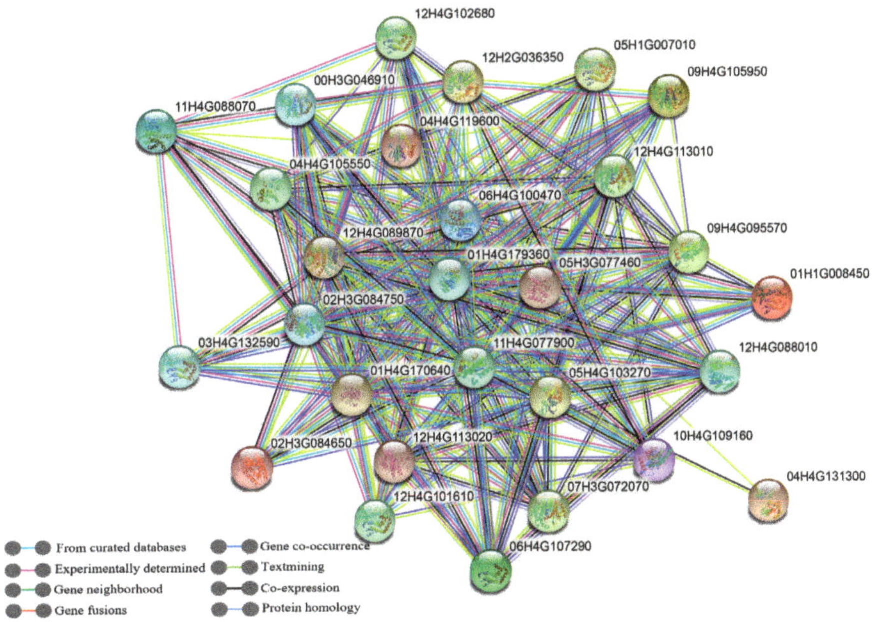

Figure 3. *StGST* protein interaction network. Nodes represent proteins and edges indicate protein–protein associations. Edge colors denote different types of interaction evidence. Minimum interaction confidence score: 0.400 (medium confidence).

3.8. Expression Analysis of StGSTs Under Drought Stress and DNA Demethylation

Previous transcriptomic analyses suggested that drought-stress responses of *StGST* genes might be regulated through DNA methylation pathways [39–41]. To investigate this relationship, we analyzed the transcriptome-wide expression (FPKM values) of all *StGST* family members in drought-tolerant (QS9) and drought-sensitive (ATL) potato varieties under both drought stress and DNA demethylation treatments. The two varieties exhibited distinct transcriptional responses to both treatments (Figure 4). From the 366 *StGST* members, we identified 16 genes that responded to both drought stress and DNA demethylation. Quantitative PCR validation demonstrated that four genes—*StGSTL4*, *StGSTT6*, *StGSTU262*, and *StGSTL21*—showed robust drought-stress responses, with expression levels exceeding a 10-fold increase. These genes also displayed significant responses to DNA demethylation treatment, with markedly different expression patterns between drought-tolerant QS9 and drought-sensitive ATL (Figure 5). Therefore, we selected these four genes as candidate genes for further study and highlighted their gene names with boxes in Figure 5. These differential responses warrant further investigation into the mechanistic relationship between drought resistance and DNA methylation regulation in these candidate genes.

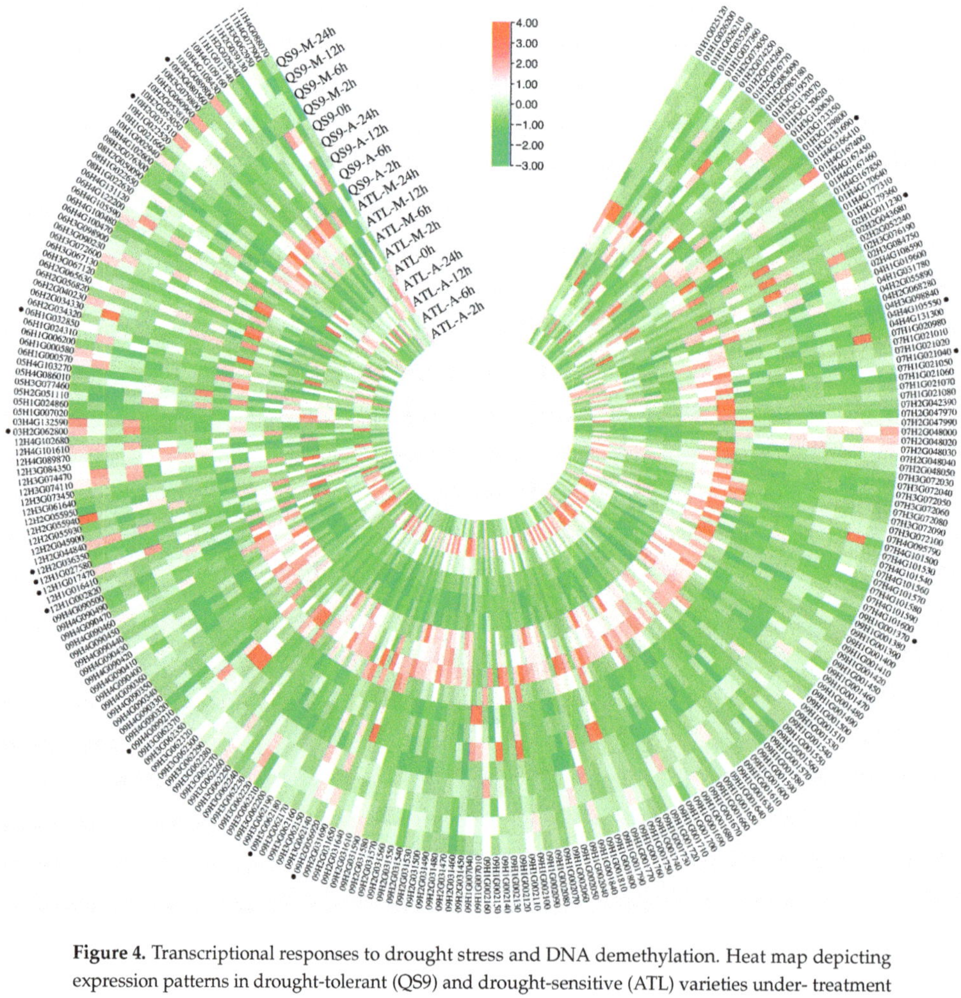

Figure 4. Transcriptional responses to drought stress and DNA demethylation. Heat map depicting expression patterns in drought-tolerant (QS9) and drought-sensitive (ATL) varieties under- treatment

(M: Drought Stress; A: DNA Demethylation) after 0, 2, 6, and 12 h. Expression levels (FPKM values) were Z-score normalized. Color scale: green represents decreased expression; red represents increased expression. Black dots indicate the 16 genes that respond to both drought treatment and DNA demethylation.

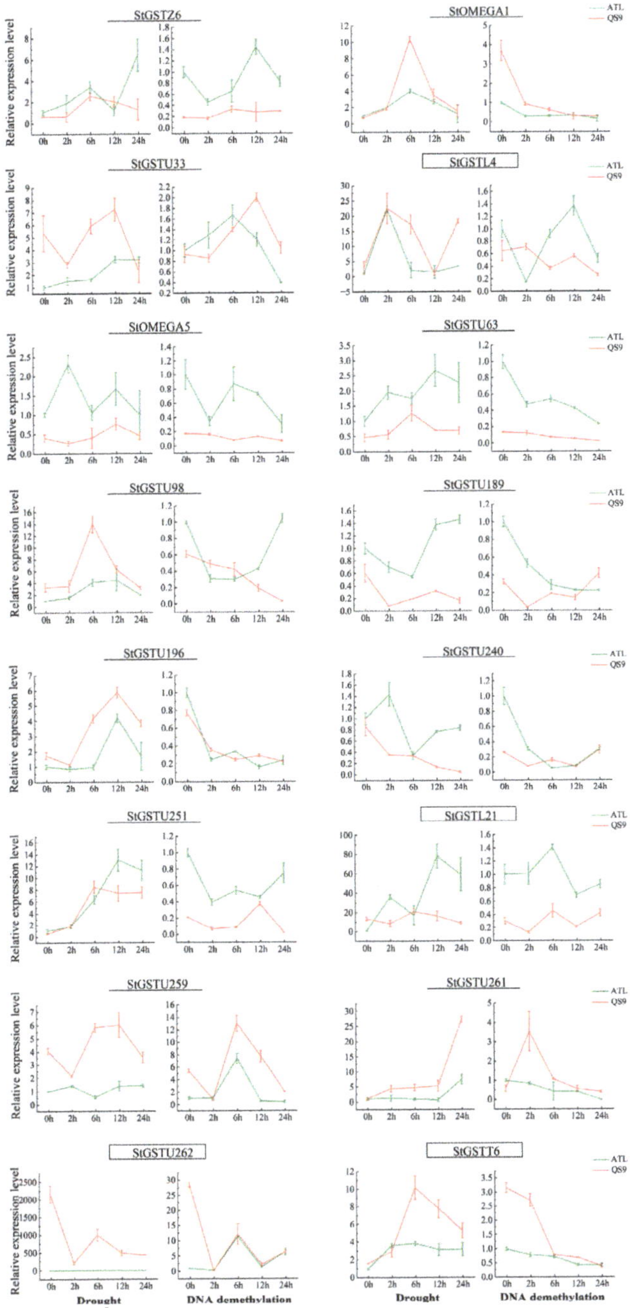

Figure 5. Gene-specific expression analysis of drought and DNA demethylation responsive *StGSTs*. Quantitative PCR analysis of 16 *StGST* genes showing dual responsiveness to drought stress and DNA demethylation in drought-tolerant (QS9) and drought-sensitive (ATL) varieties.

3.9. Subcellular Localization of Candidate Genes

To determine the subcellular localization of four candidate genes, overexpression vectors *pC2300s-StGSTL4-GFP*, *pC2300s-StGSTT6-GFP*, *pC2300s-StGSTU262-GFP*, and *pC2300s-StGSTL21-GFP*, as well as the empty vector *pC2300s-GFP*, were introduced into the lower epidermis of tobacco leaves. Protein localization was observed using laser confocal microscopy. The empty pC2300s-GFP-fusion proteins vector localized to the nucleus, cell membrane, and cytoplasm. In contrast, pC2300s-StGSTL4-GFP-fusion proteins and pC2300s-StGSTT6-GFP-fusion proteins were specifically localized to chloroplasts, while pC2300s-StGSTU262-GFP-fusion proteins and pC2300s-StGSTL21-GFP-fusion proteins were detected in the cell membrane and cytoplasm (Figure 6). These results are consistent with predictions made by gene family analysis.

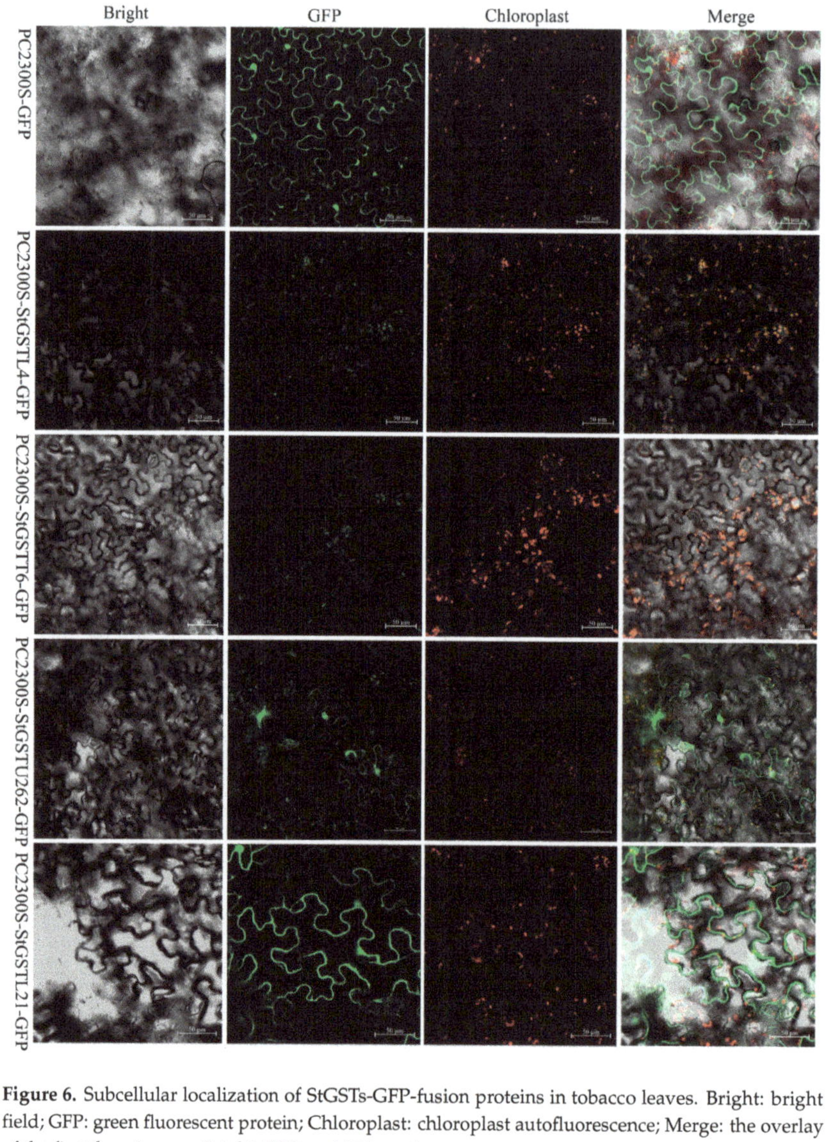

Figure 6. Subcellular localization of StGSTs-GFP-fusion proteins in tobacco leaves. Bright: bright field; GFP: green fluorescent protein; Chloroplast: chloroplast autofluorescence; Merge: the overlay of the first three images, Bright, GFP, and Chloroplast.

3.10. Candidate Genes Involved in Antioxidant Defense upon Drought Stress

To evaluate whether the four candidate genes contribute to drought resistance and DNA methylation regulation, their coding sequences (CDS) were cloned into vectors driven by the 35S promoter and transiently transformed into tobacco plants. Plants were treated with 200 mmol·L^{-1} mannitol or 60 μmol·L^{-1} DNA methylation inhibitor (5-Aza-dC) for 6 h. Stress-related physiological indices, including catalase (CAT), peroxidase (POD), superoxide dismutase (SOD) activities, and proline (Pro) content, were measured.

Under drought treatment, the CAT activity of tobacco transformed with the empty *pC2300s* vector increased by 9.39% ($p < 0.05$). In comparison, CAT activity significantly increased by 51.45% and 234.30% in *StGSTL4* and *StGSTL21*-transformed plants, respectively, but decreased by 55.71% in *StGSTU262*-transformed plants. *StGSTT6* showed no significant change ($p < 0.05$) (Figure 7A). For POD activity, the empty vector increased activity by 10.87% under drought stress. *StGSTT6* and *StGSTL21* significantly increased POD activity by 90.10% and 391.86%, respectively, while *StGSTU262* showed a 43.74% decrease ($p < 0.05$) (Figure 7B). Similarly, SOD activity increased by 24.32% in the empty vector, with significant increases of 34.20%, 105.55%, and 230.86% in *StGSTL4*, *StGSTT6*, and *StGSTL21*-transformed plants, respectively. In contrast, *StGSTU262*-transformed plants showed a significant decrease of 47.83% ($p < 0.05$) (Figure 7C). Proline content in empty-vector-transformed plants increased by 330.82% under drought stress. *StGSTL4* and *StGSTU262* significantly increased proline levels by 629.72% and 517.53%, respectively, while *StGSTL21* decreased by 80.19% ($p < 0.05$) (Figure 7D).

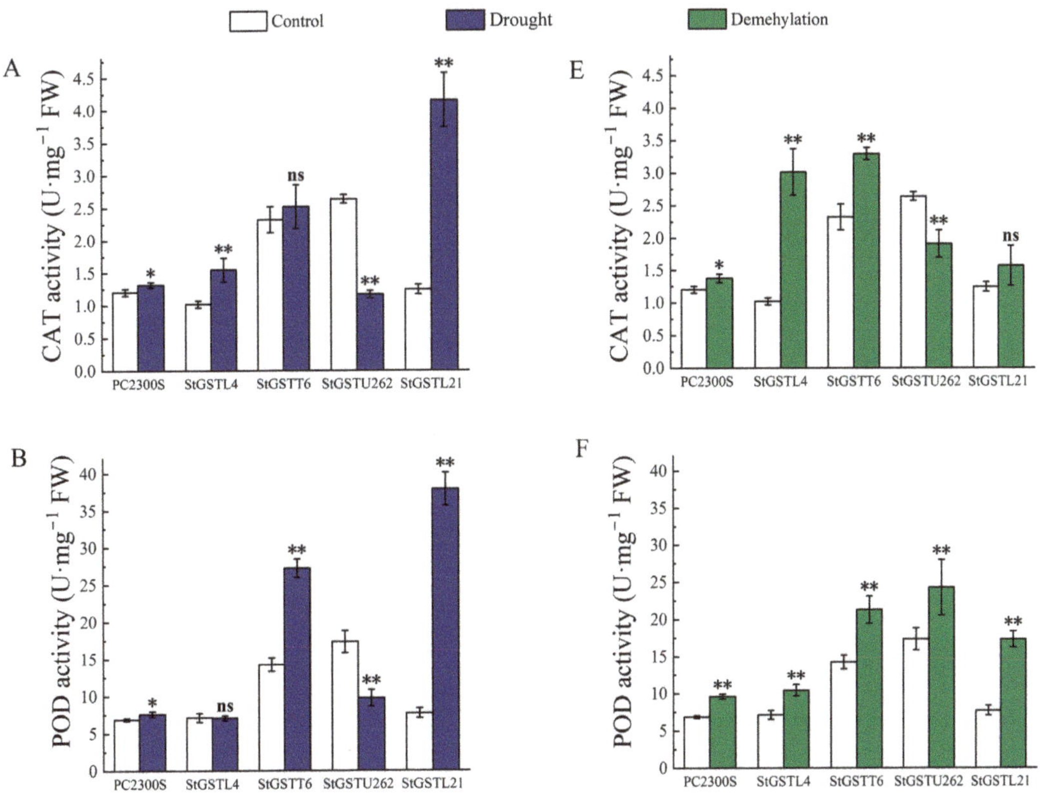

Figure 7. *Cont.*

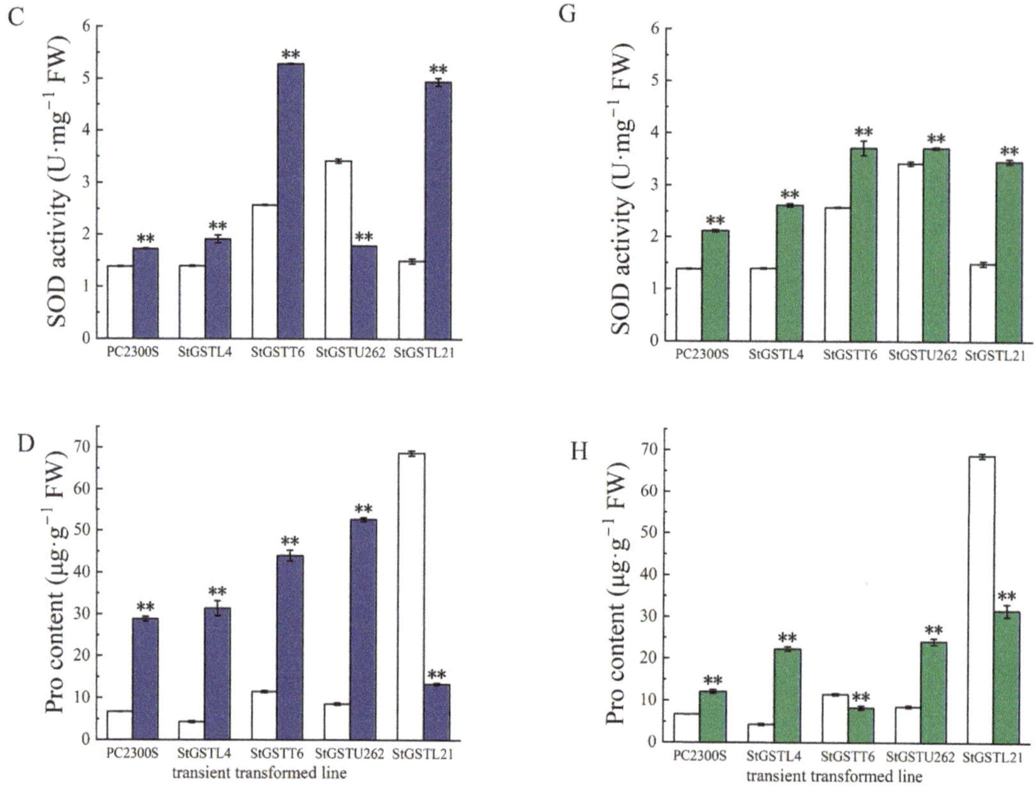

Figure 7. Physiological characteristics of transiently transformed tobacco under drought and demethylation treatments. (**A**,**E**) Catalase activity; (**B**,**F**) peroxidase activity; (**C**,**G**) superoxide dismutase activity; (**D**,**H**) proline content. Significant differences from the control are indicated, "ns" represents no significant difference. (* $p < 0.05$; ** $p < 0.01$, *t*-test).

Under DNA demethylation treatment, CAT activity increased by 14.42% in the empty vector. *StGSTL4*, *StGSTT6*, and *StGSTL21* increased CAT activity by 198.15%, 41.84%, and 26.34%, respectively, while *StGSTU262* decreased by 27.63% ($p < 0.05$) (Figure 7E). POD activity increased by 39.78% in empty vector plants. Significant increases of 46.23%, 49.42%, and 150.84% were observed in *StGSTL4*, *StGSTT6*, and *StGSTL21*-transformed plants, respectively ($p < 0.05$) (Figure 7F). For SOD activity, the empty vector showed a 52.41% increase, with *StGSTL4* and *StGSTL21* showing increases of 87.52% and 131.56%, respectively. *StGSTT6* showed no significant change, while *StGSTU262* decreased by 20.65% ($p < 0.05$) (Figure 7G). Proline content in empty vector plants increased by 80.39%. *StGSTL4* and *StGSTU262* significantly increased proline levels by 417.51% and 182.68%, respectively, while *StGSTT6* and *StGSTL21* decreased by 27.53% and 54.08% ($p < 0.05$) (Figure 7H).

Overall, the results indicate that the four candidate genes enhance antioxidant enzyme activities and osmotic regulation under drought and demethylation treatments, suggesting a positive role in drought resistance potentially mediated by DNA methylation. Further studies are required to elucidate the underlying mechanisms.

4. Discussion

4.1. Identification of Gene Family Members

The glutathione S-transferase (GST) family plays a crucial role in plant responses to abiotic stress [42,43]. Using comprehensive bioinformatics analyses, we identified 366 *StGST* gene family members in the potato C88 genome and characterized their physicochemical properties (Table S2). The encoded proteins range from 100 to 902 amino acids (09H1G010720–09H3G070630), with an average length of 246 amino acids. Their relative molecular weights vary from 11.304 to 103.295 kDa (05H2G041460–09H3G070630), averaging 28.36 kDa. These structural variations suggest evolutionary divergence in response to environmental pressures. Most *StGST* proteins exhibit a pI < 7, indicating enrichment in acidic amino acids—a characteristic shared with other dicotyledonous plants but contrasting with monocotyledonous plants, reflecting their evolutionary divergence. Through transient expression in tobacco, we demonstrated that StGSTL4 and StGSTT6 fusion proteins localize to chloroplasts, while StGSTU262 and StGSTL21 fusion proteins localize to cell membranes and cytoplasm (Figure 6), consistent with previous findings [44,45].

4.2. Gene Structure Analysis

Following the *Arabidopsis GST* classification system, we categorized the *366 StGSTs* into ten subfamilies: Phi, TCHQD, Theta, OMEGA, Zeta, MAPEG, DHAR, EF1By, Lambda, and Tau (Figure 1). Phylogenetic analysis revealed close clustering between multiple *StGSTs* and *AtGSTs*, suggesting high homology, functional similarity, and evolutionary conservation. While physicochemical properties vary considerably among *StGST* family proteins, gene structures and amino acid motifs remain relatively conserved within subfamilies (Figure S1). Gene structure analysis provides crucial insights into evolution and amplification patterns [46]. Chromosomal rearrangements and duplications can lead to exon/intron gains or losses, contributing to gene family diversification [47,48]. The identified *StGSTs* are rich in coding sequences (CDS) and untranslated regions (UTRs), with subfamily members sharing similar conserved motifs, domains, and exon–intron structures. However, these features differ substantially between subfamilies, indicating distinct functional roles [49,50]. Some subfamily members, such as 09H3G070630 (Tau) and 08H1G022650 (Theta), possess unique structural elements, suggesting functional specialization. Based on subfamily size and intron patterns, we hypothesize that the Tau class represents the ancestral *GST* subfamily in potatoes. The widespread distribution of motif 1 across *StGSTs* suggests it may be a defining feature of this gene family.

4.3. Chromosomal Distribution and Comparative Genomic Analysis

The variation in gene family size across species likely results from evolutionary processes including gene recombination, duplication events, and species-specific differentiation [51,52]. Plant *GST* family expansion primarily occurs through serial duplication of tau and phi classes, making these plant-specific subfamilies the largest [48,53,54]. Our analysis revealed that the 366 *StGST* genes are unevenly distributed across the 12 potato chromosomes, with subfamily members dispersed randomly. Chromosome 9 harbors the highest concentration of *StGST* genes, indicating chromosome-specific expression patterns (Figure S3). We identified 13 tandem duplication events involving 25 *StGST* genes on chromosomes 1, 3, 5, 6, 7, and 9. The majority of these duplicated genes belong to the Tau subfamily, suggesting coordinated protein expression and regulatory functions.

Comparative genomic analysis between potato C88 (autotetraploid) and *Arabidopsis*, rice, potato DM (Doubled Monoploid), tomato, and pepper revealed stronger syntenic relationships among dicotyledonous plants, reflecting their evolutionary history [51]. The limited collinearity between potato and rice (monocot) suggests that most *GST* genes emerged after dicot–monocot divergence. However, we observed over 100 syntenic gene pairs between autotetraploid potato C88 and its diploid relatives (potato DM, tomato, and pepper) (Figure S4), indicating high *GST* family conservation within the Solanaceae. These patterns of genome similarity and divergence illuminate species relationships and genome evolution, while suggesting strong selective pressure on the *StGST* gene family.

4.4. Promoter Cis-Regulatory Element and Protein Interaction Analyses

Promoter regions contain multiple cis-regulatory elements that serve as binding sites for transcription factors, playing crucial roles in environmental response regulation [55,56]. Analysis of the 366 *StGST* promoter sequences revealed diverse response elements related to abiotic stress and phytohormones, with meristem-specific elements being most abundant. The widespread distribution of hormone- and stress-responsive elements suggests extensive regulatory networks (Figure S2). Gene ontology (GO) enrichment analysis showed significant enrichment in biological processes (BP) and molecular functions (MF). KEGG pathway analysis revealed 265 genes significantly enriched in glutathione metabolic pathways associated with drought-stress response. GO analysis of GST-interacting proteins showed significant enrichment in stimulus response, cellular metabolic processes, cellular processes, and toxic substance response pathways, indicating coordinated roles in drought-stress resistance. These findings suggest that modulating these genes and their regulatory pathways could enhance plant drought tolerance.

4.5. Functional Characterization Under Drought Stress and DNA Demethylation

Previous studies have demonstrated the regulatory importance of *StGSTs* in abiotic stress responses. Transgenic studies have shown that overexpression of various *GST* genes (*OsGSTU4, LeGSTU2, ThGSTZ1*) enhances drought tolerance in *Arabidopsis* [57–59]. Loss of *AtGSTU17* function increases drought and salt tolerance [60]. Overexpression studies of Tau class *GSTs* from various species in tobacco have demonstrated enhanced *GST* activity and improved drought tolerance through multiple physiological mechanisms [58,61–63]. *JrGSTU1* gene in walnut (*Juglans regia* L.) can enhance GST activity to regulate the expression of other stress-related genes, thereby reducing the level of reactive oxygen species [64]. Overexpression of *MruGSTU39* in alfalfa (*Medicago sativa*) significantly improves the tolerance of transgenic plants to drought stress [65].

Abiotic stress can trigger DNA methylation changes, affecting stress-related gene expression [66]. For example, *TaGAPC1* DNA demethylation enhances drought stress response in resistant wheat varieties, while DNA methylation of *ZmNAC111* modulates maize drought tolerance [67,68]. In screening for osmotic stress-induced DNA methylation changes in maize using methylation sensitive amplification polymorphism (MSAP), a *ZmGST* was found to be induced by salt stress and undergo DNA demethylation, which resulted in up-regulation of its expression [69]. Our previous transcriptome analysis indicated potential regulation of *GST* expression through DNA demethylation under drought stress [40]. In this study, we identified 16 *StGST* genes showing significant responses to both drought and demethylation treatments, with differential expression between extremes of drought tolerance. Transient expression of four selected *StGST* genes in tobacco demonstrated that their overexpression significantly enhanced drought resistance. This enhancement operated through increased antioxidant defense and ROS-scavenging activity, suggesting positive roles in drought resistance potentially regulated through DNA methylation pathways, though specific molecular mechanisms require further investigation.

5. Conclusions

In this study, we conducted the first systematic identification of 366 *StGST* genes from the autotetraploid potato genome. These genes were classified into ten subfamilies (Phi, TCHQD, Theta, OMEGA, Zeta, MAPEG, DHAR, EF1By, Lambda, and Tau). While the physicochemical properties of the *StGST* family members showed considerable variation, their gene structures and protein motifs demonstrated high conservation within subfamilies. The *StGST* genes were distributed across all 12 potato chromosomes, with 13 tandem duplication events identified. Comparative genomic analysis revealed highest homology with other dicotyledonous plants. Cis-element analysis showed *StGST* genes particularly respond to abiotic stress. Subcellular localization studies demonstrated chloroplast localization for two StGST fusion proteins and cytoplasmic/membrane localization for two others. RNA-seq analysis revealed that most *StGSTs* responded to both drought stress and DNA demethylation treatments. Quantitative PCR validation of 16 selected *StGSTs* identified four members that showed strong responses to both treatments, with distinct expression patterns between drought-tolerant (QS9) and drought-sensitive (ATL) varieties. Functional characterization identified four *StGST* genes that responded strongly to both drought stress and DNA demethylation treatment, suggesting their positive role in stress tolerance. The underlying functional mechanisms still need to be verified and further analyzed in potato. These results provide a foundation for future investigations into the functions of *GSTs* in potato.

Supplementary Materials: The following supporting information can be downloaded at: https://www.mdpi.com/article/10.3390/antiox14020239/s1; Table S1: Primer sequences, Table S2: Characteristics of *GST* gene family members in potato (C88) genome, Table S3: Cis-element analyses of the *GST* gene promoter regions, Table S4: Covariant gene pairs, Table S5: Enrichment results statistics; Figure S1: *StGSTs* family member motif, domain, and gene structure; Figure S2: Cis-regulatory Elements in *StGST* Promoter Regions; Figure S3: Chromosomal Distribution of *StGST* Genes; Figure S4: Syntenic Relationships of *GST* Genes Between Potato C88 and Five Species.

Author Contributions: Conceptualization, N.S. and C.S.; methodology, N.S., Y.F., W.Z. and Z.Z.; software, Y.L. and Z.L. (Zhen Liu); formal analysis, Z.P., Z.L. (Zhongrun Li) and L.H.; writing—original draft preparation, N.S. and C.S.; writing—review and editing, C.S., Z.B., P.Y. and J.B.; funding, C.S. All authors have read and agreed to the published version of the manuscript.

Funding: This research was funded by the National Natural Science Foundation of China (32060502), Gansu Agricultural University Fuxi Talent Funding (Gaufx-05Y03); the State Key Laboratory of Aridland Crop Science of China (GSCS-2024-08); Gansu Provincial University Industry Support Program (2024CYZC-29).

Institutional Review Board Statement: Not applicable.

Informed Consent Statement: Not applicable.

Data Availability Statement: All data are available within the manuscript.

Conflicts of Interest: The authors declare no conflicts of interest.

References

1. Lallement, P.-A.; Meux, E.; Gualberto, J.M.; Prosper, P.; Didierjean, C.; Saul, F.; Haouz, A.; Rouhier, N.; Hecker, A. Structural and enzymatic insights into Lambda glutathione transferases from Populus trichocarpa, monomeric enzymes constituting an early divergent class specific to terrestrial plants. *Biochem. J.* **2014**, *462*, 39–52. [CrossRef] [PubMed]
2. Jens, H.; Paul, T. Reactive Oxygen Species in Phytopathogenic Fungi: Signaling, Development, and Disease. *Annu. Rev. Phytopathol.* **2011**, *49*, 369–390.
3. Michael, W.; Mikael, B.; Jaakko, K. ROS signaling loops—Production, perception, regulation. *Curr. Opin. Plant Biol.* **2013**, *16*, 575–582.

4. Chi, Y.; Cheng, Y.; Vanitha, J.; Kumar, N.; Ramamoorthy, R.; Ramachandran, S.; Jiang, S.-Y. Expansion mechanisms and functional divergence of the glutathione s-transferase family in sorghum and other higher plants. *DNA Res. Int. J. Rapid Publ. Rep. Genes Genomes* **2011**, *18*, 1–16. [CrossRef]
5. Chronopoulou, E.G.; Labrou, N.E. Glutathione transferases: Emerging multidisciplinary tools in red and green biotechnology. *Recent Pat. Biotechnol.* **2009**, *3*, 211–223. [CrossRef] [PubMed]
6. Edwards, R.; Dixon, D.P.; Cummins, I.; Brazier-Hicks, M.; Skipsey, M. New perspectives on the metabolism and detoxification of synthetic compounds in plants. In *Organic Xenobiotics and Plants: From Mode of Action to Ecophysiology*; Springer: Dordrecht, The Netherlands, 2011; pp. 125–148.
7. Cummins, I.; Wortley, D.J.; Sabbadin, F.; He, Z.; Coxon, C.R.; Straker, H.E.; Sellars, J.D.; Knight, K.; Edwards, L.; Hughes, D.; et al. Key role for a glutathione transferase in multiple-herbicide resistance in grass weeds. *Proc. Natl. Acad. Sci. USA* **2013**, *110*, 5812–5817. [CrossRef]
8. Dixon, D.P.; Cole, D.J.; Edwards, R. Dimerisation of maize glutathione transferases in recombinant bacteria. *Plant Mol. Biol.* **1999**, *40*, 997–1008. [CrossRef]
9. Axarli, I.; Dhavala, P.; Papageorgiou, A.C.; Labrou, N.E. Crystallographic and Functional Characterization of the Fluorodifen-inducible Glutathione Transferase from Glycine Max Reveals an Active Site Topography Suited for Diphenylether Herbicides and a Novel L-site. *J. Mol. Biol.* **2008**, *385*, 984–1002. [CrossRef]
10. Dixon, D.P.; Sellars, J.D.; Edwards, R. The *Arabidopsis* phi class glutathione transferase At GSTF2: Binding and regulation by biologically active heterocyclic ligands. *Biochem. J.* **2011**, *438*, 63–70. [CrossRef] [PubMed]
11. Chronopoulou, E.; Madesis, P.; Asimakopoulou, B.; Platis, D.; Tsaftaris, A.; Labrou, N.E. Catalytic and structural diversity of the fluazifop-inducible glutathione transferases from Phaseolus vulgaris. *Planta* **2012**, *235*, 1253–1269. [CrossRef] [PubMed]
12. Jain, M.; Ghanashyam, C.; Bhattacharjee, A. Comprehensive expression analysis suggests overlapping and specific roles of rice glutathione S-transferase genes during development and stress responses. *BMC Genom.* **2010**, *11*, 73. [CrossRef] [PubMed]
13. Chronopoulou, E.; Madesis, P.; Tsaftaris, A.; Labrou, N.E. Cloning and characterization of a biotic-stress-inducible glutathione transferase from Phaseolus vulgaris. *Appl. Biochem. Biotechnol.* **2014**, *172*, 595–609. [CrossRef] [PubMed]
14. Wagner, U.; Edwards, R.; Dixon, D.P.; Mauch, F. Probing the diversity of the *Arabidopsis* glutathione S-transferase gene family. *Plant Mol. Biol.* **2002**, *49*, 515–532. [CrossRef] [PubMed]
15. Lan, T.; Yang, Z.-L.; Yang, X.; Liu, Y.-J.; Wang, X.-R.; Zeng, Q.-Y. Extensive functional diversification of the Populus glutathione S-transferase supergene family. *Plant Cell* **2009**, *21*, 3749–3766. [CrossRef]
16. Pang, S.; Duan, L.; Liu, Z.; Song, X.; Li, X.; Wang, C. Co-induction of a glutathione-S-transferase, a glutathione transporter and an ABC transporter in maize by xenobiotics. *PLoS ONE* **2017**, *7*, e40712. [CrossRef]
17. Rahantaniaina, M.S.; Tuzet, A.; Mhamdi, A.; Noctor, G. Missing links in understanding redox signaling via thiol/disulfide modulation: How is glutathione oxidized in plants? *Front. Plant Sci.* **2013**, *4*, 477. [CrossRef]
18. Dixon, D.P.; Lapthorn, A.; Madesis, P.; Mudd, E.A.; Day, A.; Edwards, R. Binding and glutathione conjugation of porphyrinogens by plant glutathione transferases. *J. Biol. Chem.* **2008**, *283*, 20268–20276. [CrossRef] [PubMed]
19. Thom, R.; Dixon, D.P.; Edwards, R.; Cole, D.J.; Lapthorn, A.J. The structure of a zeta class glutathione S-transferase from *Arabidopsis thaliana*: Characterisation of a GST with novel active-site architecture and a putative role in tyrosine catabolism. *J. Mol. Biol.* **2001**, *308*, 949–962. [CrossRef] [PubMed]
20. Norrgård, M.A.; Hellman, U.; Mannervik, B. Cys-X scanning for expansion of active-site residues and modulation of catalytic functions in a glutathione transferase. *J. Biol. Chem.* **2011**, *286*, 16871–16878. [CrossRef]
21. Oakley, A. Glutathione transferases: A structural perspective. *Drug Metab. Rev.* **2011**, *43*, 138–151. [CrossRef]
22. Singh, S. Cytoprotective and regulatory functions of glutathione S-transferases in cancer cell proliferation and cell death. *Cancer Chemother. Pharmacol.* **2015**, *75*, 1–15. [CrossRef] [PubMed]
23. Singhal, S.S.; Singh, S.P.; Singhal, P.; Horne, D.; Singhal, J.; Awasthi, S. Antioxidant role of glutathione S-transferases: 4-Hydroxynonenal, a key molecule in stress-mediated signaling. *Toxicol. Appl. Pharmacol.* **2015**, *289*, 361–370. [CrossRef] [PubMed]
24. Islam, M.S.; Choudhury, M.; Majlish, A.-N.K.; Islam, T.; Ghosh, A. Comprehensive genome-wide analysis of Glutathione S-transferase gene family in potato (*Solanum tuberosum* L.) and their expression profiling in various anatomical tissues and perturbation conditions. *Gene* **2018**, *639*, 149–162. [CrossRef] [PubMed]
25. Bao, Z.; Li, C.; Li, G.; Wang, P.; Peng, Z.; Cheng, L.; Li, H.; Zhang, Z.; Li, Y.; Huang, W.; et al. Genome architecture and tetrasomic inheritance of autotetraploid potato. *Mol. Plant* **2023**, *15*, 1211–1226. [CrossRef] [PubMed]
26. Mistry, J.; Chuguransky, S.; Williams, L.; Qureshi, M.; Salazar, G.A.; Sonnhammer, E.L.L.; Tosatto, S.C.; Paladin, L.; Raj, S.; Richardson, L.J.; et al. Pfam: The protein families database in 2021. *Nucleic Acids Res.* **2020**, *49*, D412–D419. [CrossRef] [PubMed]
27. Pagnuco, I.A.; Revuelta, M.V.; Bondino, H.G.; Brun, M.; Ten Have, A. HMMER Cut-off Threshold Tool (HMMERCTTER): Supervised classification of superfamily protein sequences with a reliable cut-off threshold. *PLoS ONE* **2018**, *13*, e0193757. [CrossRef] [PubMed]

28. Chen, C.; Chen, H.; Zhang, Y.; Thomas, H.R.; Frank, M.H.; He, Y.; Xia, R. TBtools: An integrative toolkit developed for interactive analyses of big biological data. *Mol. Plant* **2020**, *13*, 1194–1202. [CrossRef] [PubMed]
29. Duvaud, S.; Gabella, C.; Lisacek, F.; Stockinger, H.; Ioannidis, V.; Durinx, C. Expasy, the Swiss Bioinformatics Resource Portal, as designed by its users. *Nucleic Acids Res.* **2021**, *49*, W216–W227. [CrossRef] [PubMed]
30. Horton, P.; Park, K.-J.; Obayashi, T.; Fujita, N.; Harada, H.; Adams-Collier, C.J.; Nakai, K. WoLF PSORT: Protein localization predictor. *Nucleic Acids Res.* **2007**, *35*, W585–W587. [CrossRef] [PubMed]
31. Bailey, T.L.; Boden, M.; Buske, F.A.; Frith, M.; Grant, C.E.; Clementi, L.; Ren, J.; Li, W.W.; Noble, W.S. MEME SUITE: Tools for motif discovery and searching. *Nucleic Acids Res.* **2009**, *37*, W202–W208. [CrossRef]
32. Marchler-Bauer, A.; Derbyshire, M.K.; Gonzales, N.R.; Lu, S.; Chitsaz, F.; Geer, L.Y.; Geer, R.C.; He, J.; Gwadz, M.; Hurwitz, D.I.; et al. CDD: NCBI's conserved domain database. *Nucleic Acids Res.* **2015**, *43*, D222–D226. [CrossRef]
33. Lescot, M.; Déhais, P.; Thijs, G.; Marchal, K.; Moreau, Y.; Van de Peer, Y.; Rouzé, P.; Rombauts, S. PlantCARE, a database of plant cis-acting regulatory elements and a portal to tools for in silico analysis of promoter sequences. *Nucleic Acids Res.* **2002**, *30*, 325–327. [CrossRef] [PubMed]
34. Livak, K.J.; Schmittgen, T.D. Analysis of relative gene expression data using real-time quantitative PCR and the 2(-Delta Delta C(T)) Method. *Methods* **2001**, *25*, 402–408. [CrossRef]
35. Beauchamp, C.; Fridovich, I. Superoxide dismutase: Improved assays and an assay applicable to acrylamide gels. *Anal. Biochem.* **1971**, *44*, 276–287. [CrossRef] [PubMed]
36. Upadhyaya, A.; Sankhla, D.; Davis, T.D.; Sankhla, N.; Smith, B. Effect of paclobutrazol on the activities of some enzymes of activated oxygen metabolism and lipid peroxidation in senescing soybean leaves. *J. Plant Physiol.* **1985**, *121*, 453–461. [CrossRef]
37. Maehly, A. The assay of catalases and peroxidases. In *Methods of Biochemical Analysis*; Willey: Hoboken, NJ, USA, 1954; pp. 357–424.
38. Xue, L.; Ren, H.; Long, W.; Leng, X.; Wang, J.; Yao, X.; Li, S. Ecophysiological responses of calcicole *Cyclobalanopsis* gliauca (Thunb.) oerst. to drought stress and calcium supply. *Forests* **2018**, *9*, 667. [CrossRef]
39. Li, P.; Bi, Z.; Sun, C.; Qin, T.; Liang, J.; Wang, Y.; Xu, D.; Liu, Y.; Zhang, J.; Bai, J. Key genes mining of DNA methylation involved in regulating drought stress response in potato. *Acta Agron. Sin.* **2021**, *47*, 599–612. [CrossRef]
40. Li, P.; Bi, Z.; Liang, J.; Sun, C.; Zhang, J.; Bai, J. DNA methylation involved in regulating drought stress response of potato. *Acta Agron. Sin.* **2019**, *45*, 1595–1603.
41. Zhenzhen, B.; Yihao, W.; Pengcheng, L.; Chengju, L.; Yindu, L.; Chao, S.; Panfeng, Y.; Yuhui, L.; Zhen, L.; Jiangping, B. Transcriptomics Analysis Reveals a More Refined Regulation Mechanism of Methylation in a Drought-Tolerant Variety of Potato. *Genes* **2022**, *13*, 2260. [CrossRef] [PubMed]
42. Labrou, N.E.; Papageorgiou, A.C.; Pavli, O.; Flemetakis, E. Plant GSTome: Structure and functional role in xenome network and plant stress response. *Curr. Opin. Biotechnol.* **2015**, *32*, 186–194. [CrossRef] [PubMed]
43. Nianiou-Obeidat, I.; Madesis, P.; Kissoudis, C.; Voulgari, G.; Chronopoulou, E.; Tsaftaris, A.; Labrou, N.E. Plant glutathione transferase-mediated stress tolerance: Functions and biotechnological applications. *Plant Cell Rep.* **2017**, *36*, 791–805. [CrossRef] [PubMed]
44. YuWen, Z.; ChuKun, W.; XiaoYu, H.; DaGang, H. Genome-Wide Analysis of the Glutathione S-Transferase (GST) Genes and Functional Identification of MdGSTU12 Reveals the Involvement in the Regulation of Anthocyanin Accumulation in Apple. *Genes* **2021**, *12*, 1733. [CrossRef] [PubMed]
45. Gu, K.-D.; Wang, C.-K.; Hu, D.-G.; Hao, Y.-J. How do anthocyanins paint our horticultural products? *Sci. Hortic.* **2019**, *249*, 257–262. [CrossRef]
46. Frova, C. Glutathione transferases in the genomics era: New insights and perspectives. *Biomol. Eng.* **2006**, *23*, 149–169. [CrossRef] [PubMed]
47. Xu, G.; Guo, C.; Shan, H.; Kong, H. Divergence of duplicate genes in exon-intron structure. *Proc. Natl. Acad. Sci. USA* **2012**, *109*, 1187–1192. [CrossRef]
48. Islam, S.; Sajib, S.D.; Jui, Z.S.; Arabia, S.; Islam, T.; Ghosh, A. Genome-wide identification of glutathione S-transferase gene family in pepper, its classification, and expression profiling under different anatomical and environmental conditions. *Sci. Rep.* **2019**, *9*, 9101. [CrossRef] [PubMed]
49. Hu, X.; Zheng, T.; Chen, W.; Duan, H.; Yuan, Z.; An, J.; Zhang, H.; Liu, X. Genome-wide identification and expression analysis of the GST gene family of Betula platyphylla. *J. For. Res.* **2024**, *35*, 123. [CrossRef]
50. Yuan, S.; Yang, C.; Zheng, B.; Ni, J.; Zhou, K.; Qian, M.; Wu, H. Genome-Wide Identification and Expression Analysis of GST Genes during Light-Induced Anthocyanin Biosynthesis in Mango (*Mangifera indica* L.). *Plants* **2024**, *13*, 2726. [CrossRef]
51. Vision, T.J.; Brown, D.G.; Tanksley, S.D. The Origins of Genomic Duplications in *Arabidopsis*. *Science* **2000**, *290*, 2114–2117. [CrossRef] [PubMed]

52. Huang, F.; Ye, X.; Wang, Z.; Ding, Y.; Cai, X.; Yu, L.; Waseem, M.; Abbas, F.; Ashraf, U.; Chen, X.; et al. The prohibitins (PHB) gene family in tomato: Bioinformatic identification and expression analysis under abiotic and phytohormone stresses. *GM Crops Food* **2021**, *12*, 530–550. [CrossRef]
53. Islam, S.; Rahman, I.A.; Islam, T.; Ghosh, A. Genome-wide identification and expression analysis of glutathione S-transferase gene family in tomato: Gaining an insight to their physiological and stress-specific roles. *PLoS ONE* **2017**, *12*, e0187504. [CrossRef] [PubMed]
54. Ma, L.; Zhang, Y.; Meng, Q.; Shi, F.; Liu, J.; Li, Y. Molecular cloning, identification of GSTs family in sunflower and their regulatory roles in biotic and abiotic stress. *World J. Microbiol. Biotechnol.* **2018**, *34*, 109. [CrossRef] [PubMed]
55. Hernandez-Garcia, C.M.; Finer, J.J. Identification and validation of promoters and cis -acting regulatory elements. *Plant Sci.* **2014**, *217*, 109–119. [CrossRef] [PubMed]
56. Chen, L.-Y.; Xin, Y.; Wai, C.M.; Liu, J.; Ming, R. The role of cis-elements in the evolution of crassulacean acid metabolism photosynthesis. *Hortic. Res.* **2020**, *7*, 5. [CrossRef]
57. Sharma, R.; Sahoo, A.; Devendran, R.; Jain, M. Over-expression of a rice tau class glutathione s-transferase gene improves tolerance to salinity and oxidative stresses in *Arabidopsis*. *PLoS ONE* **2014**, *9*, e92900. [CrossRef] [PubMed]
58. Xu, J.; Xing, X.-J.; Tian, Y.-S.; Peng, R.-H.; Xue, Y.; Zhao, W.; Yao, Q.-H. Transgenic *Arabidopsis* Plants Expressing Tomato Glutathione S-Transferase Showed Enhanced Resistance to Salt and Drought Stress. *PLoS ONE* **2017**, *10*, e0136960. [CrossRef] [PubMed]
59. Yang, G.; Wang, Y.; Xia, D.; Gao, C.; Wang, C.; Yang, C. Overexpression of a GST gene (ThGSTZ1) from Tamarix hispida improves drought and salinity tolerance by enhancing the ability to scavenge reactive oxygen species. *Plant Cell Tissue Organ Cult.* **2014**, *117*, 99–112. [CrossRef]
60. Chen, J.-H.; Jiang, H.-W.; Hsieh, E.-J.; Chen, H.-Y.; Chien, C.-T.; Hsieh, H.-L.; Lin, T.-P. Drought and Salt Stress Tolerance of an *Arabidopsis* Glutathione S-Transferase U17 Knockout Mutant Are Attributed to the Combined Effect of Glutathione and Abscisic Acid. *Plant Physiol.* **2012**, *158*, 340–351. [CrossRef] [PubMed]
61. George, S.; Venkataraman, G.; Parida, A. A chloroplast-localized and auxin-induced glutathione S-transferase from phreatophyte *Prosopis juliflora* confer drought tolerance on tobacco. *J. Plant Physiol.* **2009**, *167*, 311–318. [CrossRef]
62. Ji, W.; Zhu, Y.; Li, Y.; Yang, L.; Zhao, X.; Cai, H.; Bai, X. Over-expression of a glutathione S-transferase gene, GsGST, from wild soybean (Glycine soja) enhances drought and salt tolerance in transgenic tobacco. *Biotechnol. Lett.* **2010**, *32*, 1173–1179. [CrossRef]
63. Liu, D.; Liu, Y.; Rao, J.; Wang, G.; Li, H.; Ge, F.; Chen, C. Overexpression of the Glutathione S-Transferase Gene from *Pyrus pyrifolia* Fruit Improves Tolerance to Abiotic Stress in Transgenic Tobacco Plants. *Mol. Biol.* **2013**, *47*, 515–523. [CrossRef]
64. Yang, G.; Xu, Z.; Peng, S.; Sun, Y.; Jia, C.; Zhai, M. In planta characterization of a tau class glutathione S-transferase gene from *Juglans regia* (JrGSTTau1) involved in chilling tolerance. *Plant Cell Rep.* **2016**, *35*, 681–692. [CrossRef] [PubMed]
65. Wang, T.; Zhang, D.; Chen, L.; Wang, J.; Zhang, W.-H. Genome-wide analysis of the Glutathione S-Transferase family in wild *Medicago ruthenica* and drought-tolerant breeding application of MruGSTU39 gene in cultivated alfalfa. *Theor. Appl. Genet.* **2021**, *135*, 863–864. [CrossRef]
66. Lu, X.; Liu, Y.; Xu, J.; Liu, X.; Chi, Y.; Li, R.; Mo, L.; Shi, L.; Liang, S.; Yu, W.; et al. Recent progress of molecular mechanisms of DNA methylation in plant response to abiotic stress. *Environ. Exp. Bot.* **2024**, *218*, 105599. [CrossRef]
67. Fei, Y.; Xue, Y.; Du, P.; Yang, S.; Deng, X. Expression analysis and promoter methylation under osmotic and salinity stress of TaGAPC1 in wheat (*Triticum aestivum* L). *Protoplasma* **2017**, *254*, 987–996. [CrossRef]
68. Mao, H.; Wang, H.; Liu, S.; Li, Z.; Yang, X.; Yan, J.; Li, J.; Tran, L.S.P.; Qin, F. A transposable element in a NAC gene is associated with drought tolerance in maize seedlings. *Nat. Commun.* **2015**, *6*, 8326. [CrossRef] [PubMed]
69. Tan, M.P. Analysis of DNA methylation of maize in response to osmotic and salt stress based on methylation-sensitive amplified polymorphism. *Plant Physiol. Biochem.* **2010**, *48*, 21–26. [CrossRef] [PubMed]

antioxidants

MDPI

Article

Deciphering Antioxidant Responses in Tomato Autografts

Carlos Frey [1,2], Andrés Hernández-Barriuso [1], José Luis Acebes [1,3] and Antonio Encina [1,2,*]

[1] Área de Fisiología Vegetal, Facultad de Ciencias Biológicas y Ambientales, Universidad de León, 24007 León, Spain; cfred@unileon.es (C.F.); ahernb00@estudiantes.unileon.es (A.H.-B.); jl.acebes@unileon.es (J.L.A.)

[2] Instituto de Biología Molecular, Genómica y Proteómica, Universidad de León, 24007 León, Spain

[3] Instituto de la Viña y el Vino, Universidad de León, 24009 León, Spain

* Correspondence: a.encina@unileon.es

Abstract: Grafting is a horticultural technique that involves a healing process that requires grafted plants to develop physiological responses to overcome oxidative stress. In this study, oxidative damage, total antioxidant capacity and antioxidant enzymatic activities were analysed in functional and non-functional tomato autografts for eight days after grafting, considering scion and rootstock tissues separately. The results showed that oxidative damage, measured as lipid peroxidation, was controlled, especially in functional grafts. Scion tissues showed significant increases in total antioxidant capacity and activities of key antioxidant enzymes, including superoxide dismutase and catalase. Non-functional grafts showed elevated levels of class III peroxidase, potentially related to defensive suberisation and lignification. Principal component analysis revealed that antioxidant activities correlated dynamically with grafting stages, highlighting their critical role in stress mitigation. These results suggest that an efficient and asymmetric antioxidant response is essential for successful graft healing in tomato plants. Furthermore, different patterns in non-functional grafts underline the importance of redox balance in determining graft success.

Keywords: antioxidant capacity; catalase; lipid peroxidation; oxidative stress; redox balance; *Solanum lycopersicum*

1. Introduction

Grafting is a horticultural method in which two parts of a plant, typically a stem (called a scion) and the root-bearing part (called a rootstock), are cut and joined to create a single plant with combined characteristics. This technique allows desirable traits, such as disease resistance from the rootstock and superior fruit quality from the scion, to be combined [1]. Grafting is widely used in woody crops and also in herbaceous crops, such as cucurbits (melon, watermelon and cucumber) and solanaceous (pepper, eggplant and tomato) [2–4]. Particularly, tomato plants are grafted onto resistant rootstocks to improve their resilience under saline conditions [5,6], drought [7,8] and some diseases [9,10], among other uses.

When two stems are joined under the right conditions, the graft healing process begins, and a series of changes modulate the cut zone to form a solid graft union, a bridge between the scion and the rootstock. Previous studies have shown that graft union formation is a dynamic and continuous process involving a complex set of physiological and molecular mechanisms for successful rootstock–scion fusion [11–16]. However, despite their widespread use, some of the underlying mechanisms of graft formation remain the focus of research.

https://doi.org/10.3390/antiox14020234

Grafting involves a stressful recovery period for the plant, and one of the challenges faced by many nurseries specialising in this production is that a significant proportion of grafted plants fail during the recovery period of healing. How plants overcome and resist this highly stressful event is complex. Cutting causes a severe injury to the stem, characterised by tissue disruption and the triggering of a wound-healing response in the close vicinity of the cutting edge. Plants have evolved mechanisms that allow them, firstly, to sense the wound-induced damage and, secondly, to activate wound protection mechanisms [17]. Much is known about the molecular mechanisms underlying wound-induced responses, including the perception of cell integrity via cell wall sensing and wound signalling via ion fluxes (Ca^{2+} or H^+), the activation of MAPKs phosphorylation cascades, the de novo synthesis of hormones such as jasmonic acid and the production of reactive oxygen species (ROS) [17,18]. Indeed, wound response and oxidative stress-related genes are significantly correlated with grafting at very early time points (1 h after grafting, 1 HAG) [12]. Furthermore, the discontinuity created between the scion and the rootstock during grafting leads to a disruption in sap transport along the stem, drastically reducing water availability to the scion. As a result, the stomata close to optimise water use [19–21], leading to a decrease in CO_2 content and detrimental effects on the photosynthetic process [20,22]. The excess photon energy beyond the photosynthetic capacity of the plant leads to the inhibition and dysfunction of photosystem II and photosystem I, disrupting the electron flow path and potentially triggering the generation of highly toxic ROS within the plant [23–25]. The healing conditions (high relative humidity and low light intensity) in the recovery chambers help to mitigate the effects of water stress but are not sufficient for the plant to show no change in turgor. Grafted plants are then exposed to water stress, which has been shown to induce ROS accumulation [26,27]. In addition, increased accumulation of hydrogen peroxide (H_2O_2) leading to oxidative damage and increased activity of some antioxidant enzymes, such as peroxidase, associated with increased accumulation of phenolic compounds have been reported in tomato incompatible plants [25].

It is hypothesised that the stress during graft healing (mainly caused by wound syndrome and the onset of water deficit in the scion) converges in a generalised oxidative stress, particularly at the site of graft union [14,28]. To buffer this, the plant promotes an antioxidant response to prevent oxidative damage throughout the healing process. This response can be mediated by antioxidant molecules, antioxidant enzymes or a combination of both (Figure 1). The most prominent antioxidant molecules are ascorbic acid and glutathione. Both are part of the ascorbate-glutathione cycle, a biochemical process known to buffer ROS [29,30]. It is regulated by enzymes including ascorbate peroxidase (A-POX) and glutathione reductase (GR). Superoxide dismutase (SOD), catalase (CAT), class III peroxidase (CIII-POX) and malate dehydrogenase (MD) are other key enzymes in the regulation of redox homeostasis.

Considering the above, the aim of this study was to use tomato autografts to shed light on the following: (1) Whether oxidative damage occurs in the tissues adjacent to the cut area during graft healing. To this purpose, the levels of lipid peroxidation (LIPOX) in the stem tissues involved in the grafting process were monitored throughout the graft healing period (0–8 days after grafting, DAG). (2) The functional dynamics of the antioxidant machinery in the union tissues during the graft healing process. For this purpose, the total antioxidant capacity (TAC) and the activity of several antioxidant enzymes (SOD, CAT, CIII-POX, A-POX and GR) were evaluated in the same tissues as in point 1. (3) Whether there are differences in oxidative damage and antioxidant response between functional and non-functional grafts.

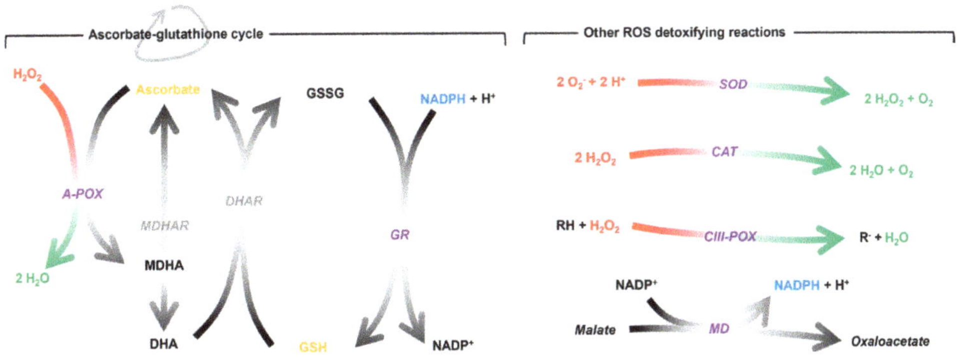

Figure 1. Schematic representation of the antioxidant mechanisms mentioned above and the reactions catalysed by the enzymes. The stoichiometry of the reactions catalysed by the grey enzymes has been omitted for simplicity. The enzymatic activities studied in this work are shown in **purple**. A-POX, ascorbate peroxidase; CAT, catalase; CIII-POX, class III peroxidase; DHA, dehydroascorbate; DHAR, dehydroascorbate reductase; GR, glutathione reductase; GSH, reduced glutathione; GSSG, oxidised glutathione; MD, malate dehydrogenase; MDHA, monodehydroascorbate; MDHAR, monodehydroascorbate reductase; NADPH, nicotinamide adenine dinucleotide phosphate; SOD, superoxide dismutase.

2. Materials and Methods

2.1. Plant Growth

Seeds of tomato (*Solanum lycopersicum* "Minibel", Mascarell Semillas S.L., Benissoda, Spain) were imbibed for 24 h and sown in individual 200 mL containers with a moist substrate based on peat and wood fibres. The containers were covered with a plastic film to keep the humidity for four days and placed in a growth chamber at 24 $\pm$ 1 °C under light ($\approx$40 µmol m^{-2} s^{-1}) with 16/8 photoperiodic conditions and over 50–60% relative humidity. The substrate was irrigated twice a week to field capacity with Hoagland solution.

2.2. Grafting Technique and Healing Conditions

When the stems were over 4–5 mm in diameter (about one month old), a splice autograft with a 45° angle was made under the cotyledonary leaves: scion and rootstock were brought together, realigning vascular bundles, and tied with a graft clip (Toogoo®). A total of 120 grafted plants were grouped in batches of 12 plants and distributed in 10 trays (56 cm $\times$ 42 cm $\times$ 8 cm). The trays were placed under a plastic cover constituting the graft healing chamber and maintained in a growth chamber at 24 $\pm$ 1 °C. During the initial grafting phase, high relative humidity conditions were maintained (90–100%) to reduce wilting. Light conditions and relative humidity were modified during graft healing as described in [21].

2.3. Sampling

On the different sampling days (0–8 DAG), five plants were randomly selected. From each plant, a 5 mm long stem segment was collected above and below the cut line (Figure 2). The 0 DAG samples corresponded to 5 mm long stem segments from non-grafted plants taken at a similar position to the grafting site on grafted plants. Non-functional grafts were identified by careful visual inspection, including the signs of lack of graft union and scion wilting.

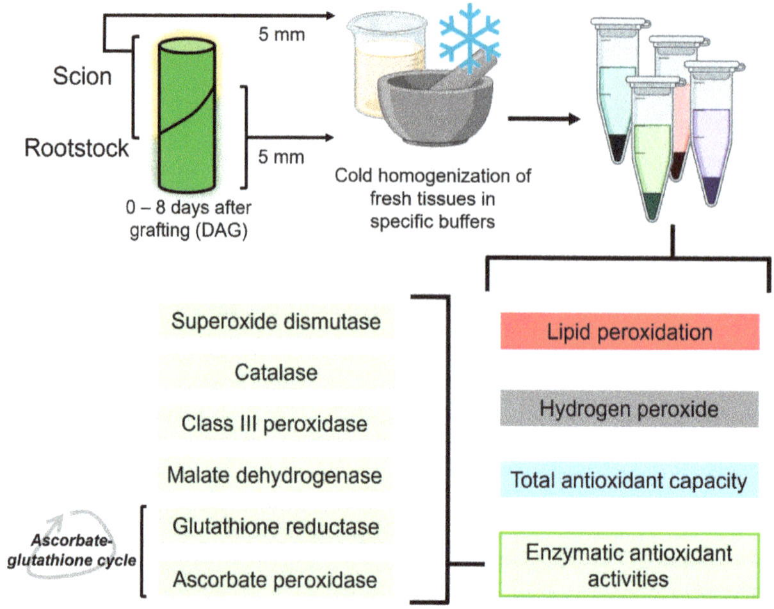

Figure 2. Schematic view of the experimental procedure, showing all the variables determined from the sample extracts. Created using BioRender.

Stem segments were homogenised at 4 °C (1 g of fresh weight/6 mL) using a cold mortar and pestle and appropriate extraction buffer. For LIPOX, TAC, SOD, CAT, MD and GR activity assays, a buffer consisting of 50 mM Tris-HCl; 0.1 mM EDTA; 0.1% Triton X-100; 2 mM DTT and 10% glycerol was used. For A-POX and CIII-POX activity assays, an extraction buffer consisting of 40 mM Tris-HCl; 1 mM EDTA and 5% glycerol was used. For the determination of H_2O_2, a special buffer consisting of K-phosphate buffer pH 6.4 and 5 mM KCN was used. Homogenates were incubated for 1 h at room temperature and then stored at −80 °C until use.

2.4. Measurement of Lipid Peroxidation

LIPOX levels were measured from malondialdehyde (MDA, a product of LIPOX) content determined by the thiobarbituric acid reaction. The extracts were centrifuged (15,000× *g*, 2 min), and 50 µL of the supernatant were transferred to a test tube, to which 50 µL of 20% trichloroacetic acid and 25 µL of 0.675% thiobarbituric acid were added. The mixture was heated to 100 °C for 20 min, rapidly cooled in a cold-water bath and centrifuged again (15,000× g, 10 min). The samples were loaded into a 96-well plate and the absorbance was measured at 600, 532 and 440 nm in a plate reader. The MDA assay required a sucrose standard line to account for the interference of soluble sugars in the assay. In this way, the concentration of sugars measured in the samples was proportional to the absorbance of sucrose at 440 nm. Thus, the concentrations of thiobarbituric acid-reactive substances) are equivalent to the concentration of MDA expressed in nmol/mL according to the following formula [31]:

$$\text{MDA equivalent}\left(\frac{\text{nmol}}{\text{mL}}\right) = \left(\frac{\left(\text{A532} - \text{A600} - \left((\text{A440} - \text{A600}) \times \left(\frac{\text{MA of sucrose at 532 nm}}{\text{MA of sucrose at 440 nm}}\right)\right)\right)}{157000}\right) \times 10^6$$

MA: molar absorbance. MA of sucrose at 532 nm = 8.4; MA of sucrose at 440 nm = 147.

2.5. Measurement of H_2O_2 Content

To determine the concentration of H_2O_2 present in the samples, 30 µL of the extract was mixed with 200 µL of H_2O_2 reaction buffer (100 µM xylenol orange, 100 µM D-sorbitol, 250 µM $FeSO_4$, 250 µM $(NH_4)_2SO_4$ and 1% ethanol in 25 mM H_2SO_4). The samples were incubated in a 96-well plate for 40 min in the dark with gentle shaking. Absorbance was measured at 550 nm, and H_2O_2 was determined using a standard line of H_2O_2 (Sigma) according to the method described by Cheeseman [32].

2.6. Measurement of Total Antioxidant Capacity (TAC)

Total antioxidant capacity was measured using an e-BQC lab device (BioQuoChem, Oviedo, Spain), an electrochemical method based on the measurement of a redox potential [33]. The device measures the resistance of the sample to be oxidised, which is translated into antioxidant capacity. e-BQC lab distinguishes between fast (Q1) and slow (Q2) acting antioxidants, covering compounds with a higher or lower rate of free radical reduction. The instrument also calculates a QTotal value, which is the sum of both. Results were expressed in charge units (µC).

2.7. Measurement of Antioxidant Enzymatic Activities

The different levels of antioxidant activities in the grafted tomato stems were determined using specific methods. The frozen extracts were thawed and centrifuged (15,000× *g*, 2 min), and the supernatant was used to determine the different activities.

SOD activity was determined using the SOD assay kit-WST (Sigma, St. Louis, MO, USA), based on the measurement of the decrease in superoxide anions generated by the xanthine oxidase enzyme. CAT activity was quantified according to the method [34], based on the decrease in absorbance at 240 nm due to the decomposition of H_2O_2. The reaction buffer used consisted of 50 mM K-phosphate buffer pH 7 and 37.5 mM H_2O_2. CIII-POX was evaluated by the technique described by [35], where the absorbance of guaiacol oxidation products was measured at 470 nm. The reaction buffer used consisted of 100 mM sodium acetate pH 5.5, 0.94 mM guaiacol and 1.3 mM H_2O_2. MD activity was determined by the method described in [36] based on the increase in absorbance due to the conversion of $NADP^+$ to NADPH in the presence of malate. The reaction buffer used consisted of 50 mM HEPES-NaOH pH 7.6, 1 mM $MgCl_2$, 0.8 mM $NADP^+$ and 500 mM L-malic acid. A-POX activity was quantified by the method described by [37] based on the reduction of absorbance at 290 nm due to ascorbate oxidation. The reaction buffer used consisted of 50 mM HEPES-NaOH pH 7.6, 0.2 mM ascorbate and 1.3 mM H_2O_2. GR activity was evaluated by the method described in [38] based on the decrease in the absorbance due to the oxidation of NADPH to $NADP^+$ by conversion of oxidised glutathione to reduced glutathione. The reaction buffer used consisted of 0.1 M HEPES-NaOH pH 7.8, 4 M NaOH, 1 mM EDTA, 3 mM $MgCl_2$, 0.5 mM oxidised glutathione and 0.2 mM NADPH.

2.8. Statistical Analyses

All data were subjected to Shapiro–Wilk and Levene tests for normal distribution and homoscedasticity. For physiological and redox parameters, all values were subjected to one-way ANOVA using IBM SPSS Statistics v. 25 software. In addition, Tukey's test was used to compare the means of the different replicates. The statistical significance level was set at $p \leq 0.05$. All results were expressed as mean ± standard deviation (SD). Principal component analysis (PCA) was performed with the antioxidant variables by RStudio (v. 2024.09.0+375 "Cranberry Hibiscus") using the packages Factoextra [39], FactoMineR [40] and Gplots [41].

3. Results

3.1. Oxidative Damage Measured as Lipid Peroxidation

To monitor the effect of putative oxidative stress in grafted tissues, LIPOX was measured in both the scion and rootstock sampling tissues in the close vicinity of the graft zone throughout grafting. Tomato autografts are a highly compatible grafting model that has allowed the establishment of a time course of graft healing in tomato, with a wound response phase at 1–2 DAG, a callus formation and adhesion phase at 3–6 DAG and vascular reconnection at 8 DAG [42].

As shown in Figure 3, the level of LIPOX did not change significantly during grafting and was not significantly different from the LIPOX measured in the non-grafted plants. At very early times after grafting (1 HAG), an apparent increase in the level of LIPOX of the scion tissues was observed. However, the results obtained did not allow significant differences to be established with respect to the non-grafted controls (0 DAG) due to the high variability.

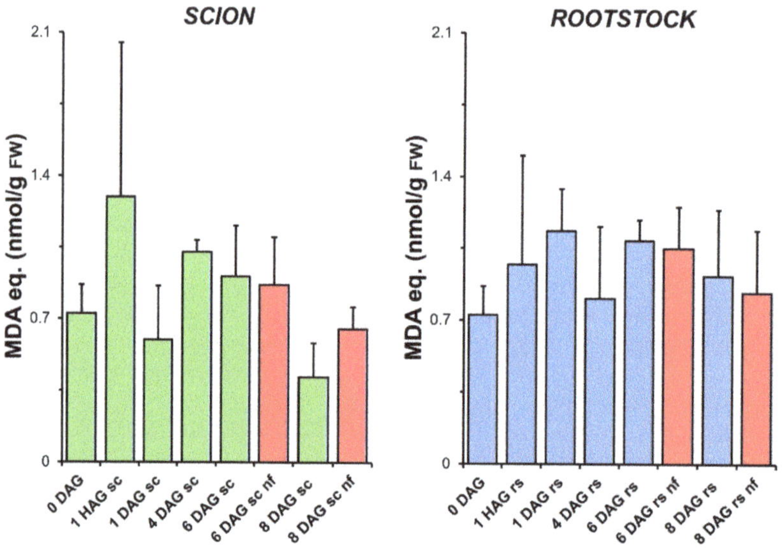

Figure 3. Variation in the level of lipid peroxidation measured as the production of malondialdehyde (MDA) equivalents relative to fresh weight (FW) in scion (sc, green) and rootstock (rs, blue) from 0 to 8 DAG throughout the grafting process. Red columns show non-functional grafts (nf). Mean ± SD (n = 5). No significant differences at the *p*-value < 0.05 level were detected after ANOVA analysis of variance followed by Tukey's test.

In addition, no higher levels of LIPOX were found in the scion and rootstock tissues corresponding to the non-functional grafts compared to the non-grafted plants or grafted plants at the same stage of grafting. The levels of H_2O_2 detected in the tissues also decreased during the healing process, although the non-functional grafts had higher levels than the functional grafts (Figure S1).

3.2. Total Antioxidant Capacity in Grafted Tissues

TAC in the scion tissues increased sharply soon after grafting (Figure 4). At 4 DAG, the antioxidant capacity of the scion tissues was significantly higher than that of the non-grafted plants (0 DAG). Thereafter, a decrease in TAC was measured up to 8 DAG. Both short-term (Q1) and long-term (Q2) antioxidant capacities, representing the contribution of fast- and slow-acting antioxidants, respectively, reflected the changes observed in TAC.

However, our results showed that the antioxidant capacity of the slow-acting antioxidants (Q2) was greater than that of the fast-acting antioxidants (Q1).

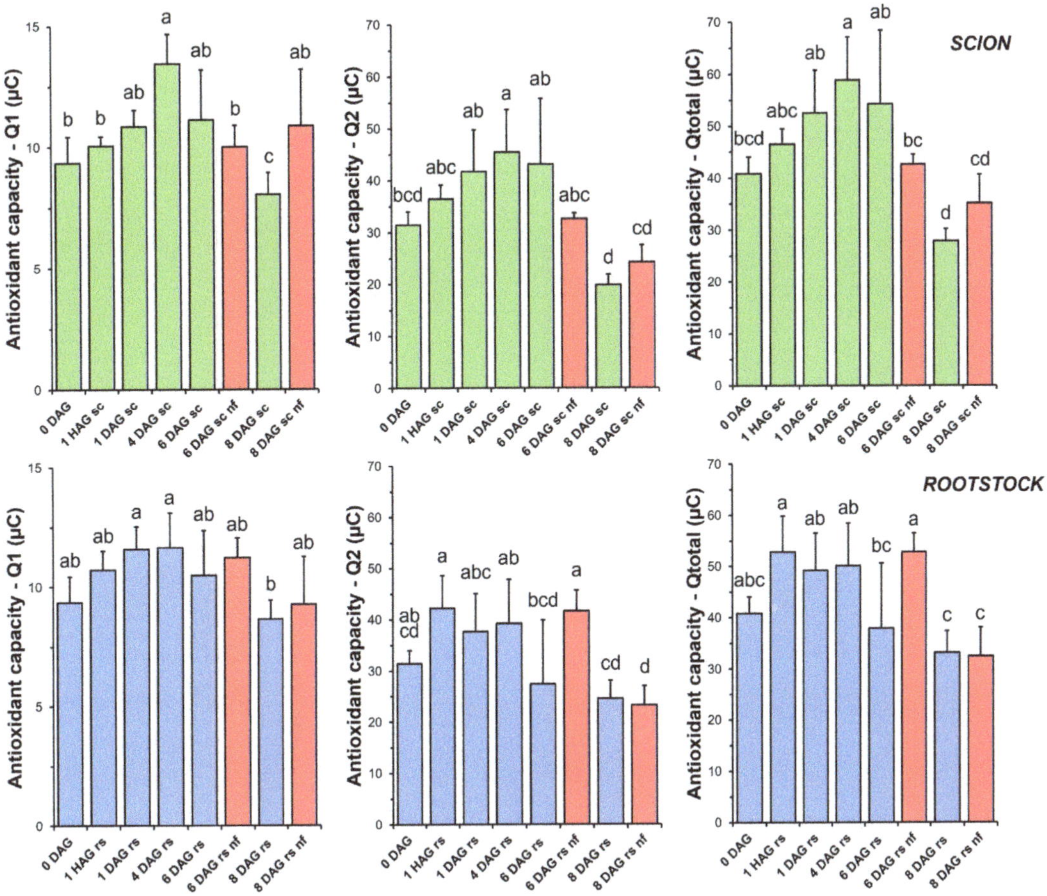

Figure 4. Total antioxidant capacity (TAC) in scion (sc, green) and rootstock (rs, blue) from 0 to 8 DAG throughout the grafting process. Q1 indicates the short-term TAC and Q2 the long-term TAC. Qtotal is the sum of Q1 and Q2. **Red** columns indicate non-functional grafts (nf). Mean ± SD (n = 5). Different letters indicate significant differences at the p-value < 0.05 level after ANOVA analysis of variance followed by Tukey's test.

TAC analysis on the rootstock samples showed less variation throughout the grafting process (Figure 4). Overall, no significant differences were found when comparing the non-functional grafts with the non-grafted or grafted plants at the same stage of grafting.

3.3. Assay of Antioxidant Enzymes Activities

In order to gain insight into the contribution of the enzymatic machinery to the antioxidant status of the grafted tissues, a time course of changes in activities of antioxidant enzymes (SOD, CAT, CIII-POX, MD, A-POX and GR) was carried out in both the scion and rootstock tissues.

3.4. Superoxide Dismutase

Concurrent with the increase in TAC observed in the scion tissues (Figure 4), a significant increase in SOD activity (Figure 5) was measured during the sampling period. Interestingly, in the scion tissue, SOD activity at 8 DAG was significantly reduced in the non-functional grafts compared to the functional grafts. In the case of the rootstock, SOD activity did not change significantly after grafting. Similarly, no significant differences were found when comparing the non-functional grafts with the non-grafted or grafted plants at the same stage of grafting.

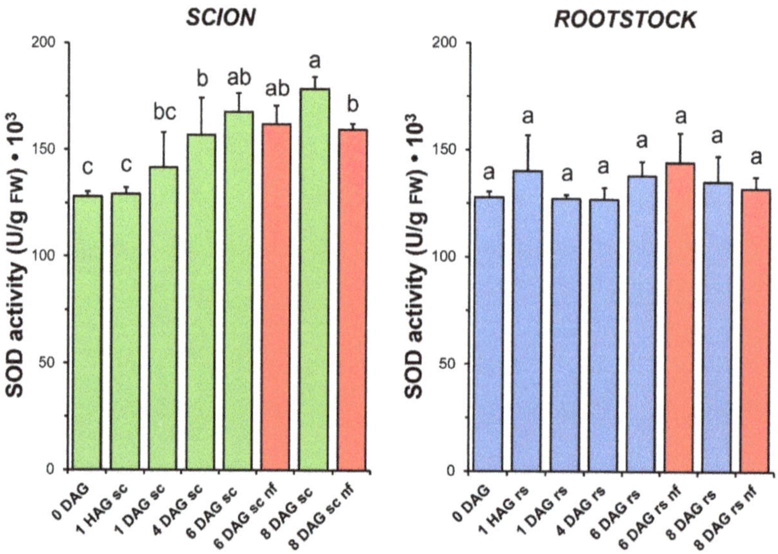

Figure 5. Superoxide dismutase (SOD) activity in scion (sc, green) and rootstock (rs, blue) from 0 to 8 DAG throughout the grafting process. FW, fresh weight. Red columns show non-functional grafts (nf). Mean ± SD (n = 5). Different letters indicate significant differences at the p-value < 0.05 level after ANOVA analysis of variance followed by Tukey's test.

3.5. Catalase

Although an increasing trend was observed, no statistical differences in CAT activity (Figure 6) were found between the non-grafted and grafted tissues (both scion and rootstock) over the time window when a measurable increase in TAC was observed (1 HAG to 4 DAG in Figure 4). At 8 DAG, a significant decrease in CAT activity was observed in the scion tissues. Interestingly, this sharp decrease in CAT activity coincided in time with a significant decrease in TAC in the same tissues. In the rootstocks, no significant differences were found when comparing the non-functional grafts with the non-grafted or grafted plants at the same stage of grafting.

3.6. Class III Peroxidase

CIII-POX activity remained unchanged throughout the graft healing process in both the scions and rootstocks in the functional grafts. In the case of the scions, a non-significant increase in this activity was observed at 6 DAG (Figure 7). Interestingly, a significant increase in CIII-POX activity was measured at 8 DAG in the scion and rootstock tissues from non-functional grafts compared to both the functional grafts at 8 DAG and the non-grafted controls (Figure 7).

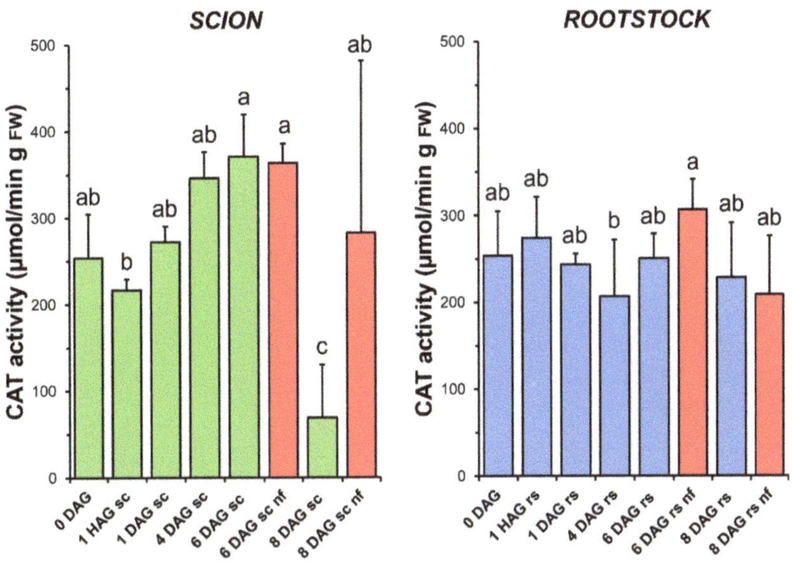

Figure 6. Catalase (CAT) activity in scion (sc, green) and rootstock (rs, blue) from 0 to 8 DAG throughout the grafting process. FW, fresh weight. Red columns show non-functional grafts (nf). Mean ± SD (n = 5). Different letters indicate significant differences at the p-value < 0.05 level after ANOVA analysis of variance followed by Tukey's test.

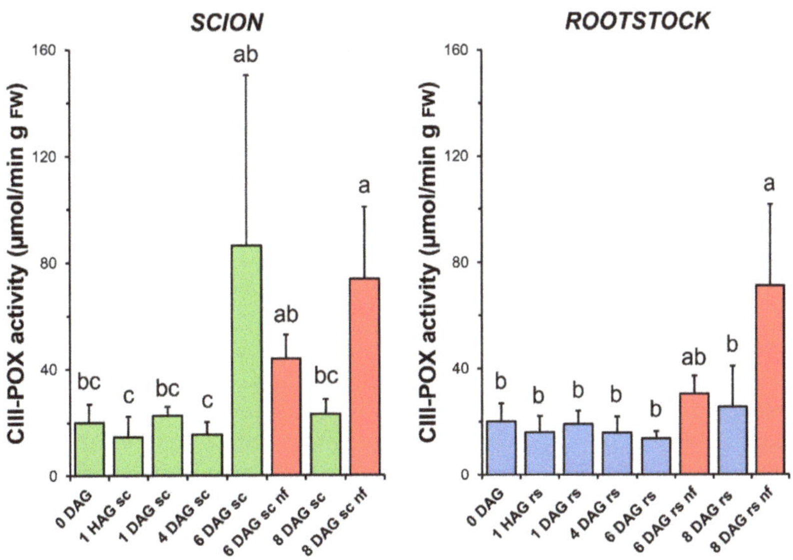

Figure 7. Class III peroxidases (CIII-POX) activity in scion (sc, green) and rootstock (rs, blue) from 0 to 8 DAG throughout the grafting process. FW, fresh weight. Red columns show non-functional grafts (nf). Mean ± SD (n = 5). Different letters indicate significant differences at the p-value < 0.05 level after ANOVA analysis of variance followed by Tukey's test.

3.7. Malate Dehydrogenase

The enzymatic activity of MD underwent some variations throughout the graft healing process in the scions, first an increase at 1 DAG and then a decrease at 4 DAG. It was also found that the specific activity of MD decreased significantly at 6 DAG with respect to the non-grafted plants (Figure 8). In the rootstocks, the level of MD activity did not vary significantly throughout the graft healing process. This activity was only slightly increased in the non-functional rootstocks compared to the non-grafted plants or to the functional grafts at the same stage of grafting (Figure 8).

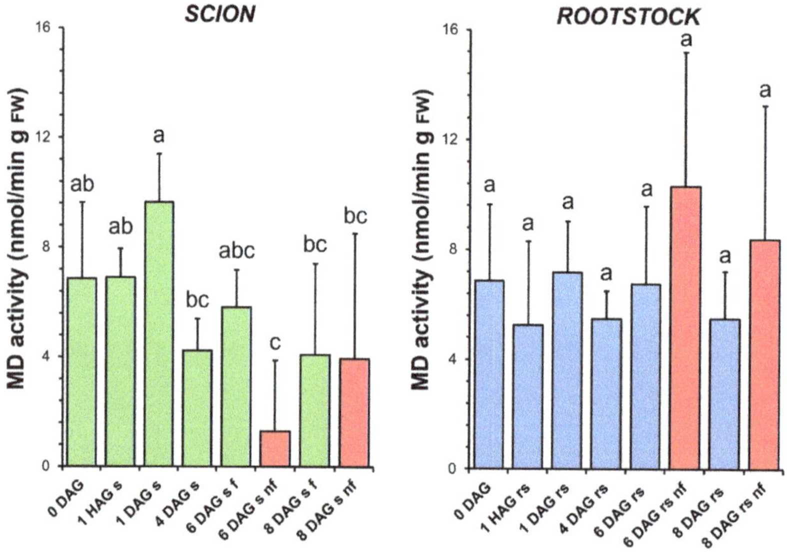

Figure 8. Malate dehydrogenase (MD) activity in scion (sc, green) and rootstock (rs, blue) from 0 to 8 DAG throughout the grafting process. FW, fresh weight. Red columns show non-functional grafts (nf). Mean ± SD (n = 5). Different letters indicate significant differences at the *p*-value < 0.05 level after ANOVA analysis of variance followed by Tukey's test.

3.8. Ascorbate Peroxidase and Glutathione Reductase

As shown in Figure 9a, A-POX activity decreased immediately after grafting and then recovered at early DAG in both the scions (4 DAG) and rootstocks (1 DAG). However, due to the variability of the measurements, no significant differences were found in most of the comparisons. In addition, no significant differences were found when comparing the non-functional grafts with the non-grafted or grafted plants at the same stage of grafting.

No significant variations in GR activity (Figure 9b) were observed during the healing process in either functional scion or rootstock samples. However, in the non-functional scions, a significant increase in the activity of this enzyme was observed at 8 DAG compared to the non-grafted or grafted plants at the same stage of grafting. On the other hand, the response of the rootstocks was moderate and similar to the A-POX activity in this type of samples.

3.9. Multivariate Analysis: Principal Component Analysis

Principal component analysis (PCA) was performed to better interpret the changes in antioxidant activities during the graft healing process. As shown in Figure 10, the clustering of the scions was found to be associated with a timeline of the graft reconnection along PC2 (or even distributed in a counter clockwise pattern starting on the negative side of PC2 with the non-grafted samples and the early times after grafting). Samples corresponding to the initial phase after grafting (0 DAG, 1 HAG and 1 DAG) were located on the negative side of PC2, which showed a correlation with MD activity rather than with SOD, GR, APOX or CIII-POX. Subsequent intermediate times after grafting were located on the positive side of

PC2 and on the right side of PC1, with a predominant correlation observed with TAC and CAT. Finally, the last time evaluated of the graft healing process (8 DAG) was positioned away from the intermediate times and in contraposition to antioxidant activities and TAC, suggesting that these are not so necessary when the graft is consolidated.

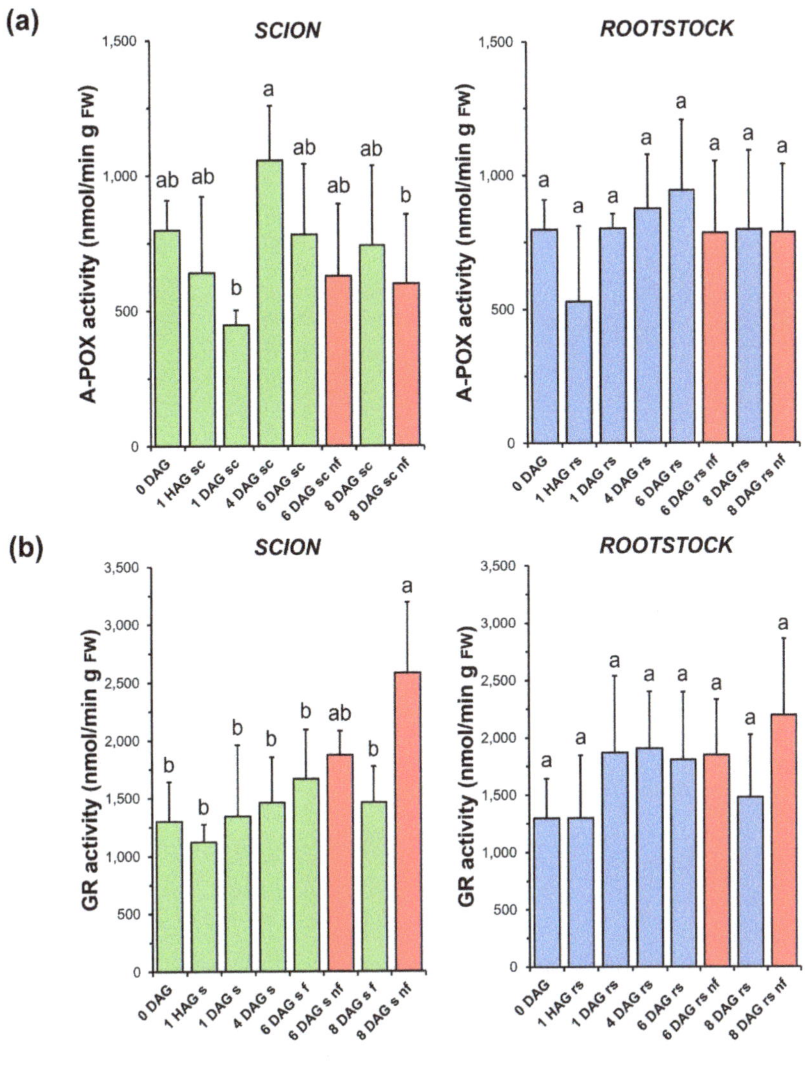

Figure 9. Ascorbate peroxidase (A-POX) (**a**) and glutathione reductase (GR) (**b**) activity in scion (sc, green) and rootstock (rs, blue) from 0 to 8 DAG throughout the grafting process. FW, fresh weight. Red columns show non-functional grafts (nf). Mean ± SD (n = 5). Different letters indicate significant differences at the *p*-value < 0.05 level after ANOVA analysis of variance followed by Tukey's test.

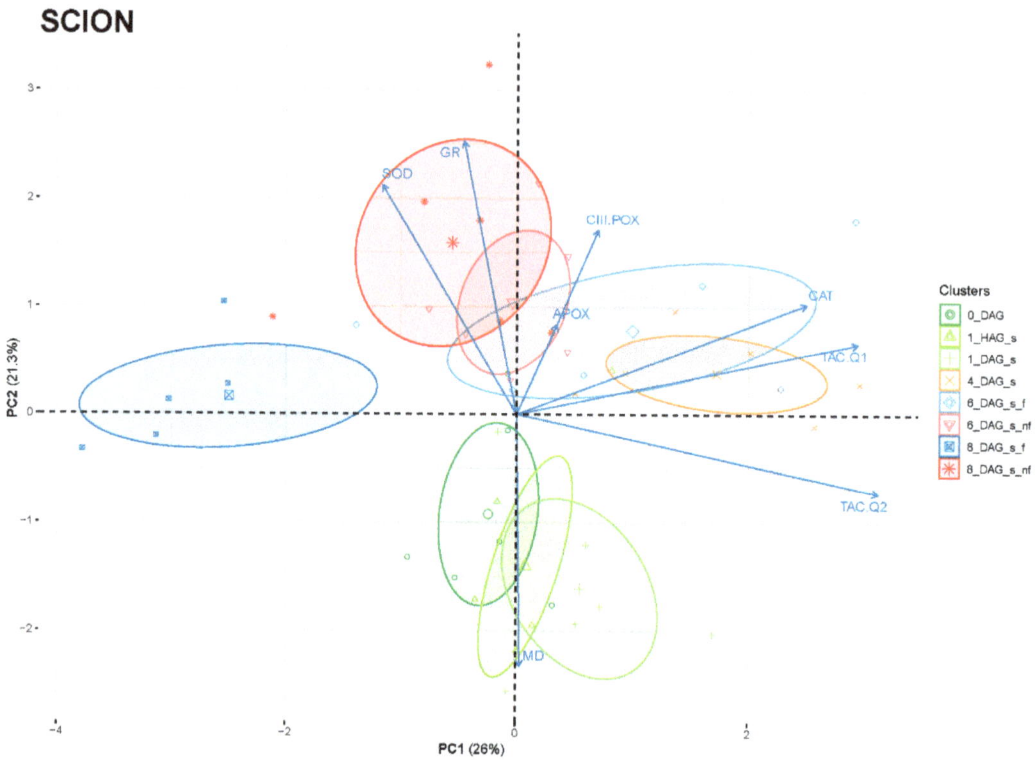

Figure 10. Principal component analysis (PCA) of scion tissues of evaluated variables related to the variation of antioxidant power during the healing process of grafting: total antioxidant capacity (TAC), ascorbate peroxidase (A-POX), glutathione reductase (GR), superoxide dismutase (SOD), catalase (CAT), class III peroxidase (CIII-POX) and malate dehydrogenase (MD) activities. PC1 and PC2 accounted for 47.3% of the variance (26 and 21.3%, respectively). Variables are shown as vectors (arrows). DAG, days after grafting; nf, non-functional; s, scion.

PCA analysis of rootstock clustering did not show a clear trend, consistent with a high degree of group overlapping (Figure S2).

4. Discussion

During graft healing, the wound syndrome together with the onset of water deficit in the scion causes the release of ROS and the activation of antioxidant mechanisms dedicated to controlling oxidative damage [43]. In fact, Cui et al. [12] demonstrated the simultaneous early overexpression of genes related to wound response, oxidative stress and response to water deficit in tomato grafting. If there is an imbalance between the production and scavenging of ROS, the consequences for the success of the graft could be dire. On the one hand, the activation of antioxidant mechanisms is necessary, since the presence of excessive levels of ROS is toxic to the plant, as mentioned above. Nevertheless, ROS contribute to the lignification and/or suberisation of the cutting zone and are used as a signalling/defence mechanism in response to stress and pathogens that attack grafted plants through injured tissues [44,45].

4.1. Oxidative Damage Appears to Be Controlled During Grafting in Tissues Close to the Cut Zone

LIPOX is a reliable measure of oxidative stress in grafted plants as it indirectly reflects the formation of hydroxyls by ROS [46]. The results presented here indicate that oxidative damage remained controlled throughout the graft healing process (Figure 3), as no significant differences in LIPOX were found between the non-grafted and grafted plants.

It has previously been shown that highly compatible grafting models have a much greater antioxidant response to ROS than less compatible grafting systems [25,43]. Antioxidant defences help to inactivate the cytotoxic ROS compounds and minimise their ability to diffuse into the intracellular space [24,43]. Therefore, it is possible to suggest that the success of a graft is closely linked to the efficiency of the antioxidant mechanisms and their ability to generate an early response to the wound damage situation [43].

4.2. Total Antioxidant Capacity Increases in Tissues Close to the Cut Zone, Likely Controlling Oxidative Damage

In this experimental model, a significant increase in TAC (peaking at 4 DAG) was found in the scion tissues during graft healing (Figure 4). In the case of the rootstock tissues, a non-statistically significant trend towards an increase in TAC was also observed. These results may indicate a link between the induction of a non-enzymatic antioxidant response (Figure 4) and the control of oxidative damage (Figure 3) in successfully autografted plants. These results would be consistent with previous findings indicating a positive relationship between grafting success and an efficient antioxidant response [43].

The eBQC-based-method for TAC measurement assesses the oxidant-buffering capacity of a liquid sample by electrochemical methods. TAC is related to the presence of non-enzymatic antioxidants such as ascorbic acid or glutathione, and, indeed, antioxidant activity is often expressed as ascorbic acid equivalents [47]. One of the main advantages of the eBQC assay for TAC is that it provides information on the contribution of fast (Q1) and slow (Q2) antioxidants [33]. In this regard, our results indicated that slow-acting antioxidants (Q2), which could include compounds such as phenylpropanoids and flavonoids, contributed more to TAC (Figure 4) than fast-acting antioxidants (Q1) such as ascorbic acid [48].

4.3. During Graft Healing, Changes in Antioxidant Activities Have Been Observed to Be Associated with Scion Tissues

Although TAC is a rapid and reliable method for monitoring changes in the antioxidant status of a tissue, it does not provide information on the contribution of enzymatic antioxidants. Therefore, an analysis of changes in enzymatic antioxidants during graft healing was performed.

The results presented here highlight the differential contribution of antioxidant activities in the scion and rootstock, emphasising their biochemical diversity during grafting. In functional tomato grafts, it can be concluded that the main variations in antioxidant activities were restricted to the scion tissue (Figures 5–9). In addition, principal component analysis revealed a dynamic correlation between antioxidant activities and grafting stages (Figure 10), highlighting their critical role in mitigating oxidative stress.

None of the antioxidant activities evaluated in the rootstocks (SOD, CAT, CIII-POX, MD, A-POX and GR) showed differences with respect to the non-grafted controls during the healing process. A possible explanation for this result is the difference in the level of stress experienced by the rootstock and the scion: the rootstock, which is in contact with the soil and has a source of water and nutrients, would be less stressed than the scion, which is more vulnerable in the phases before the graft is fully healed.

4.4. SOD and CAT Activities Increase During Graft Healing in Scion Tissues

The relationship between graft healing and enzymatic antioxidant capacity was particularly well followed in the scion by changes in SOD (Figure 5) and CAT (Figure 6) activities. In both cases, an increase in enzymatic activity was observed over time, with a maximum at 6 DAG. A similar relationship between increases in SOD and CAT and the graft healing process has previously been reported in functional tomato autografts [25].

In this work, oxidative damage was measured by MDA production, a good marker of LIPOX by O_2^- [49,50]. O_2^- is one of the most reactive molecules with the greatest capacity to cause not only lipid degradation but also tissue damage [51]. Furthermore, it should be noted that SODs are the enzymes responsible for the reduction of O_2^- to H_2O_2, controlling the accumulation of O_2^- in both the cytoplasm and the cell wall [45,52,53]. Therefore, SOD activity may be responsible for controlling oxidative damage during the graft healing process. O_2^- and H_2O_2 are naturally intertwined, as H_2O_2 is produced to a greater extent by the SOD-catalysed dismutation of O_2^- [45]. Consequently, an increase in SOD activity is expected to correlate with the accumulation of H_2O_2, which is also more stable than another ROS [45]. In this respect, the data show that H_2O_2 levels were highly variable immediately after grafting and then decreased to non-grafted levels at late DAG (Figure S1).

H_2O_2 is scavenged by peroxisomal/glyoxysomal CAT, among others [45]. The key role of CAT in H_2O_2 control has already been demonstrated by using mutants in *Arabidopsis* [54] and in heat-tolerant tomato scions grafted onto a commercial rootstock [55]. As in the case of SOD (Figure 5), the results indicated here point to an increase in CAT activity (Figure 6) up to 6 DAG, also suggesting a role for this enzyme activity in controlling oxidative damage during graft healing, as previously reported for tomato heterografts [25,56]. Similarly, previous reports on tomato grafting have reported an increase in CAT activity

An unexpected finding was observed for catalase activity at 8 DAG, as it was significantly lower in the grafted plants than in the non-grafted plants. In this respect, it is interesting to note that CAT activity is often associated with the accumulation of antioxidant molecules (such as glutathione) during stress situations [57]. The lower catalase activity found in the functional grafted plants compared to the non-grafted plants may therefore indicate an inhibition of activity once oxidative stress has been overcome.

In addition to CAT, other peroxidases such as organellar A-POX and cell wall CIII-POX are involved in controlling H_2O_2 levels [45,58]. However, the results obtained here do not allow us to establish a relationship between the variation of CIII-POX (Figure 7) and A-POX (Figure 9a) activities with graft healing in successful grafts.

4.5. What About Non-Functional Grafts?

At 6–8 DAG, functional and non-functional grafts could be distinguished morphologically. Taking advantage of this fact, we set out to assess oxidative damage, TAC and enzymatic antioxidant activities in non-functional grafts to evaluate the possible relationship between graft failure and exacerbated oxidative damage.

No differences in the level of oxidative damage (Figure 3) were observed between the scion and rootstock tissues when comparing the non-functional grafts with functional grafts of the same age, although several studies have previously reported an accumulation of H_2O_2 causing oxidative damage in incompatible grafts [25,28]. In our work, we also observed few significant differences in TAC (scion at 8 DAG and rootstock at 6 DAG, Figure 4), as well as in SOD (scion at 8 DAG, Figure 5), CAT (scion at 8 DAG, Figure 6) and GR (Figure 9b) activities. These results would suggest that, although there is no causal relationship between graft failure and uncontrolled oxidative stress in our tomato autograft experimental model, changes in antioxidant activities could be a cause or consequence of autograft failure.

CIII-POX is a class of secreted peroxidases involved in cell wall stiffening by oxidatively polymerising lignin, suberin, structural proteins and polysaccharides using H_2O_2 as an oxidant [59]. Interestingly, their activity was significantly increased in non-functional grafts (Figure 7). Previous findings allowed us to demonstrate the deposition of defensive cell wall materials, such as suberin and lignin, on the cut surface of non-functional tomato autografts at 20 DAG [60]. It is therefore tempting to hypothesise that the increased CIII-POX activity found here may be related to suberification and lignification during autograft failure.

5. Conclusions

The results presented here show that oxidative damage in the stem tissue surrounding the cut surface of autografted tomato plants was not statistically different from that observed in non-grafted control plants during the period up to 8 DAG. This time frame is sufficient for graft healing and consolidation, probably due to the activation of antioxidant mechanisms. An asymmetry in the antioxidant response was reported between the scion and the rootstock, with the scion being the tissue that showed more changes in each of the variables studied. There was a consistent increase in of TAC and CAT and SOD activities as the graft healed. These results suggest that there was an increase in the antioxidant capacity to deal with reactive oxygen species. On the other hand, the non-functional grafts were characterised by an increase in CIII-POX and GR activities. Taken together, these results support the conclusion that the antioxidant enzymatic machinery is altered and enhanced during tomato graft healing and that these changes are also partially different in non-functional grafts. The results presented here suggest that an efficient antioxidant response in scion tissue may be critical for the development of a successful graft.

Supplementary Materials: The following supporting information can be downloaded at: https://www.mdpi.com/article/10.3390/antiox14020234/s1, Figure S1. Hydrogen peroxide (H_2O_2) concentration in scion and rootstock tissues, including non-functional grafts (nf), expressed as percentage of non-grafted plants at 0 days after grafting. Times after grafting evaluated were: 0 days after grafting, 1 h after grafting, 1, 4, 6 and 8 days after grafting. Figure S2. Principal component analysis (PCA) of rootstock tissues of evaluated variables related to the antioxidant power variation during the healing grafting process: total antioxidant capacity (TAC), ascorbate peroxidase (A-POX), glutathione reductase (GR), superoxide dismutase (SOD), catalase (CAT), class III peroxidase (CIII-POX) and malate dehydrogenase (MD) activities. PC1 and the PC2 represented 43.9 % of the variance (23.7 and 20.2 % respectively). Variables are represented as vectors (arrows). DAG, days after grafting; nf, non-functional; rs, rootstock.

Author Contributions: J.L.A. and A.E. conceived, designed and supervised the research; C.F., with the help of A.H.-B., conducted the experiments and analysed the data; C.F. wrote the manuscript draft; J.L.A. and A.E. contributed to analysing the data and supervised the manuscript. All authors have read and agreed to the published version of the manuscript.

Funding: This research was funded by the Vice-Rectorate for Research and Transfer of the Universidad de León through the financial support of the "Fisiología y Biotecnología de las Plantas" Research Group. C.F. acknowledges the PhD grant from the FPU programme of the Spanish Ministry of Universities (Ref. FPU18/04934).

Institutional Review Board Statement: Not applicable.

Informed Consent Statement: Not applicable.

Data Availability Statement: The original contributions presented in this study are included in the article and Supplementary Materials. Further inquiries can be directed to the corresponding author.

Acknowledgments: The authors acknowledge Julio Martín Mateos for his valuable collaboration in the initial approaches to the subject of this study.

Conflicts of Interest: The authors declare no conflicts of interest.

Abbreviations

A-POX	ascorbate peroxidase
CIII-POX	class III peroxidase
CAT	catalase
DAG	days after grafting
GR	glutathione reductase
HAG	hours after grafting
LIPOX	lipid peroxidation
MD	malate dehydrogenase
MDA	malondialdehyde
PCA	principal component analysis
POX	peroxidase
ROS	reactive oxygen species
SOD	superoxide dismutase
TAC	total antioxidant capacity

References

1. Feng, M.; Augstein, F.; Kareem, A.; Melnyk, C.W. Plant grafting: Molecular mechanisms and applications. *Mol. Plant* **2024**, *17*, 75–91. [CrossRef] [PubMed]
2. Kyriacou, M.C.; Rouphael, Y.; Colla, G.; Zrenner, R.; Schwarz, D. Vegetable grafting: The implications of a growing agronomic imperative for vegetable fruit quality and nutritive value. *Front. Plant Sci.* **2017**, *8*, 741. [CrossRef] [PubMed]
3. Argento, S.; Treccarichi, S.; Melilli, M.G.; Branca, F. Grafting compatibility and environmental conditions on soilless eggplant (*Solanum melongena*) grown in the Mediterranean greenhouse. *Horticulturae* **2023**, *9*, 1060. [CrossRef]
4. Bento da Silva, E.P.P.; Mendonça, S.R.; de Moraes, M.G. Trends and gaps in tomato grafting literature: A systematic approach. *Span. J. Agric. Res.* **2023**, *21*, e0904. [CrossRef]
5. Singh, H.; Kumar, P.; Kumar, A.; Kyriacou, M.C.; Colla, G.; Rouphael, Y. Grafting tomato as a tool to improve salt tolerance. *Agronomy* **2020**, *10*, 263. [CrossRef]
6. Zeist, A.R.; Henschel, J.M.; Silva Júnior, A.D.; Oliveira, G.J.A.; Neto, J.G.; Beauboeuf, C.B.; Parthasarathi, T.; de Resendee, J.T.V. Responses of rootstocks variability to tolerate salinity in tomato. *S. Afr. J. Bot.* **2023**, *153*, 280–289. [CrossRef]
7. Zhang, Z.; Liu, Y.; Cao, B.; Chen, Z.; Xu, K. The effectiveness of grafting to improve drought tolerance in tomato. *Plant Growth Regul.* **2020**, *91*, 157–167. [CrossRef]
8. Khapte, P.S.; Kumar, P.; Wakchaure, G.C.; Jangid, K.K.; Colla, G.; Cardarelli, M.; Rane, J. Application of phenomics to elucidate the influence of rootstocks on drought response of tomato. *Agronomy* **2022**, *12*, 1529. [CrossRef]
9. Nakano, K. Mechanisms of resistance to *Ralstonia solanacearum* in tomato rootstocks and integrated management of bacterial wilt using high grafting. *J. Gen. Plant Pathol.* **2021**, *87*, 398–402. [CrossRef]
10. Saman, P.; Kawicha, P.; Sangdee, A.; Wongpakdee, S.; Rattanapolsan, L.; Ponpang-Nga, P.; Suwor, P.; Thanyasiriwat, T. Grafting compatibility, scion growth, and fusarium wilt disease incidence of intraspecific grafted tomato. *J. Hortic. Res.* **2022**, *30*, 95–104. [CrossRef]
11. Nanda, A.K.; Melnyk, C.W. The role of plant hormones during grafting. *J. Plant Res.* **2018**, *131*, 49–58. [CrossRef] [PubMed]
12. Cui, Q.; Xie, L.; Dong, C.; Gao, L.; Shang, Q. Stage-specific events in tomato graft formation and the regulatory effects of auxin and cytokinin. *Plant Sci.* **2021**, *304*, 110803. [CrossRef] [PubMed]
13. Frey, C.; Manga-Robles, A.; Acebes, J.L.; Encina, A. The graft framework: Quantitative changes in cell wall matrix polysaccharides throughout the tomato graft union formation. *Carbohydr. Polym.* **2022**, *276*, 118781. [CrossRef]
14. Loupit, G.; Brocard, L.; Ollat, N.; Cookson, S.J. Grafting in plants: Recent discoveries and new applications. *J. Exp. Bot.* **2023**, *74*, 2433–2447. [CrossRef] [PubMed]
15. Duan, Y.; Zhang, F.; Meng, X.; Shang, Q. Spatio-temporal dynamics of phytohormones in the tomato graft healing process. *Hortic. Plant J.* **2024**, *10*, 1362–1370. [CrossRef]

16. Frey, C.; Saez-Aguayo, S.; Encina, A.; Acebes, J.L. Deepening the role of pectin in the tissue assembly process during tomato grafting. *Plants* **2024**, *13*, 3519. [CrossRef] [PubMed]
17. Savatin, D.V.; Gramegna, G.; Modesti, V.; Cervone, F. Wounding in the plant tissue: The defence of a dangerous passage. *Front. Plant Sci.* **2014**, *5*, 470. [CrossRef]
18. Vega-Muñoz, I.; Duran-Flores, D.; Fernández-Fernández, Á.D.; Heyman, J.; Ritter, A.; Stael, S. Breaking bad news: Dynamic molecular mechanisms of wound response in plants. *Front. Plant Sci.* **2020**, *11*, 610445. [CrossRef]
19. Chen, G.; Lips, S.H.; Sagi, M. Biomass production, transpiration rate and endogenous abscisic acid levels in grafts of flacca and wild-type tomato (*Lycopersicon esculentum*). *Funct. Plant Biol.* **2002**, *29*, 1329–1335. [CrossRef] [PubMed]
20. Correia, M.J.; Osório, M.L.; Osório, J.; Barrote, I.; Martins, M.; David, M.M. Influence of transient shade periods on the effects of drought on photosynthesis, carbohydrate accumulation and lipid peroxidation in sunflower leaves. *Environ. Exp. Bot.* **2006**, *58*, 75–84. [CrossRef]
21. Frey, C.; Hernández-Barriuso, A.; Encina, A.; Acebes, J.L. Non-invasive monitoring of tomato graft dynamics using thermography and fluorescence quantum yields measurements. *Physiol. Plant.* **2023**, *175*, e13935. [CrossRef]
22. Aroca, R.; Ruiz-Lozano, J.M.; Zamarreño, A.M.; Paz, J.A.; García-Mina, J.M.; Pozo, M.J.; López-Ráez, J.A. Arbuscular mycorrhizal symbiosis influences strigolactone production under salinity and alleviates salt stress in lettuce plants. *J. Plant Physiol.* **2013**, *170*, 47–55. [CrossRef]
23. Havaux, M. Rapid photosynthetic adaptation to heat stress triggered in potato leaves by moderately elevated temperatures. *Plant. Cell Environ.* **1993**, *16*, 461–467. [CrossRef]
24. Mittler, R. ROS are good. *Trends Plant Sci.* **2017**, *22*, 11–19. [CrossRef]
25. Babar, M.; Afzal, N.; Siddiqui, K.; Azhar, A.; Galani, S. Exploring graft incompatibility markers: Intraspecific and interspecific grafts of tomato (*Solanum lycopersicum* L.). *Sci. Hort.* **2023**, *310*, 111762. [CrossRef]
26. Miller, G.; Shulaev, V.; Mittler, R. Reactive oxygen signaling and abiotic stress. *Physiol. Plant.* **2008**, *33*, 481–489. [CrossRef]
27. Baxter, A.; Mittler, R.; Suzuki, N. ROS as key players in plant stress signalling. *J. Exp. Bot.* **2014**, *65*, 1229–1240. [CrossRef]
28. Aloni, B.; Karni, L.; Deventurero, G.; Levin, Z.; Cohen, R.; Katzir, N.; Lotan-Pompan, M.; Edelstein, M.; Aktas, H.; Turhan, E.; et al. Physiological and biochemical changes at the rootstock–scion interface in graft combinations between *Cucurbita* rootstocks and a melon scion. *J. Hortic. Sci. Biotechnol.* **2008**, *83*, 777–783. [CrossRef]
29. Noctor, G.; Foyer, C.H. Ascorbate and glutathione: Keeping active oxygen under control. *Annu. Rev. Plant Physiol. Mol. Biol.* **1998**, *49*, 249–279. [CrossRef] [PubMed]
30. Foyer, C.H.; Kunert, K. The ascorbate–glutathione cycle coming of age. *J. Exp. Bot.* **2024**, *75*, 2682–2699. [CrossRef]
31. Du, Z.; Bramlage, W.J. Modified thiobarbituric acid assay for measuring lipid oxidation in sugar-rich plant tissue extracts. *J. Agric. Food Chem.* **1992**, *40*, 1566–1570. [CrossRef]
32. Cheeseman, J.M. Hydrogen peroxide concentrations in leaves under natural conditions. *J. Exp. Bot.* **2006**, *57*, 2435–2444. [CrossRef] [PubMed]
33. Aranaz, M.; Costas-Rodríguez, M.; Lobo, L.; González-Iglesias, H.; Pereiro, R.; Vanhaecke, F. Homeostatic alterations related to total antioxidant capacity, elemental concentrations and isotopic compositions in aqueous humor of glaucoma patients. *Anal. Bioanal. Chem.* **2022**, *414*, 515–524. [CrossRef] [PubMed]
34. Droillard, M.J.; Paulin, A.; Massot, J.C. Free radical production, catalase and superoxide dismutase activities and membrane integrity during senescence of petals of cut carnations (*Dianthus caryophyllus*). *Physiol. Plant.* **1987**, *71*, 197–202. [CrossRef]
35. Ádám, A.; Bestwick, C.; Barna, B.; Mansfield, J. Enzymes regulating the accumulation of active oxygen species during the hypersensitive reaction of bean to *Pseudomonas syringae* pv. *phaeolicola*. *Planta* **1995**, *197*, 240–249.
36. Corpas, J.F.; Barroso, B.F.; Sandalio, M.L.; Distefano, S.; Palma, M.J.; Lupiáñez, J.A.; del Río, A.L. A dehydrogenase-mediated recycling system of NADPH in plant peroxisomes. *Biochem. J.* **1998**, *330*, 777–784. [CrossRef]
37. Hossain, M.A.; Asada, K. Inactivation of ascorbate peroxidase in spinach chloroplasts on dark addition of hydrogen peroxide: Its protection by ascorbate. *Plant Cell Physiol.* **1984**, *25*, 1285–1295.
38. Edwards, E.A.; Rawsthorne, S.; Mullineaux, P.M. Subcellular distribution of multiple forms of glutathione reductase in leaves of pea (*Pisum sativum* L.). *Planta* **1990**, *180*, 278–284. [CrossRef] [PubMed]
39. Kassambara, A.; Mundt, F. *Factoextra: Extract and Visualize the Results of Multivariate Data Analyses*; R Package Version 1; 2020; p. 7. Available online: https://cran.r-project.org/web/packages/factoextra/index.html (accessed on 12 December 2024).
40. Lê, S.; Josse, J.; Husson, F. FactorMineR: An R package for multivariate analysis. *J. Stat. Softw.* **2008**, *25*, 1–18. Available online: https://www.jstatsoft.org/article/view/v025i01 (accessed on 12 December 2024). [CrossRef]
41. Warnes, G.; Bolker, B.; Bonebakker, L.; Gentleman, R.; Liaw, W.H.A.; Lumley, T.; Magnusson, A.; Moeller, S.; Schwartz, M.; Venables, B. *Gplots: Various R Programming Tools for Plotting Data*; R package version 3; 2020; p. 1. Available online: https://cran.r-project.org/web/packages/gplots/gplots.pdf (accessed on 12 December 2024).
42. Frey, C.; Martínez-Romera, N.; Encina, A.; Acebes, J.L. Immunohistochemical dynamics of cell wall matrix polymers during tomato autograft healing. *Plant Mol. Biol.* **2023**, *113*, 353–365. [CrossRef]

43. Irisarri, P.; Binczycki, P.; Errea, P.; Juel, H.; Pina, A. Oxidative stress associated with rootstock–scion interactions in pear/quince combinations during early stages of graft development. *J. Plant Physiol.* **2015**, *176*, 25–35. [CrossRef] [PubMed]
44. Minibayeva, F.; Beckett, R.P.; Kranner, I. Roles of apoplastic peroxidases in plant response to wounding. *Phytochemistry* **2015**, *112*, 122–129. [CrossRef] [PubMed]
45. Smirnoff, N.; Arnaud, D. Hydrogen peroxide metabolism and functions in plants. *New Phytol.* **2019**, *221*, 1197–1214. [CrossRef]
46. Pina, A.; Cookson, S.J.; Calatayud, A.; Trinchera, A.; Errea, P. Physiological and molecular mechanisms underlying graft compatibility. In *Vegetable Grafting: Principles and Practices*; Colla, G., Pérez-Alfocea, F., Schwarz, D., Eds.; CABI Publishing: Wallingford, UK, 2017; pp. 132–154.
47. Pisoschi, A.M.; Negulescu, G.P. Methods for total antioxidant activity determination: A review. *Biochem. Anal. Biochem.* **2011**, *1*, 106. [CrossRef]
48. Schauss, A.G.; Wu, X.; Prior, R.L.; Ou, B.; Huang, D.; Owens, J.; Agarwal, A.; Jensen, G.S.; Hart, A.N.; Shanbrom, E. Antioxidant capacity and other bioactivities of the freeze-dried amazonian palm berry, *Euterpe oleraceae* Mart. (Acai). *J. Agric. Food Chem.* **2006**, *54*, 8604–8610. [CrossRef] [PubMed]
49. Del Río, D.; Stewart, A.J.; Pellegrini, N. A review of recent studies on malondialdehyde as toxic molecule and biological marker of oxidative stress. *Nutr. Metab. Cardiovasc. Dis.* **2005**, *15*, 316–328. [CrossRef] [PubMed]
50. Tsikas, D. Assessment of lipid peroxidation by measuring malondialdehyde (MDA) and relatives in biological samples: Analytical and biological challenges. *Anal. Biochem.* **2017**, *524*, 13–30. [CrossRef]
51. Del Río, L.A. ROS and RNS in plant physiology: An overview. *J. Exp. Bot.* **2015**, *66*, 2827–2837. [CrossRef]
52. Apel, K.; Hirt, H. Reactive oxygen species: Metabolism, oxidative stress, and signal transduction. *Annu. Rev. Plant Biol.* **2004**, *55*, 373–399. [CrossRef] [PubMed]
53. Gill, S.S.; Anjum, N.A.; Gill, R.; Yadav, S.; Hasanuzzaman, M.; Fujita, M.; Mishra, P.; Sabat, S.C.; Tuteja, N. Superoxide dismutase—Mentor of abiotic stress tolerance in crop plants. *Environ. Sci. Pollut. Res.* **2015**, *22*, 10375–10394. [CrossRef] [PubMed]
54. Queval, G.; Issakidis-Bourguet, E.; Hoeberichts, F.A.; Vandorpe, M.; Gakiere, B.; Vanacker, H.; Miginiac-Maslow, M.; Van Breusegem, F.; Noctor, G. Conditional oxidative stress responses in the Arabidopsis photorespiratory mutant *cat2* demonstrate that redox state is a key modulator of daylength dependent gene expression, and define photoperiod as a crucial factor in the regulation of H_2O_2-induced cell death. *Plant J.* **2007**, *52*, 640–657. [PubMed]
55. Lee, C.; Harvey, J.T.; Qin, K.; Leskovar, D. Physio-biochemical responses of grafted tomatoes differing in thermotolerance to heat stress and recovery. *Sci. Hortic.* **2023**, *308*, 111546. [CrossRef]
56. Fernández-García, N.; Carvajal, M.; Olmos, E. Graft union formation in tomato plants: Peroxidase and catalase involvement. *Ann. Bot.* **2004**, *93*, 53–60. [CrossRef] [PubMed]
57. Noctor, G.; Mhamdi, A.; Chaouch, S.; Han, Y.; Neukermans, J.; Marquez-Garcia, B.; Queval, G.; Foyer, C.H. Glutathione in plants: An integrated overview. *Plant Cell Environ.* **2012**, *35*, 454–484. [CrossRef] [PubMed]
58. Wang, Y.; Wang, Q.; Zhao, Y.; Han, G.; Zhu, S. Systematic analysis of maize class III peroxidase gene family reveals a conserved subfamily involved in abiotic stress response. *Gene* **2015**, *566*, 95–108. [CrossRef] [PubMed]
59. Francoz, E.; Ranocha, P.; Nguyen-Kim, H.; Jamet, E.; Burlat, V.; Dunand, C. Roles of cell wall peroxidases in plant development. *Phytochemistry* **2015**, *112*, 15–21. [CrossRef] [PubMed]
60. Frey, C.; Álvarez, R.; Encina, A.; Acebes, J.L. Tomato graft union failure is associated with alterations in tissue development and onset of cell wall defense responses. *Agronomy* **2021**, *11*, 1197. [CrossRef]

antioxidants

MDPI

Article

Exogenous GABA-Ca Alleviates Growth Inhibition Induced by a Low-P Environment in Peanuts (*Arachis hypogaea*)

Zhiyu Sun [1], Mingzhu Ma [1], Huan Liu [1], Dongbing Tao [2], Shaikh Amjad Salam [1], Xiaori Han [1], Yifei Liu [1,3,4,*] and Jean Wan Hong Yong [5,*]

1 China-Australia Joint Laboratory for Plant Nutrition and Germplasm Resources Innovation, College of Land and Environment, National Engineering Research Center for Efficient Utilization of Soil and Fertilizer Resources, Northeast China Plant Nutrition and Fertilization Scientific Observation and Research Center for Ministry of Agriculture and Rural Affairs, Shenyang Agricultural University, Shenyang 110866, China
2 College of Food Science, Shenyang Agricultural University, Shenyang 110866, China
3 Institute of Agriculture, The University of Western Australia, Perth, WA 6009, Australia
4 School of Biological Sciences, The University of Western Australia, Perth, WA 6009, Australia
5 Department of Biosystems and Technology, Swedish University of Agricultural Sciences, 23456 Alnarp, Sweden
* Correspondence: liuyifei@syau.edu.cn (Y.L.); jean.yong@slu.se (J.W.H.Y.)

Abstract: Phosphorus (P) deficiency is a major global factor constraining peanut production. Exogenous γ-aminobutyric acid (GABA) and Ca^{2+} are essential to improve stress resilience in peanuts growing under low-P conditions. This study therefore examined the detailed physiological effects of GABA-Ca on restoring peanut growth under low-P conditions. These included the root–shoot ratio, leaf nutrients, photochemical activity, reactive oxygen species (ROS), cyclic electron flow (CEF), ATP synthase activity, and the proton gradient (ΔpH), all of which were measured under low-P (LP, 0.5 mM) and optimized-P (1 mM) conditions. Specifically, supplying GABA-Ca under LP conditions regulated the ΔpH by causing adjustments in CEF and ATP synthase activities, buffering the photosystems' activities, restoring the antioxidant enzyme system, and lowering ROS production. Interestingly, exogenous GABA-Ca restored peanut growth under low-P conditions, possibly by the putative signaling crosstalk between GABA and Ca^{2+}. The plausible signal amplification between GABA and Ca^{2+} suggested that the combination of GABA and Ca, may offer an effective strategy for enhancing peanut adaptation to low-P conditions. Moving forward, the strategic supplementation of GABA-Ca, either during cultivation or through the formulation of novel fertilizers, opens up many possibilities for better and more resilient plant production in soils with low P.

Keywords: gamma-aminobutyric acid; calcium; signal crosstalk effect; p deficiency; cyclic electron flow; ROS toxicity

1. Introduction

Phosphorus (P) is a critical major limiting nutrient for plant growth; in particular, its availability is restricted in tropical regions, with a resulting impact on global food production [1,2]. Varying degrees of P limitation are common in terrestrial ecosystems, and the extent of phosphorus limitation varies in different climate zones and regions [1]. P has low mobility and a high fixation rate in soil [2,3] resulting in less than 25% of phosphatic fertilizer being utilized during the growing period. The high reliance of conventional and often intensive farming activities on phosphatic fertilizers is inevitable for maintaining high production in order to cope with increasing food demands [4,5]. The long-term and excessive use of phosphatic fertilizer has caused negative environmental issues such as the eutrophication of water bodies and poor drinking water quality. In China, about 131 major lakes are classified as highly eutrophic [6]. Furthermore, the intense and unsustainable use of phosphatic rocks could deplete the finite resource of naturally occurring phosphatic rocks by 2050 [7,8]. Moreover, the effective management of phosphate-based fertilizers to improve agricultural productivity while maintaining environmental sustainability is indeed a very challenging task globally. Specifically for China, a country with a large population, the efficient management of P resources, P utilization, sustainable food production, and food security requires strategic, science-based, multi-disciplinary, and multi-stakeholder policies [9].

P is an essential element for plant growth and production [10,11]. It also regulates nucleic acid, protein, ATP, and various P-containing enzymes in plants. It is involved in important physiological and biochemical processes in plants [10,11]. Even a slight P deficiency will lower photosynthesis and cause growth inhibition in peanuts [12]. Low P triggers ROS accumulation and damages the photosynthetic membranes of peanut plants [13]. Additionally, an increase in the ΔpH of peanut thylakoids under low-P stress induces changes in photosynthetic control and strengthens the non-photochemical quenching (NPQ) photoprotection mechanism, along with cyclic electron flow (CEF) around PSI.

Peanut is an important oilseed and leguminous crop that is highly sensitive to variation in P availability in soils [14]. In China, and globally, peanut production in low-yielding fields is often associated with cultivation in P-limiting environments [15]. Peanuts are physiologically sensitive to the availability of P, and minor phosphorus deficiency will lower photosynthesis and increase ROS toxicity, ultimately reducing growth [12,13]. As low-P-induced ROS stress is linked to photosynthetic perturbations, efforts to reduce ROS and enhance photosynthesis are crucial for restoring peanut production in P-deficient soils.

GABA (γ-aminobutyric acid) is well known and proven for physiological improvement, particularly for mental and physical relief. Because of its non-toxic and health-beneficial properties, it is widely used in the food industry as a natural edible additive [16–19]. In recent years, GABA has been gradually applied in agricultural production due to the large-scale industrial production of bio-based green pathways [20–22]. Supplemental GABA accumulates rapidly in plants and enhances their resistance to abiotic stress [21]. A stress-induced increase in GABA synthesis also leads to changes in the intracellular calcium flux [23]. Glutamate decarboxylase (GAD), a key enzyme involved in the GABA shunt pathway, was identified as a calmodulin (CaM)-binding protein that can be stimulated by Ca^{2+} and CaM [24]. Stress in plants generally elevates the levels of Ca^{2+} and GABA simultaneously; thus, the application of both supplements may offer an avenue to help plants adapt to unfavorable conditions [25]. Based on the literature, Ca^{2+} is an important second messenger in plants [26,27]. Under abiotic stress, increasing the levels of free Ca^{2+} in the cytoplasm and binding with calmodulin will initiate various physiological and biochemical processes [26,28]. GABA is known to be involved in regulating plant metabolism through signaling molecules, especially during unfavorable conditions [21]. Ma et al. [29] found that the interplay of Ca^{2+} and GABA improved the stress resilience of barley, particularly during exposures to high NaCl levels. Interestingly, Ca^{2+} signaling is involved in the accumulation of GABA under stress; GABA also transmits Ca^{2+} signaling through receptors to trigger downstream signaling cascades [30]. There is plausible evidence of a complex system of crosstalk between GABA and Ca^{2+} signaling. Xie et al. [31] also suggested that utilization of GABA and Ca^{2+} interactions possibly enhanced the resilience of soybeans to high NaCl levels.

From earlier studies, GABA effectively alleviated the growth inhibition and photoinhibition of peanuts grown under low-P stress [13]. Exogenous applied Ca^{2+} treatments were shown to enhance the photosynthesis of peanuts under P-limited environments [15]. However, it remains unclear whether the combination of GABA and Ca can deliver positive effects on enhancing peanuts' resilience to low P. Thus, this study aimed to evaluate the physiological and biochemical effects of using Ca^{2+}, GABA, and GABA-Ca treatments on peanuts, measured under low-P (LP, 0.5 mM) and optimized-P (1 mM) conditions.

2. Materials and Methods

The Chinese peanut (*Arachis hypogaea* L.) cultivar "Xiaobaisha" was used in the study. Uniformly sized peanut seeds were selected and transplanted into 60 pots (11.4 cm height, 13.6 cm diameter, 1 seed per pot), each containing an equal amount of sand. Each treatment had 12 pots of peanut plants to provide sufficient biological materials for the different analyses, and there were 3 independent biological replicates for each measurement. Peanut seeds were cultured in an incubator (Conviron, Winnipeg, MB, Canada). Peanuts were grown on a 12/12 h photoperiod (light intensity, 1000 μmol quanta m^{-2} s^{-1}), 28/23 °C day/night temperature, and 60% relative air humidity. Prior to treatment, Hoagland's nutrient solution was applied every 3 days. After 21 days of planting, pure water was added to flush away the local residual nutrients and salt patches from the pots.

The pots with uniformly sized peanut seedlings were then divided into five treatments: (1) control plants (CK) (normal P level + foliar spray of ultrapure water); (2) LP (low P level + foliar spray of type 1 ultrapure water); (3) LP + Ca (low P level + foliar spray of 15 mM $CaCl_2$); (4) LP + G (low P level + foliar spray of 10 mM GABA); and (5) LP + GCa (low-P stress + foliar spray of GABA-Ca solution contained 10 mM GABA and 15 mM $CaCl_2$). Hoagland's nutrient solution was prepared in accordance with the previous studies [12,13]. Starting 27 days after planting, the leaves of the peanut plants were sprayed twice a day (at 8:00 and 16:00) for three days (at 27, 28, and 29 days after planting) according to different treatments. With reference to Hoagland's nutrient solution, it was determined that the low-P-concentration solution (−P, 0.5 mM) corresponded approximately to having 15.5 mg/kg ± 3 mg/kg of soil available P, typically found in soils of medium-yielding to low-yielding peanut fields; the normal-P-level solution (CK, 1 mM) was equivalent to that of high-yielding peanut fields with soil available phosphorus exceeding 30 mg/kg.

Two types of Hoagland's nutrient solution were prepared in the form of KH_2PO_4 for simulating normal CK plants (1 mM) and low P for LP plants (0.5 mM) 22 days after planting. The levels of nitrogen (N) and potassium (K) in the two P levels of Hoagland's nutrient solution were balanced by altering the amounts of KNO_3 and NH_4NO_3. Every two days, 25 mL of the nutrient solution for each treatment was supplied.

At 30 days after planting, chlorophyll fluorescence and P700 parameters were measured on the third youngest fully expanded leaf of peanut plants using a Dual-PAM 100 (Heinz Walz, Effeltrich, Germany) according to the method of [12,26]. The light intensity gradient setting of the RLCs was set to 0, 22, 48, 71, 144, 226, 284, 513, 771, 1190, 1467, and 1821 μmol quanta·m^{-2}·s^{-1}. The CEF was determined by the formula CEF = ETR (I) − ERT (II), and the ratio of the quantum yield of CEF was Y (II) [Y (CEF)/Y (II)] = [Y (I) − Y (II)]/Y (II) [13,26]. ΔpH and membrane potential (ΔΨ) were measured simultaneously using a Dual-PAM 100 with a P515/535 module, following the method of [13,26].

The net photosynthetic rate (Pn) was monitored on the third youngest fully expanded leaf with an open gas exchange system (GFS 3000; Heinz Walz, Effeltrich, Germany) at 30 days after planting. The setting parameters of the instrument program were chosen with reference to the previous method [12,13]. Relative chlorophyll concentration was determined with a chlorophyll meter (SPAD-502 plus, Tokyo, Japan) at 31 days after planting.

At 32 days after planting, the superoxide radical ($O_2^{\bullet-}$) generation rate, leaf hydrogen peroxide (H_2O_2) concentration, and MDA concentration of peanut leaves were measured. Malondialdehyde (MDA) in peanut leaves was assayed following the method of Heath and Packer [32,33] and our modified version for peanuts [13]. Peanut leaf samples (0.2 g) frozen in liquid nitrogen were homogenized with trichloroacetic acid and centrifuged. The absorbance of thiobarbituric acid-reactive substances in the supernatant was measured at 450, 532, and 600 nm for the calculation of MDA content. H_2O_2 in peanut leaves was assayed following the method of [13] with modifications. Peanut leaf samples (0.2 g) frozen in liquid nitrogen were homogenized with acetone and centrifuged. The supernatant was mixed with 5% TiCl4 and 2 mol H_2SO_4 and centrifuged at 3000 r/min for 10 min. The supernatant was taken, and its absorbance at 415 nm was measured; the H_2O_2 content was then calculated. The generation rate of $O_2^{\bullet-}$ was assayed following the method of [34,35] and the modified version of Sun et al. [13]. Peanut leaf samples (0.2 g) frozen in liquid nitrogen were homogenized with 50 mM phosphate buffer (PBS, pH 7.8) and centrifuged at 10,000 r/min for 20 min at 4 °C. The supernatant was mixed with 1 mM hydroxylamine hydrochloride and stored at 25 °C for 60 min. The mixture was supplemented with 17 mM p-aminobenzene sulfonic acid and 7 mM 1-naphthylamine and stored at 25 °C for 20 min. The absorbance of the mixture at 530 nm was measured, and the $O_2^{\bullet-}$ content (μmol/L) was calculated.

Antioxidant enzymes—superoxide dismutase (SOD), peroxidase (POD), ascorbate peroxidase (APX), and catalase (CAT)—were extracted and their activity assayed according to the previous methods [36–39]. Peanut leaf samples (0.5 g) frozen in liquid nitrogen were homogenized with 50 mM PBS (pH 7.8) and centrifuged at 10,000 r/min for 20 min at 4 °C. The supernatant was taken and mixed with PBS containing nitroblue tetrazolium, EDTA-Na_2, methionine, and riboflavin. After light treatment, its absorbance at 560 nm was measured, and the SOD activity was calculated. Peanut leaf samples (0.5 g) frozen in liquid nitrogen were homogenized with 200 mM PBS (pH 6.0) and centrifuged at 10,000 r/min for 20 min at 4 °C. The supernatant was taken and mixed with PBS containing guaiacol and H_2O_2. Its absorbance at 470 nm was measured at 1 min intervals for a total of 3 min. The POD activity was calculated based on the change in absorbance values. Peanut leaf samples (0.5 g) frozen in liquid nitrogen were homogenized with 0.15 M PBS (pH 7.8) and centrifuged at 10,000 r/min for 20 min at 4 °C. The supernatant was taken and mixed with PBS containing H_2O_2. The absorbances at 240 nm were measured at 1 min intervals for a total of 4 min. The CAT activity was calculated based on the decrease in absorbance values. Peanut leaf samples (0.5 g) frozen in liquid nitrogen were homogenized with 0.05 M PBS (pH 7.0) and centrifuged at 10,000 r/min for 20 min at 4 °C. The supernatant was taken and mixed with PBS containing EDTA-Na_2, PBS containing ascorbic acid, and PBS containing H_2O_2. Its absorbance at 290 nm was measured over a period of 40 s, and the APX activity was calculated.

At 33 days after planting, leaf area and plant height were measured. Peanut leaves (0.2 g DW) were digested with a mixture of H_2SO_4 and H_2O_2. Leaf N concentrations were determined using the micro-Kjeldahl method [40]. Leaf P concentrations were determined using the vanadomolybdate method [41]. The mixture contained peanut leaves (0.1 g DW), 5 mL HNO_3 (65%, guaranteed reagent), 2 mL H_2O (type 1 ultrapure water), and 1 mL H_2O_2 (30%, guaranteed reagent); the leaves in the mixture were digested with a microwave system (Milestone Ethos one, Italy). Program settings followed previous studies [13]. The amounts of potassium (K), calcium (Ca), magnesium (Mg), iron (Fe), manganese (Mn), copper (Cu), and zinc (Zn) were determined by flame atomic absorption spectrometry (AA-4002400304, PerkinElmer, Waltham, MA, USA). Ca, Mg, Fe, Mn, Cu, and Zn content were obtained as metal concentrations and DW Values.

The experiment adopted a random block design. The results presented the mean values and standard errors of three biological replicates of a final experiment. A one-way ANOVA model including a homogeneity-of-variance test and Fisher's LSD test in Origin 2024 was used for statistical analysis. The results were presented as mean values and standard errors of three biological replicates. Different letters above columns in the figures indicate significant differences among the variants ($p \leq 0.05$). All graphs were plotted using GraphPad Prism 9.5 software.

3. Results

3.1. Growth

The peanut plants' growth and physiological parameters were tested through the utilization of GABA and calcium supplementation. Compared with CK, plants that had received LP treatment showed lower plant height, plant dry weight, SPAD value, and Pn (Figure 1a–c,e). The treatment LP + GCa enhanced the plant height, plant dry weight, SPAD value, and Pn compared with LP. LP treatment significantly increased the root–shoot ratio of the peanut (Figure 1d), and LP + GCa significantly decreased the root–shoot ratio

(Figure 1c). The LP + G group showed greater plant height than the LP + Ca group; the LP + Ca group showed greater plant dry weight and SPAD values than the LP + G group (Figure 1a–c).

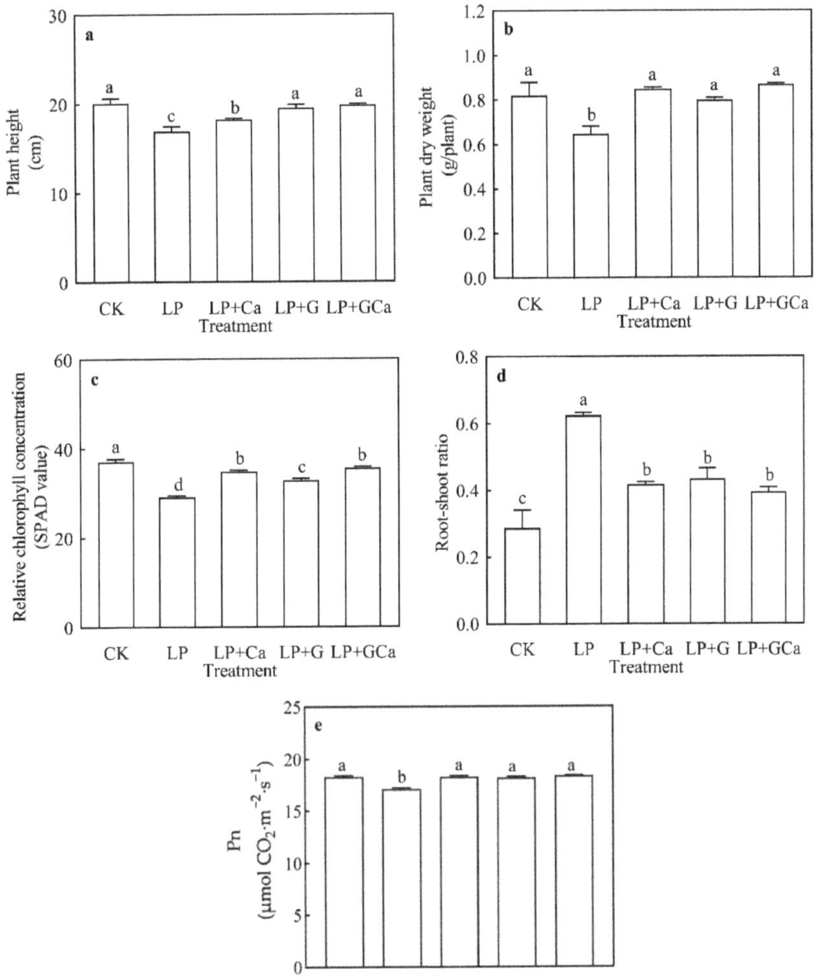

Figure 1. Effects of exogenous GABA-Ca on growth and development in peanuts under low-P stress. Plant height (**a**), Plant dry weight (**b**), Relative chlorophyll concentration (**c**), Root-shoot ratio (**d**), and Pn (**e**). Values are means of three biological replicates ± SE (n = 3). Different and unshared letters above columns in the figures indicate significant differences among the 5 variants ($p \leq 0.05$).

3.2. *Mineral Element*

Compared with CK, LP treatment resulted in a lower P concentration in leaves. According to the obtained results, LP + G and LP + GCa significantly increased the observed nutrient levels in peanut plants. LP + G treatment showed higher P concentration in leaves than LP + Ca (Figure 2a). LP + G and LP + GCa treatments significantly increased the N concentration of peanut leaves, LP + Ca, but the effect of the applied treatment was not obvious (Figure 2b).

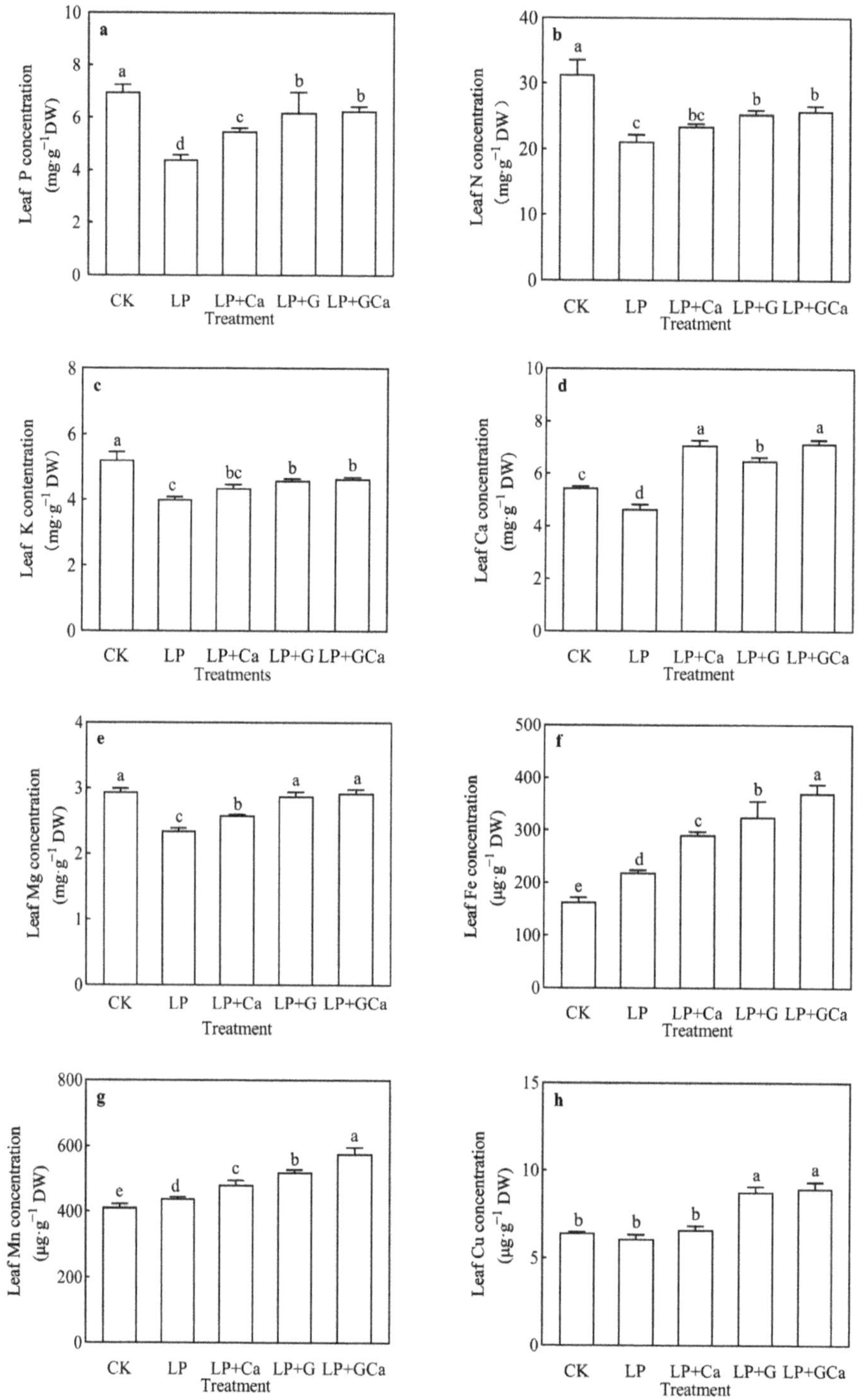

Figure 2. *Cont.*

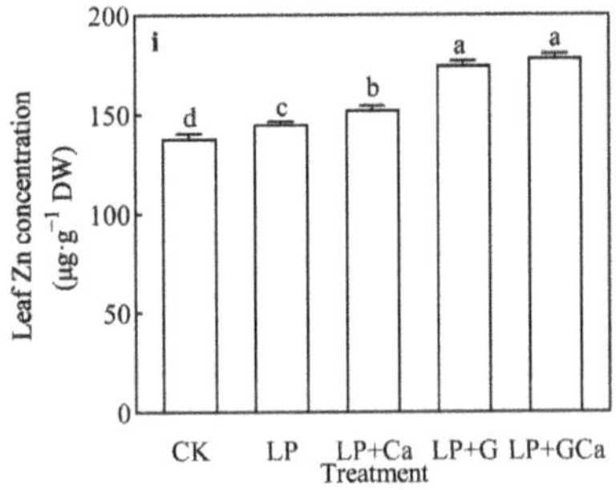

Figure 2. Effects of exogenous GABA-Ca on P (**a**), N (**b**), K (**c**), Ca (**d**), Mg (**e**), Fe (**f**), Mn (**g**), Cu (**h**), and Zn (**i**) levels in peanut leaves under low-P stress. Values are means of three biological replicates ± SE (n = 3). There are different and unshared letters above columns in the figures which indicate significant differences among the 5 variants ($p \leq 0.05$).

3.3. ROS

The LP treatment caused a higher leaf $O_2^{\bullet-}$ generation rate and H_2O_2 level compared with CK (Figure 3a,b). Furthermore, exogenous GABA-Ca alleviated oxidative damage in peanut leaves under LP stress, as indicated by a significant reduction in MDA content (Figure 3c). LP + Ca, LP + G, and LP + GCa treatments significantly reduced the ROS in peanut leaves under low P. The combined application of GABA and Ca proved more effective than the sole application of GABA or Ca.

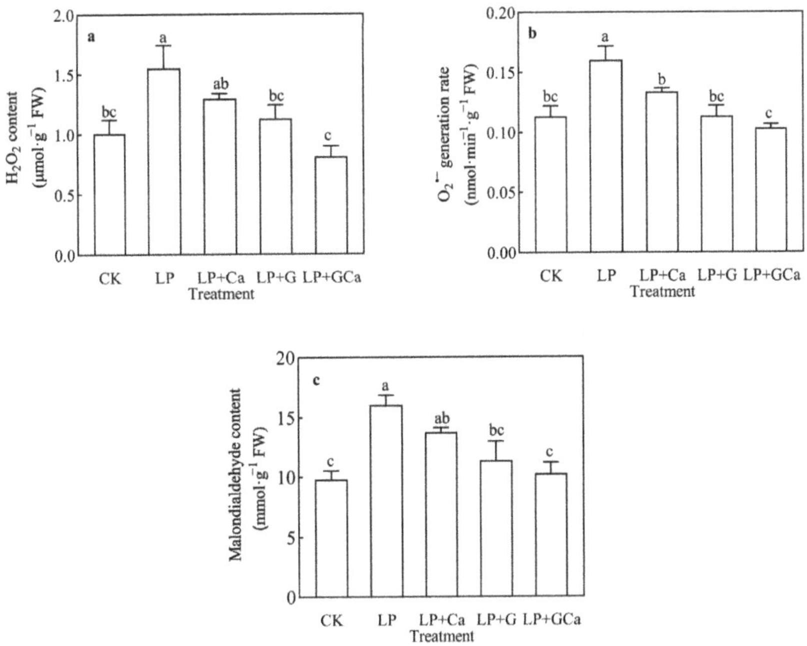

Figure 3. Effects of exogenous GABA-Ca on H_2O_2 content (**a**), $O_2^{\bullet-}$ generation rate (**b**), and MDA content (**c**) in peanut leaves under low-P stress. Values are means of three biological replicates ± SE (n = 3). Different and unshared letters above columns in the figures indicate significant differences among the 5 variants ($p \leq 0.05$).

3.4. Antioxidant Enzyme Activity

Compared with CK, LP treatment induced higher antioxidant enzymatic activity in leaves. The application of GABA and Ca^{2+} significantly enhanced the antioxidant activity of the peanut plants except for control or CK plants (Figure 4a,b,d). The LP + GCa treatment showed a significant difference from all other remaining treatments. The LP + GCa treatment significantly enhanced SOD, POD, CAT, and APX enzyme activities (Figure 4a–d). However, the LP + G treatment showed higher antioxidant enzyme activities compared with LP + Ca.

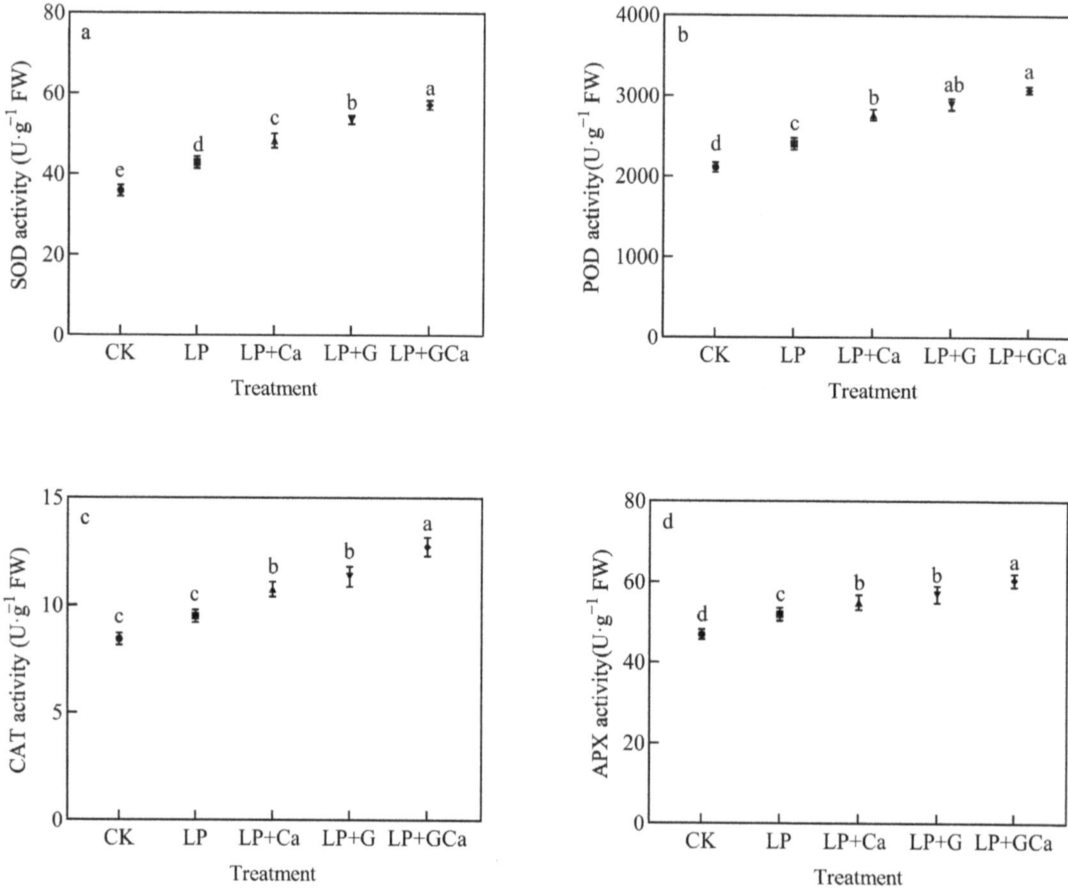

Figure 4. Effects of exogenous GABA-Ca on SOD (**a**), POD (**b**), CAT (**c**), and APX (**d**) activity in peanut leaves under low-P stress. Values are means of three biological replicates ± SE ($n = 3$). Different and unshared letters above columns in the figures indicate significant differences among the 5 variants ($p \leq 0.05$).

3.5. Photosystem I and II Activity

Compared with CK, LP treatment was associated with lower Y (I), Y (II), and F_v/F_m in leaves. Compared with LP, LP + Ca, LP + G, and LP + GCa treatments significantly increased the F_v/F_m in peanut leaves under low-P stress (Figure 5a). Compared with LP, LP + Ca, LP + G, and LP + GCa treatments significantly decreased the NPQ in peanut leaves under low-P stress (Figure 5b). CK and LP + GCa had the highest Y (I), followed by LP + Ca and LP + GABA, with the lowest values observed in LP (Figure 5c). CK and

LP + GCa treatments had the highest Y (II), followed by LP + Ca, while the lowest was observed in LP treatment (Figure 5d).

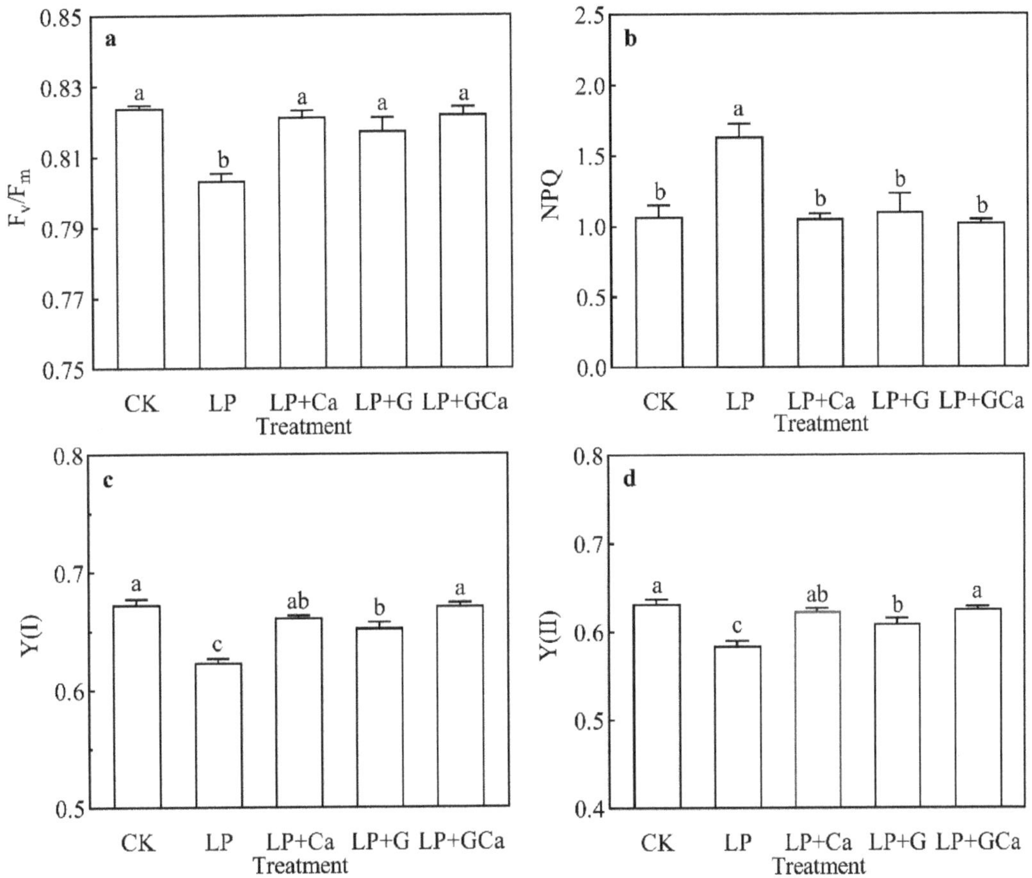

Figure 5. Effects of exogenous GABA-Ca on F_v/F_m (**a**), NPQ (**b**), Y (I) (**c**), and Y (II) (**d**) in peanut leaves under low-P stress. Values are means of three biological replicates ± SE (n = 3). Different and unshared letters above columns in the figures indicate significant differences among the 5 variants ($p \leq 0.05$).

3.6. ETR and CEF

The light response curve indicated the relative electron flow rate of the photosystem under different light intensities (Figure 6a,b). The difference between ETR (II) and ETR (I) was cyclic electron flow (Figure 6c). Compared with CK, LP treatment was associated with lower ETR (II) and ETR (I) in leaves (Figure 6a,b). LP + GCa treatment enhanced both ETR (II) and ETR (I) under high-light conditions compared with LP (Figure 6a,b). Compared with CK, LP treatment was associated with higher CEF in leaves (Figure 6c). LP + Ca was associated with higher ETR (II), ETR (I), and CEF compared with LP + G (Figure 6a–c).

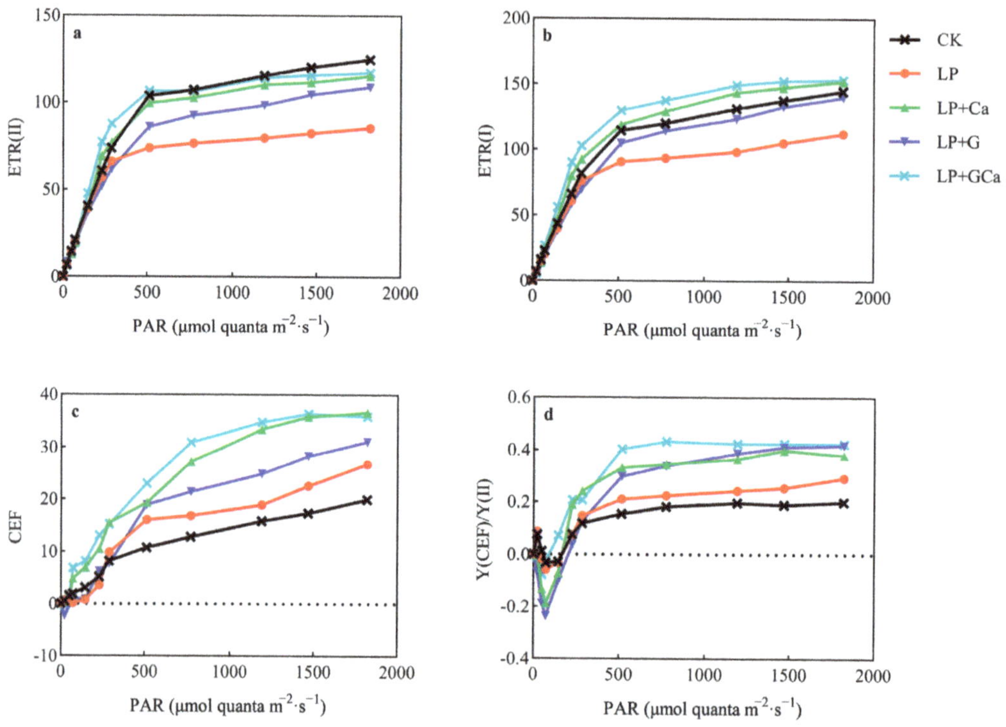

Figure 6. Effects of exogenous GABA-Ca on ETR (II) (**a**), ETR (I) (**b**), CEF (**c**), and Y (CEF)/Y (II) (**d**) in peanut leaves under low-P stress. Values are means of three biological replicates ± SE (n = 3).

3.7. ΔpH and ΔΨ

The increased ΔpH of the thylakoid indicated thylakoid lumen acidification. Compared with CK, LP treatment was associated with a higher ΔpH of thylakoids in leaves (Figure 7a,b). Compared with LP treatment, LP + Ca, LP + G, and LP + GCa treatments significantly reduced the ΔpH of thylakoids in peanut leaves under low-P stress (Figure 7b).

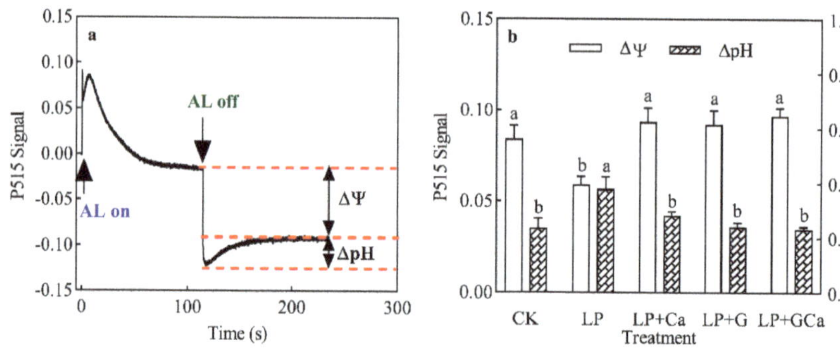

Figure 7. Effects of exogenous GABA-Ca on the proton gradient (ΔpH) and membrane potential (ΔΨ) (**a**,**b**) in peanut leaves under low-P stress. Values are means of three biological replicates ± SE (n = 3). Different and unshared letters above columns in the figures indicate significant differences among the 5 variants ($p \leq 0.05$).

3.8. Thylakoid Membrane Integrity and ATP Synthase Activity

The faster decay of the P515 signal in the dark indicated that the thylakoid membrane integrity was impaired (Figure 8a). The lower decay of the P515 signal after light exposure indicated that ATP synthase activity was inhibited (Figure 8b). Compared with CK, LP treatment was associated with lower thylakoid membrane integrity and ATP synthase activity in leaves (Figure 8a,b). LP + GCa treatment significantly enhanced thylakoid membrane integrity and ATP synthase activity (Figure 8a,b).

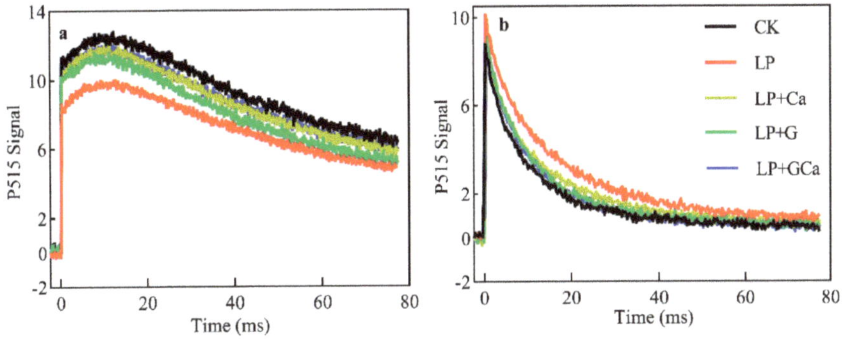

Figure 8. Effects of exogenous GABA-Ca on rapid P515 kinetics (**a**,**b**) in peanut leaves under low-P stress. Rapid kinetics over 1 h in darkness: thylakoid membrane integrity (**a**). Rapid kinetics under light exposure for 10 min at 1000 μmol photons·m^{-2}·s^{-1} and darkness for 4 min: ATP synthase activity (**b**) in peanut leaves under low-P stress. Values are means of three biological replicates ± SE ($n = 3$).

4. Discussion

Low availability of P in soils is one of the key factors affecting crop production [1], and the cultivation of peanuts is highly sensitive to P supply [12]. A search of the literature revealed that there are limited studies pertaining to the improvement of growth and physiological resilience to low-P conditions for peanuts. Guo et al. [30] and Wu et al. [42] demonstrated that exogenous GABA application was effective in improving plants' resilience to a variety of abiotic stresses, and Sun et al. [13] recently confirmed this statement. Exogenous Ca^{2+} is also beneficial for mitigating low-P stress in peanuts [15]. Additionally, there may be complex signal crosstalk between GABA and Ca^{2+} [30]. At present, there is limited information regarding the effects of GABA-Ca on strengthening plants' resilience to low-P conditions. Our study demonstrated that a combination of GABA-Ca, containing 10 mM GABA and 15 mM Ca^{2+}, alleviated the sub-optimal physiological status of peanut plants growing in low-P conditions.

4.1. Effects of GABA-Ca on Growth, Leaf Nutrients, and ROS in Peanut Under Low-P Stress

Exogenous GABA-Ca enhanced the height (Figure 1a), dry matter weight (Figure 1b), and chlorophyll concentration (Figure 1c) of peanut plants under low-P stress and alleviated the influence of P deficiency on peanut growth. Previous studies have shown that GABA and Ca^{2+} restored the growth inhibition of peanuts under low-P stress [13,15]. This study further demonstrated that the combination of GABA and Ca was more effective (Figure 1) in enhancing P (Figure 2a), N (Figure 2b), K (Figure 2c), Ca (Figure 2d), Mg (Figure 2e), Fe (Figure 2f), Mn (Figure 2g), Cu (Figure 2h), and Zn (Figure 2i) concentrations in peanut leaves under low-P stress, while P deficiency lowered the N, P, K, Ca, and Mg concentrations in peanut leaves. GABA generally regulates the activity of plant-specific anion transporters and is also recognized as a plant ion regulator [43,44]. GABA delivered a better effect on mineral element accumulation in leaves than Ca under LP. However, Ca had a better effect on SPAD and plant dry weight than GABA under LP.

Based on our analyses, the $O_2^{\bullet -}$ generation rate (Figure 3b), H_2O_2 content (Figure 3a), MDA content (Figure 3c), antioxidant enzyme activity (Figure 4), and thylakoid membrane integrity (Figure 8a) demonstrated that exogenous GABA-Ca alleviated the photooxidative stress induced in peanut leaves by low-P stress. P deficiency induces oxidative stress and damages cell membranes [13]. The application of GABA has been found to enhance the antioxidant enzyme activity of maize under salt stress and reduce the levels of $O_2 \cdot^-$ and MDA in leaves [45]. GABA also enhances the antioxidant capacity of seedlings under stress and significantly decreases ROS and MDA accumulation [46]. Exogenous GABA improved the activity of antioxidant enzymes such as POD, CAT, and SOD, and it diminished the steady-state level of ROS. Existing study results suggest that GABA application, which reduces the levels of accumulated $O_2 \cdot^-$, H_2O_2, and MDA, results in alleviation of oxidative damage in plants under drought and salt stress [47]. Similarly, Ca^{2+} alleviates oxidative damage in crops under stress by enhancing the activity of enzymes such as SOD and CAT, which are crucial for the detoxifying action of the ROS scavenging system in plants [48]. Sun et al. [15] found that exogenous Ca^{2+} reduces the ROS levels in peanut plants under low-P stress, thereby mitigating oxidative stress damage. GABA also delivered a better effect on antioxidant enzyme activities than Ca under LP. GABA and Ca^{2+} are closely involved in antioxidant defense mechanisms, and combined GABA-Ca treatment further enhances the antioxidant capacity of peanuts under low-P stress and enhances their resistance to low-P stress.

4.2. GABA-Ca Enhances CEF and Alleviates Photoinhibition in Peanut Leaves Under Low-P Stress

Exogenous GABA has been proven to regulate the photosynthetic electron transport chain (PETC) in plants by reducing the stress induced by ROS. The excess electrons in the PETC may increase ROS production, thereby reducing photosynthetic activity. Based on our analyses, exogenous GABA-Ca decreased the acidity of the thylakoid lumen (Figure 8a) and increased linear electron flow (LEF) (Figure 6a,b) and CEF (Figure 6c,d) in leaves of peanuts under low-P stress. The water splitting and plastoquinone (PQ) cycle within the PETC transports H^+ into the thylakoid lumen [49]. Therefore, improvements in the CEF and LEF contribute to H^+ accumulation in the thylakoid lumen [49]. Controlled H^+ efflux via ATP synthase in the thylakoid lumen is a key factor in the photoprotection mechanism of "photosynthetic control" [50]. Under low-P stress, the activity of ATP synthase (Figure 8b) in the thylakoid of leaves decreased, leading to the accumulation of protons in the thylakoid lumen (Figure 7b). This regulates photosynthetic control mechanisms aligned with the previous findings [13,15]. Photosynthetic control mechanisms are crucial for the photosystem because they adjust photosynthetic electron transport to balance energy supply and demand [51]. The increase in ATP synthase activity is the main factor through which GABA-Ca alleviates the acidification of the thylakoid lumen in peanut leaves under low-P stress.

GABA-Ca enhanced the CEF in peanut leaves under low-P stress, particularly under high light intensity (Figure 6). Under LEF-limited conditions, CEF around PSI can contribute a higher ΔpH intensity [52]. Under low-P stress, an increase in the CEF helps to achieve photosynthetic control [13]. Both GABA and Ca enhance the CEF of peanuts under low-P stress and alleviate photoinhibition [13,15]. However, it appears paradoxical that CEF transports more H^+ to the thylakoid lumen, where acidification needs to be reduced. CEF includes two pathways: proton gradient regulation 5 (PGR5)/PRG5-like photosynthetic phenotype 1 (PGRL1) and NADH dehydrogenase-like complex [53]. Lu et al. [54] showed that the inhibiting pathway of CEF significantly reduces the ΔpH in the thylakoid. The increase in CEF (Figure 6) induced by GABA-Ca under low P was not considered to conflict with the decrease in H^+ (Figure 7). The significant increase in and flexible regulation of CEF induced by GABA-Ca contributed to maintaining ΔpH stability, prevent an excessive decrease in ΔpH due to substantial H^+ efflux induced by ATP synthase activity contributing to the excessive acidification of thylakoid lumen. Improvement in CEF is necessary despite the generation of a ΔpH gradient without PSII [54]. Our results suggest that GABA-Ca-induced CEF elevation is beneficial for the regulation of thylakoid acidification.

5. Conclusions

Low-P stress in peanut plants induced severe leaf-level photoinhibition and ROS toxicity that significantly limited photosynthesis and growth. Exogenous GABA-Ca effectively improved the sub-optimal physiological status under low-P stress through various processes: it activated the antioxidant enzyme system, reduced oxidative damage to the photosynthetic membrane, and decreased the ROS toxicity induced by low P. Additionally, exogenous GABA-Ca effectively balanced the H^+ accumulation in the thylakoid lumen by enhancing the CEF and increasing ATP synthase activity. Under low P, exogenous GABA restored key physiological attributes such as plant height, foliar mineral content, and antioxidant enzyme activities. Interestingly, exogenous Ca delivered better effects on foliar chlorophyll (SPAD values), plant dry weight, and photochemical activity than GABA. Compared to the sole application of either Ca^{2+} or GABA, combined GABA-Ca application further alleviated P deficiency symptoms, including increased root–shoot ratios, photoinhibition, and ROS toxicity. The underlying mechanism of GABA-Ca interactions in plants is unclear and requires further research. We postulated that there might be a plausible signal amplification between GABA and Ca^{2+}, suggesting that the combination of GABA and Ca may offer an effective strategy for enhancing peanut adaptation to low-P conditions. Moving forward, the strategic supplementation of GABA-Ca, either during cultivation or by formulating novel fertilizers, opens up many possibilities for better and more resilient plant production.

Author Contributions: Conceptualization: Y.L., X.H. and J.W.H.Y.; methodology: Z.S., M.M. and D.T.; formal analysis: Z.S. and H.L.; data curation: Y.L.; writing—original draft preparation: Z.S.; writing—review and editing: Z.S., Y.L., S.A.S. and J.W.H.Y.; supervision: Y.L. and J.W.H.Y.; project administration: Y.L. All authors have read and agreed to the published version of the manuscript.

Funding: This research was funded by the Liaoning Province International Science and Technology Cooperation Project (2024JH2/101900004); the Liaoning Province Science and Technology Plan Project (2023JH1/10200009); the National Key R&D Program of China (2018YFD1000703, 2018YFD1000700); the National Natural Science Foundation of China (31772391, 31601627, and 31301842); the China Scholarship Council Cooperation Project with Canada, Australia, New Zealand and Latin America for Scientific Research and High-level Personnel Training (Z20230147); and the ARC Linkage Project (LP200100341).

Institutional Review Board Statement: Not applicable.

Informed Consent Statement: Not applicable.

Data Availability Statement: The data presented in this study are available on request from the corresponding author.

Acknowledgments: This study was carried out with support from the China Scholarship Council (No. 202408210273). Special thanks to Nanning Hanhe Biotechnology Co., Ltd. for providing high-quality GABA agents for this study.

Conflicts of Interest: The authors declare no conflicts of interest.

References

1. Hou, E.; Luo, Y.; Kuang, Y.; Chen, C.; Lu, X.; Jiang, L.; Luo, X.; Wen, D. Global meta-analysis shows pervasive phosphorus limitation of aboveground plant production in natural terrestrial ecosystems. *Nat. Commun.* **2020**, *11*, 637. [CrossRef] [PubMed]
2. López-Arredondo, D.L.; Leyva-González, M.A.; González-Morales, S.I.; López-Bucio, J.; Herrera-Estrella, L. Phosphate nutrition: Improving low-phosphate tolerance in crops. *Annu. Rev. Plant Biol.* **2014**, *65*, 95–123. [CrossRef] [PubMed]
3. Dong, R.; Hu, W.; Bu, L.; Cheng, H.; Liu, G. Legume cover crops alter soil phosphorus availability and microbial community composition in mango orchards in karst areas. *Agric. Ecosyst. Environ.* **2024**, *364*, 108906. [CrossRef]

4. Yu, X.; Keitel, C.; Dijkstra, F.A. Global analysis of phosphorus fertilizer use efficiency in cereal crops. *Glob. Food Sec.* **2021**, *29*, 100545. [CrossRef]
5. Roy, E.D.; Richards, P.D.; Martinelli, L.A.; Coletta, L.D.; Lins, S.R.M.; Vazquez, F.F.; Willig, E.; Spera, S.A.; VanWey, L.K.; Porder, S. The phosphorus cost of agricultural intensification in the tropics. *Nat. Plants* **2016**, *2*, 16043. [CrossRef]
6. Zhao, Y.; Deng, X.; Zhang, J.-Y.; Xi, B.; Lu, Q. Progress on preventing and controlling strategies of lake eutrophication in China. *Environ. Technol.* **2010**, *33*, 92–98.
7. Smil, V. Phosphorus in the environment: Natural flows and human interferences. *Annu. Rev. Environ. Resour.* **2000**, *25*, 53–88. [CrossRef]
8. Tilman, D.; Fargione, J.E.; Wolff, B.G.; D'Antonio, C.M.; Dobson, A.P.; Howarth, R.W.; Schindler, D.W.; Schlesinger, W.H.; Simberloff, D.; Swackhamer, D.L. Forecasting agriculturally driven global environmental change. *Science* **2001**, *292*, 281–284. [CrossRef]
9. Chowdhury, R.B.; Moore, G.A.; Weatherley, A.J.; Arora, M. Key sustainability challenges for the global phosphorus resource, their implications for global food security, and options for mitigation. *J. Clean. Prod.* **2017**, *140*, 945–963. [CrossRef]
10. Lambers, H.; Oliveira, R. *Plant Physiological Ecology*, 3rd ed.; Springer: Berlin/Heidelberg, Germany, 2019.
11. Lambers, H. Phosphorus Acquisition and Utilization in Plants. *Annu. Rev. Plant Biol.* **2022**, *73*, 17–42. [CrossRef]
12. Shi, Q.; Pang, J.; Yong, J.W.H.; Bai, C.; Pereira, C.G.; Song, Q.; Wu, D.; Dong, Q.; Cheng, X.; Wang, F.; et al. Phosphorus-fertilisation has differential effects on leaf growth and photosynthetic capacity of *Arachis hypogaea* L. *Plant Soil.* **2020**, *447*, 99–116. [CrossRef]
13. Sun, Z.; Bai, C.; Liu, Y.; Ma, M.; Zhang, S.; Liu, H.; Bai, R.; Han, X.; Yong, J.W.H. Resilient and sustainable production of peanut (*Arachis hypogaea*) in phosphorus-limited environment by using exogenous gamma-aminobutyric acid to sustain photosynthesis. *Ecotox. Environ. Safe* **2023**, *263*, 115388. [CrossRef] [PubMed]
14. Zeng, R.; Chen, T.; Li, X.; Cao, J.; Li, J.; Xu, X.; Zhang, L.; Chen, Y. Integrated physiological, transcriptomic and metabolomic analyses reveal the mechanism of peanut kernel weight reduction under waterlogging stress. *Plant Cell Environ.* **2024**, *47*, 3198–3214. [CrossRef] [PubMed]
15. Sun, Z.; Liu, X.; Zhang, S.; Ma, M.; Bai, R.; Liu, H.; Yi, B.; Han, X.; Liu, Y. Exogenous calcium alleviates phosphorus deficiency-induced photosynthetic inhibition in peanuts (*Arachis hypogaea*). *J. Plant Nutr.* **2022**, *28*, 1055–1066.
16. Foster, A.C.; Kemp, J.A. Glutamate- and GABA-based CNS therapeutics. *Curr. Opin. Pharmacol.* **2006**, *6*, 7–17. [CrossRef]
17. Sun, Y.; Mehmood, A.; Battino, M.; Xiao, J.; Chen, X. Enrichment of gamma-aminobutyric acid in foods: From conventional methods to innovative technologies. *Int. Food Res. J.* **2022**, *162*, 111801. [CrossRef]
18. Hou, D.; Tang, J.; Feng, Q.; Niu, Z.; Shen, Q.; Wang, L.; Zhou, S. Gamma-aminobutyric acid (GABA): A comprehensive review of dietary sources, enrichment technologies, processing effects, health benefits, and its applications. *Crit. Rev. Food Sci. Nutr.* **2023**, *64*, 8852–8874. [CrossRef]
19. Boonstra, E.; de Kleijn, R.; Colzato, L.S.; Alkemade, A.; Forstmann, B.U.; Nieuwenhuis, S. Neurotransmitters as food supplements: The effects of GABA on brain and behavior. *Front. Psychol.* **2015**, *6*, 01520. [CrossRef]
20. Renes, E.; Linares, D.M.; González, L.; Fresno, J.M.; Tornadijo, M.E.; Stanton, C. Production of conjugated linoleic acid and gamma-aminobutyric acid by autochthonous lactic acid bacteria and detection of the genes involved. *J. Funct. Food* **2017**, *34*, 340–346. [CrossRef]
21. Bown, A.W.; Shelp, B.J. Plant GABA: Not just a metabolite. *Trends Plant Sci.* **2016**, *21*, 811–813. [CrossRef]
22. Liu, C.; Wang, H.; Zhang, X.; Ma, F.; Guo, T.; Li, C. Activation of the ABA signal pathway mediated by GABA improves the drought resistance of apple seedlings. *Int. J. Mol. Sci.* **2021**, *22*, 12676. [CrossRef] [PubMed]
23. Knight, M.R.; Campbell, A.K.; Smith, S.M.; Trewavas, A.J. Transgenic plant aequorin reports the effects of touch and cold-shock and elicitors on cytoplasmic calcium. *Nature* **1991**, *352*, 524–526. [CrossRef] [PubMed]
24. Kinnersley, A.M.; Turano, F.J. Gamma aminobutyric acid (GABA) and plant responses to stress. *Crit. Rev. Plant Sci.* **2000**, *19*, 479–509. [CrossRef]
25. Braam, J.; Davis, R.W. Rain-, wind-, and touch-induced expression of calmodulin and calmodulin-related genes in *Arabidopsis*. *Cell* **1990**, *60*, 357–364. [CrossRef] [PubMed]
26. Wu, D.; Liu, Y.; Pang, J.; Yong, J.W.H.; Chen, Y.; Bai, C.; Han, X.; Liu, X.; Sun, Z.; Zhang, S.; et al. Exogenous calcium alleviates nocturnal chilling-induced feedback inhibition of photosynthesis by improving sink demand in peanut (*Arachis hypogaea*). *Front. Plant Sci.* **2020**, *11*, 607029. [CrossRef]
27. Song, Q.; Liu, Y.; Pang, J.; Yong, J.W.H.; Chen, Y.; Bai, C.; Gille, C.; Shi, Q.; Wu, D.; Han, X.; et al. Supplementary calcium restores peanut (*Arachis hypogaea*) growth and photosynthetic capacity under low nocturnal temperature. *Front. Plant Sci.* **2020**, *10*, 1637. [CrossRef]
28. Liu, J.; Hu, J.; Li, Y.; Li, G.; Wu, H. Chapter 10-Calcium channels and transporters in plants under salinity stress. In *Calcium Transport Elements in Plants*; Upadhyay, S.K., Ed.; Academic Press: Cambridge, MA, USA, 2021; pp. 157–169.
29. Ma, Y.; Wang, P.; Gu, Z.; Tao, Y.; Shen, C.; Zhou, Y.; Han, Y.-B.; Yang, R. Ca^{2+} involved in GABA signal transduction for phenolics accumulation in germinated hulless barley under NaCl stress. *Food Chem. X* **2019**, *2*, 100023. [CrossRef]
30. Guo, Z.; Gong, J.; Luo, S.; Zuo, Y.; Shen, Y. Role of gamma-aminobutyric acid in plant defense response. *Metabolites* **2023**, *13*, 741. [CrossRef]
31. Xie, C.; Sun, M.; Wang, P.; Yang, R. Interaction of Gamma-Aminobutyric Acid and Ca^{2+} on Phenolic Compounds Bioaccumulation in Soybean Sprouts under NaCl Stress. *Plants* **2022**, *11*, 3503. [CrossRef]

32. Heath, R.L.; Packer, L. Photoperoxidation in isolated chloroplasts: I. Kinetics and stoichiometry of fatty acid peroxidation. *Arch. Biochem. Biophys.* **1968**, *125*, 189–198. [CrossRef]
33. Gupta, P.; Srivastava, S.; Seth, C.S. 24-Epibrassinolide and Sodium Nitroprusside alleviate the salinity stress in *Brassica juncea* L. cv. Varuna through cross talk among proline, nitrogen metabolism and abscisic acid. *Plant Soil* **2017**, *411*, 483–498. [CrossRef]
34. Elstner, E.F.; Heupel, A. Inhibition of nitrite formation from hydroxylammoniumchloride: A simple assay for superoxide dismutase. *Anal. Biochem.* **1976**, *70*, 616–620. [CrossRef] [PubMed]
35. Nguyen, T.T.; Uthairatanakij, A.; Srilaong, V.; Laohakunjit, N.; Kato, M.; Jitareerat, P. Impact of electron beam irradiation on the chlorophyll degradation and antioxidant capacity of mango fruit. *Appl. Biol. Chem.* **2021**, *64*, 19. [CrossRef] [PubMed]
36. Egley, G.H.; Paul, R.N.; Vaughn, K.C.; Duke, S.O. Role of peroxidase in the development of water-impermeable seed coats in *Sida spinosa* L. *Planta* **1983**, *157*, 224–232. [CrossRef]
37. Misra, H.P.; Fridovich, I. The Role of Superoxide Anion in the Autoxidation of Epinephrine and a Simple Assay for Superoxide Dismutase. *J. Biol. Chem.* **1972**, *247*, 3170–3175. [CrossRef]
38. Nakano, Y.; Asada, K. Hydrogen Peroxide is Scavenged by Ascorbate-specific Peroxidase in Spinach Chloroplasts. *Plant Cell Physiol.* **1981**, *22*, 867–880. [CrossRef]
39. Hadwan, M. Simple spectrophotometric assay for measuring catalase activity in biological tissues. *BMC Biochem.* **2018**, *19*, 1. [CrossRef]
40. Dahnke, W.C.; Johnson, G.V. Testing Soils for Available Nitrogen. In *Soil Testing and Plant Analysis*; SSSA Book Series; SSSA: Madison, WI, USA, 1990; pp. 127–139.
41. Jones, J.B. Phosphorus toxicity in tomato plants: When and how does it occur? *Commun. Soil Sci. Plant Anal.* **1998**, *29*, 1779–1784. [CrossRef]
42. Wu, X.; Jia, Q.; Ji, S.; Gong, B.; Li, J.; Lü, G.; Gao, H. Gamma-aminobutyric acid (GABA) alleviates salt damage in tomato by modulating Na^+ uptake, the GAD gene, amino acid synthesis and reactive oxygen species metabolism. *BMC Plant Biol.* **2020**, *20*, 465. [CrossRef]
43. Ramesh, S.A.; Tyerman, S.D.; Xu, B.; Bose, J.; Kaur, S.; Conn, V.; Domingos, P.; Ullah, S.; Wege, S.; Shabala, S.; et al. GABA signalling modulates plant growth by directly regulating the activity of plant-specific anion transporters. *Nat. Commun.* **2015**, *6*, 7879. [CrossRef]
44. Ramesh, S.A.; Kamran, M.; Sullivan, W.; Chirkova, L.; Okamoto, M.; Degryse, F.; McLaughlin, M.; Gilliham, M.; Tyerman, S.D. Aluminum-activated malate transporters can facilitate GABA transport. *Plant Cell* **2018**, *30*, 1147–1164. [CrossRef] [PubMed]
45. Wang, Y.; Gu, W.; Meng, Y.; Xie, T.; Li, L.; Li, J.; Wei, S. Gamma-aminobutyric acid imparts partial protection from salt stress injury to maize seedlings by improving photosynthesis and upregulating osmoprotectants and antioxidants. *Sci. Rep.* **2017**, *7*, 43609. [CrossRef]
46. Salah, A.; Zhan, M.; Cao, C.; Han, Y.; Ling, L.; Liu, Z.; Li, P.; Ye, M.; Jiang, Y. γ-Aminobutyric Acid Promotes Chloroplast Ultrastructure, Antioxidant Capacity, and Growth of Waterlogged Maize Seedlings. *Sci. Rep.* **2019**, *9*, 484. [CrossRef] [PubMed]
47. Zhou, M.; Hassan, M.J.; Peng, Y.; Liu, L.; Liu, W.; Zhang, Y.; Li, Z. γ-Aminobutyric acid (GABA) priming improves seed germination and seedling stress tolerance associated with enhanced antioxidant metabolism, DREB expression, and dehydrin accumulation in white clover under water stress. *Front. Plant Sci.* **2021**, *12*, 776939. [CrossRef] [PubMed]
48. Hu, W.; Wu, Q.; Liu, X.; Shen, Z.; Chen, J.; Liu, T.; Chen, J.; Zhu, C.; Wu, F.; Chen, L.; et al. Comparative proteomic analysis reveals the effects of exogenous calcium against acid rain stress in *Liquidambar formosana Hance* leaves. *J. Proteome Res.* **2016**, *15*, 216–228. [CrossRef]
49. Tikkanen, M.; Mekala, N.R.; Aro, E.-M. Photosystem II photoinhibition-repair cycle protects Photosystem I from irreversible damage. *Biochim. Biophys. Acta (BBA)-Bioenerg.* **2014**, *1837*, 210–215. [CrossRef]
50. Barbato, R.; Tadini, L.; Cannata, R.; Peracchio, C.; Jeran, N.; Alboresi, A.; Morosinotto, T.; Bajwa, A.A.; Paakkarinen, V.; Suorsa, M.; et al. Higher order photoprotection mutants reveal the importance of ΔpH-dependent photosynthesis-control in preventing light induced damage to both photosystem II and photosystem I. *Sci. Rep.* **2020**, *10*, 6770. [CrossRef]
51. Foyer, C.H.; Neukermans, J.; Queval, G.; Noctor, G.; Harbinson, J. Photosynthetic control of electron transport and the regulation of gene expression. *J. Exp. Bot.* **2012**, *63*, 1637–1661. [CrossRef]
52. Yamori, W.; Makino, A.; Shikanai, T. A physiological role of cyclic electron transport around photosystem I in sustaining photosynthesis under fluctuating light in rice. *Sci. Rep.* **2016**, *6*, 20147. [CrossRef]
53. Ma, M.; Liu, Y.; Bai, C.; Yang, Y.; Sun, Z.; Liu, X.; Zhang, S.; Han, X.; Yong, J.W.H. The physiological functionality of PGR5/PGRL1-dependent cyclic electron transport in sustaining photosynthesis. *Front. Plant Sci.* **2021**, *12*, 702196. [CrossRef]
54. Lu, J.; Wang, Z.; Yang, X.; Wang, F.; Qi, M.; Li, T.; Liu, Y. Cyclic electron flow protects photosystem I donor side under low night temperature in tomato. *Environ. Exp. Bot.* **2020**, *177*, 104151. [CrossRef]

antioxidants

MDPI

Article

Physiological and Microstructure Analysis Reveals the Mechanism by Which Formic Acid Delays Postharvest Physiological Deterioration of Cassava

Yannian Che [1,2,3], Zhongping Ding [1,2,3], Chen Shen [2,3,4], Alisdair R. Fernie [5], Xiangning Tang [1,2,3], Yuan Yao [2,3], Jiao Liu [2,3], Yajie Wang [2,3], Ruimei Li [2,3,*] and Jianchun Guo [2,*]

1 School of Life Sciences, Hainan University, Haikou 570228, China; 22110710000011@hainanu.edu.cn (Y.C.); 21220951310010@hainanu.edu.cn (Z.D.); 22220860000067@hainanu.edu.cn (X.T.)
2 National Key Laboratory for Tropical Crop Breeding, Institute of Tropical Bioscience and Biotechnology, Chinese Academy of Tropical Agricultural Sciences, Haikou 571101, China; 2023301120122@webmail.hzau.edu.cn (C.S.); yaoyuan@itbb.org.cn (Y.Y.); liujiao@itbb.org.cn (J.L.); wangyajie@itbb.org.cn (Y.W.)
3 Sanya Research Institute, Chinese Academy of Tropical Agricultural Sciences, Sanya 572024, China
4 College of Plant Science & Technology, Huazhong Agricultural University, Wuhan 430070, China
5 Root Biology and Symbiosis, Max Planck Institute of Molecular Plant Physiology, Am Mühlenberg 1, 14476 Potsdam, Germany; fernie@mpimp-golm.mpg.de
* Correspondence: liruimei@itbb.org.cn (R.L.); guojianchun@itbb.org.cn (J.G.); Tel.: +86-898-6696-2953 (R.L. & J.G.)

Abstract: Formic acid is reported to act as a food preservative and feed additive, but its effects on controlling postharvest physiological deterioration (PPD) development in cassava are unclear. In this study, we assessed the effectiveness of different concentrations of formic acid in attenuating PPD occurrence in fresh-cut cassava. The results showed that the concentration of 0.1% (*v*/*v*) formic acid could significantly delay the occurrence of PPD, and that the higher the concentration of formic acid supplied, the later the occurrence of PPD symptoms. The physiological and biochemical analysis of 0.5%-formic-acid-treated cassava slices revealed that formic acid decreased the degradation of starch, inhibited the accumulation of hydrogen peroxide (H_2O_2), malondialdehyde (MDA), and water-soluble pectin in cassava slices with PPD development, and increased the activities of the antioxidant enzymes ascorbate peroxidase (APX) and glutathione reductase (GR). A microscopic observation showed that the formic acid treatment inhibited the enlargement of the intercellular space during the cassava PPD process, which suggests that the formation of an intercellular layer of the cell wall was inhibited by formic acid. This study thus revealed the mechanism used by formic acid to extend the cassava shelf life; however, a detailed evaluation of the possible side effects on, for example, the cyanide content will be needed to categorically ensure the safety of this method.

Keywords: formic acid; shelf life; cassava; postharvest; cell wall; pectin

1. Introduction

The starch-rich cassava (*Manihot esculenta* Crantz) tuberous roots are an important source of human food, animal feed and industrial bioenergy. Considering the advantages of cassava in various applications, the demand for cassava is increasing day by day. However, cassava is prone to postharvest physiological deterioration (PPD), which generally starts within 12–72 h after harvest symptoms appear, resulting in discoloration and a bad taste, with the starch quality additionally being seriously impaired [1,2]. The activation and intensity of PPD in cassava is caused by mechanical damage, as evidenced by the first occurrence of PPD symptoms at the injured site. The injuring of the cassava triggers an increase in the respiratory rate, the accumulation of reactive oxygen species (ROS), and the production of cyanide (which is toxic to animals) [3,4]. After mechanical injury, the production of hydrogen peroxide significantly increases within 15 min. Within 4 h after harvest, the malondialdehyde (MDA) content increases in the regions closest to the injury site [5]. The rapid PPD of cassava is a limiting factor in cassava production, leading to significant postharvest losses of up to 29% globally [6]. Therefore, delaying the occurrence of PPD in cassava has become an important issue that producers urgently need to solve.

Formic acid is the smallest organic acid and the precursor for amino acids; therefore, it plays significant roles in organic life. In agricultural production, formic acid is used as a food preservative, feed additive, and non-antibiotic growth promoter [7–9]. The formic acid content decreases gradually during the storage of tofu, and is used as an indicator of tofu freshness [10]. In the process of silage production, the addition of formic acid not only improves the quality of the silage, but also extends its shelf life [8,11,12]. Using feed additives containing formic acid can improve the growth performance of weaned piglets [13]. Feeding diets containing formic acid could regulate intestinal growth and improve the disease resistance of gilthead sea bream [14]. Formic acid fumigation greatly reduces the rot rate of citrus fruits [15]. Formic acid is also used for chitosan dissolution and is used to control the postharvest decay of strawberries and table grape [16,17]. Formic acid inactivates *Salmonella* and maintains the postharvest quality of cherry tomatoes [18]. Formic acid can inhibit the growth of *Bacillus subtilis*, which causes the postharvest spoilage of potatoes, thereby preventing the deterioration and discoloration of potatoes during storage [7]. A mixture of formic acid and levulinic acid is used to control postharvest citrus blueberry disease [19]. However, the effectiveness of formic acid in controlling the PPD of cassava has not been studied.

The objective of this study was to evaluate whether formic acid can control the occurrence of PPD in cassava slices. Another objective is to analyze the potential mechanisms of formic acid in controlling the PPD of cassava.

2. Materials and Methods

2.1. Plant Materials and Treatments

The cassava utilized in this study was harvested from bitter cassava variety SC8 plants grown in a field for 6 months. Immediately after harvest, the cassava plants were transported to the laboratory for experimental use. The cassavas were washed with tap water and ultrapure water in turn and then drained using tissue paper. The cleaned cassavas were evenly cut into 5 mm slices and dipped in deionized water (for control) and 0.1%, 0.3%, 0.5%, 1%, 2%, and 5% (v/v) formic acid for 2 h. Subsequently, the treated cassava slices were rinsed with deionized water, drained, and placed in a plastic storage box with a temperature of 26 °C and a relative humidity of 70%. The changes in the sections of each cassava slice were observed and photographed at 0 h, 6 h, 12 h, 24 h, and 48 h after storage to evaluate the PPD occurrence. Three slices per sample of 0.5%-formic-acid-treated cassava at the 0 h, 6 h, 12 h, 24 h, and 48 h time points were collected and rapidly frozen with liquid nitrogen and then frozen at −80 °C for subsequent utilization.

2.2. Starch Content Measurement

The starch content of the cassava slices treated with 0.5% formic acid and the control after storage for different durations (0, 6, 12, 24, and 48 h) was measured by the anthrone colorimetric method using a plant starch content kit (Comin, Suzhou, China). According to the manufacturer's instructions, the cassava samples were homogenized with 80% ethanol and extracted at 80 °C in a water bath for 30 min, and then centrifuged at 25 °C and 3000× g for 5 min. The precipitate was suspended with distilled water, gelatinized in a 95 °C water bath for 15 min, and then extracted with 9.2 mol/L $HClO_4$ and diluted with distilled water. Subsequently, the mixture of supernatant and 2% anthrone was heated in a boiling water bath for 10 min. After cooling down, the mixture was used to measure the absorbance value at a 620 nm wavelength. The starch content was expressed in mg/g. Three biological replicates were performed.

2.3. H_2O_2 Content Analysis

The hydrogen peroxide (H_2O_2) content of cassava slices treated by 0.5% formic acid and control after storage for different time points (0, 6, 12, 24, and 48 h) was determined based on the titanium sulfate method. Briefly, the cassava samples were homogenized in with 4 °C precooled acetone and centrifuged to obtain the supernatant. The supernatant was mixed with 5% titanium sulfate and concentrated ammonia to obtain the precipitate. The precipitate was then washed with acetone and dissolved with 2 mol/L sulfuric acid. After that, 200 μL liquid was used to determine the absorption value at a 415 nm wavelength. The H_2O_2 content was expressed in μmol/g FW. Three biological replicates were performed of H_2O_2 content measurement.

2.4. Determination of Malondialdehyde Content

The malondialdehyde (MDA) content in cassava was measured based on the thiobarbituric acid method. The cassava samples were homogenized in liquid nitrogen and extracted with 10% TCA solution. The supernatant was mixed with 0.5% TBA and heated at 90 °C for 30 min. The absorbance at 532 nm and 600 nm of supernatant was measured. The MDA content was expressed in nmol/g FW. Three biological replicates were performed for the MDA content measurement.

2.5. Ascorbate Peroxidase Activity

The ascorbate peroxidase (APX) activity was determined using APX activity kit (Comin, Suzhou, China), according to the manufacturer's instructions. An amount of 0.1 g of the sample was weighed and added into 1 mL of 50 mmol/L K_2HPO_4-KH_2PO_4 (pH7.0) for ice bath homogenization. The mixture was centrifuged at 4 °C and 13,000× *g* for 20 min to obtain the supernatant. The mixture of 20 μL of supernatant, 140 μL of K_2HPO_4-KH_2PO_4 (pH 7.0), 20 μL of 0.3 mmol/L AsA, and 0.06 mmol/L H_2O_2 was used to measure the 10 s absorbance (A1) and 130 s absorbance (A2) at a 290 nm wavelength. The APX activity was calculated using the following formula: APX (μmol/min/g FW (fresh weight)) = 3571 × (A1 − A2) ÷ W (sample weight).

2.6. Glutathione Reductase Activity

The glutathione reductase (GR) activity was determined using GR activity kit (Comin, Suzhou, China), according to the manufacturer's instructions. About 0.1 g of the sample was weighed and added into 10 mL of 4 °C precooled 100 mmol/L phosphoric acid buffer (including 1 mmol/L EDTA, pH 7.5) for ice bath homogenization. Subsequently, the mixture was centrifuged at 4 °C and 8000× *g* for 15 min, and the supernatant is the crude enzyme. Next, 150 μL of 100 mmol/L phosphoric acid buffer (including 1 mmol/L EDTA, pH 7.5), 20 μL of 5 mmol/L GSSG solution, 20 μL of crude enzyme liquid, and 10 μL of 4 mmol/L NADPH solution were added to 96-well plates which were mixed immediately. The initial absorbance (A1) and 180 s absorbance (A2) of supernatant were quickly determined at a 340 nm wavelength. The GR activity was calculated using the following formula: GR activity (nmol/min/g FW) = 1072 × (A1 − A2) ÷ W (sample weight).

2.7. Water-Soluble Pectin Content Measurement

To measure the water-soluble pectin (WSP) content, about 0.3 g of the sample was weighed, and 1 mL of 80% ethanol was added prior to rapid homogenization at room temperature; next, the mixture was placed in a water bath at 95 °C for 20 min, and then cooled to room temperature and centrifuged at 25 °C and 4000× *g* for 10 min, and the supernatant was then discarded. The precipitate was subsequently washed with 1.5 mL of 80% ethanol and 1.5 mL of 80% acetone one by one (these solutions were swirled for about 2 min and then centrifuged at 25 °C and 4000× *g* for 10 min, and the supernatant was then discarded). Next, 1 mL 63% ethanol was added to the precipitate (in order to remove starch) and soaked for 15 h, prior to centrifuging at 4000× *g* and 25 °C for 10 min, and then the supernatant was discarded and the precipitate was dried. Next, 3 mg of dried precipitate was fully homogenized in 1 mL of distilled water and the homogenate was centrifuged at 4 °C and 8000× *g* for 10 min with the supernatant representing WSP extract. Subsequently, 50 μL of WSP extract was mixed with 50 μL of 0.1% carbazole reagent and 400 μL of concentrated sulfuric acid. The mixture was heated at 85 °C for 5 min. Next, 200 μL of the solution was taken to measure the absorbance value at a 530 nm wavelength. The galacturonic acid was used for standard curve preparation. The mixture of distilled water, carbazole reagent, and concentrated sulfuric acid was measured as a blank. The WSP content was expressed in mg/g DW (dry weight).

2.8. Microscopic Observation

The cassava samples were stored for 6 h after treatment with (0.5%) or without formic acid, and control (fresh cassava) samples were cut into 1–2 mm^3 small blocks and fixed by microscope fixation solution (Servicebio, Wuhan, China). The tissues were embedded with resin 812, and then ultra-thin sections were produced using an ultra-microtome (Leica UC7). The microstructure were observed and photographed using a HITACHI HT7800 transmission electron microscope.

2.9. Statistical Analysis

All data are represented as means ± SEM of three replicates. A statistical analysis was conducted using GraphPad Prism 8.0 software (San Diego, CA, USA) with one-way analysis of variance (ANOVA) and Dunnett's multiple comparisons tests for measuring significance. $p < 0.05$ was considered statistically significant.

3. Results

3.1. Formic Acid Treatment Maintained the Appearance of Cassava Slices

Visual observations indicated that the CK-treated group (without formic acid) of fresh-cut cassava began to generally appear black and brown stripes within 6 h of storage at 26 °C and 70% relative humidity condition (Figure 1). Moreover, with the extension of storage time, the browning degree of the surface of the slices was deepened. The slight browning was not observed in the 0.1%-formic-acid-treated group until 24 h storage, and the browning was delayed with the increase in formic acid concentration. When the concentration of formic acid was 0.5%, no browning occurred in the root slices 48 h after treatment (Figure 1). Subsequently, we chose the samples treated with 0.5% formic acid at different time points for further study.

3.2. Formic Acid Inhibited the Degradation of Starch during Storage

Starch content is an important quality indicator of cassava after harvest. The occurrence of PPD can lead to the degradation of starch, so this study investigated whether formic acid affects the change in starch content during the occurrence of PPD in cassava. The results show that the starch content of cassava slices in control group decreased gradually with the extension of storage time. While the starch content of the 0.5%-formic-acid-treated group was significantly higher than that in the control group at 12, 24, and 48 h (Figure 2). These results indicate that formic acid reduces starch degradation in the cassava slices during storage, thereby maintaining the quality of cassava after harvest.

3.3. Formic Acid Inhibited the Accumulation of H_2O_2

Reactive oxygen species (ROS) accumulation in cassava slices of the treated group and control group during PPD was analyzed by measuring hydrogen peroxide (H_2O_2) content. As shown in Figure 3, with the extension of storage time, the H_2O_2 content of cassava slices in the control group gradually increased, while the H_2O_2 content of cassava slices in the treated group showed a slow trend of decline, and the H_2O_2 content was significantly lower than that in the control group, indicating that formic acid treatment reduced ROS accumulation in the cassava PPD process.

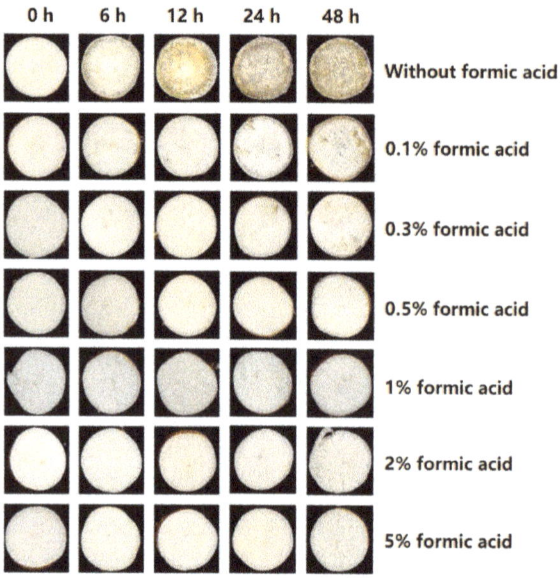

Figure 1. Effect of formic acid in delaying cassava postharvest physiological deterioration (PPD). Visual examination of cassava slices at different time points during storage (0, 6, 12, 24, and 48 h) after incubation with different formic acid solution (0.1%, 0.3%, 0.5%, 1%, 2%, and 5% (v/v)) for 2 h.

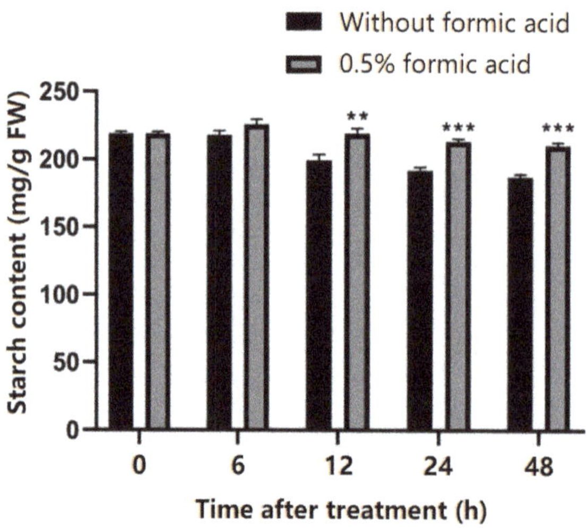

Figure 2. Effect of formic acid on starch content during storage. ** indicates the significant difference $p \leq 0.01$, *** indicates the significant difference $p \leq 0.001$.

3.4. Formic Acid Inhibited the Accumulation of MDA

The MDA content of cassava slices in the treated group and the control group was determined, and it was found that the MDA content of cassava slices in the control group gradually increased with the extension of storage time. Although the MDA content of cassava slices in the treated group also increased slowly with the extension of storage time, it was significantly lower than that in the control group at 6 h, 12 h, 24 h, and 48 h (Figure 4), indicating that the membrane lipid peroxidation degree of the slices in the treated group was lower than that in the control group during PPD, and the membrane suffered less peroxidation damage.

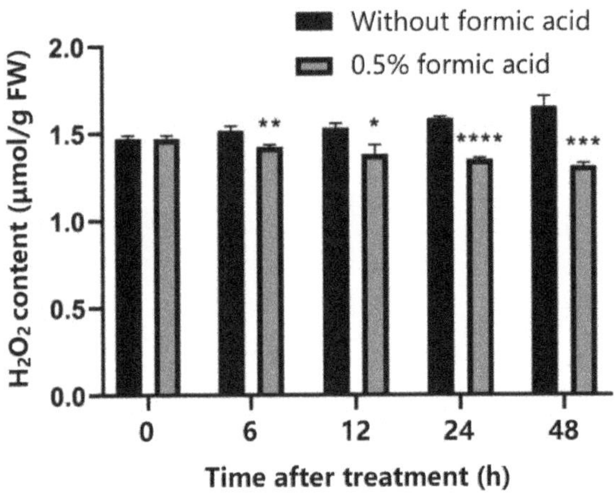

Figure 3. Effect of formic acid on reactive oxygen species (ROS) accumulation during cassava PPD occurrence. H_2O_2 content was determined in cassava slices with (0.5% v/v) or without formic acid treatment during PPD occurrence. * indicates the significant difference $p \leq 0.1$, ** indicates the significant difference $p \leq 0.01$, *** indicates the significant difference $p \leq 0.001$, **** indicates the significant difference $p \leq 0.0001$.

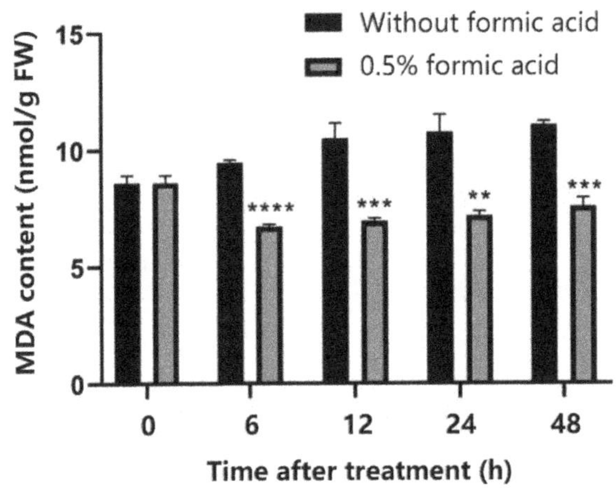

Figure 4. Effect of formic acid on membrane integrity during cassava storage. Malondialdehyde (MDA) content was determined in cassava slices with (0.5% v/v) or without formic acid treatment during cassava storage. ** indicates the significant difference $p \leq 0.01$, *** indicates the significant difference $p \leq 0.001$, **** indicates the significant difference $p \leq 0.0001$.

3.5. Formic Acid Enhance APX and GR Activities

To further clarify whether glutathione reductase (GR) and ascorbate peroxidase (APX) are directly involved in the regulation of PPD symptoms and their association with formic acid in ROS clearance, we measured the enzymatic activity of these two enzymes during PPD development in cassava slices. It was found that the activities of APX and GR decreased gradually with the extension of storage time in cassava slices of control group, while the activities of APX and GR increased slowly with the extension of storage time in the cassava slices of the treated group. Moreover, the APX enzyme activity in the 0.5%-formic-acid-treated group was significantly higher than that in the control group at 24 h and 48 h, and the GR activity in the treated group at 6 h, 12 h, 24 h, and 48 h was significantly higher than that in the control group (Figure 5). These results indicated that the activated oxygen scavenging ability of the formic acid-treated group was enhanced, and the structural integrity of the biofilm system (such as plasma, nuclear membrane, and mitochondria membrane) was better than that in the control group during PPD. The ability of defense membrane lipids to peroxidation is stronger.

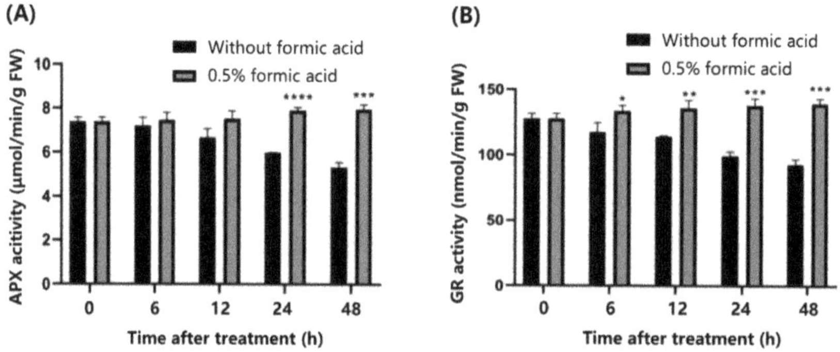

Figure 5. Effect of formic acid on the activities of ascorbate peroxidase (APX) and glutathione reductase (GR) during cassava slices storage. (**A**) APX enzyme activities in cassava slices with (0.5% v/v) or without formic acid treatment during PPD occurrences. (**B**) GR activities enzyme activities in cassava slices with (0.5% v/v) or without formic acid treatment during PPD occurrences. * indicates the significant difference $p \leq 0.1$, ** indicates the significant difference $p \leq 0.01$, *** indicates the significant difference $p \leq 0.001$, **** indicates the significant difference $p \leq 0.0001$.

3.6. Changes in Water-Soluble Pectin Content

The water-soluble pectin (WSP) content of cassava slices in the treated group and the control group was measured. It was found that the WSP content of cassava slices in the control group gradually increased with the extension of storage time, while the WSP content in the treated group began to decrease at 6 h and tended to be stable at 12 h, 24 h, and 48 h. The WSP content of cassava slices in the treated group was significantly lower than that in the control group at 6 h, 12 h, 24 h, and 48 h (Figure 6), indicating that the softening degree of cassava slices in the PPD process was lower than that in the control group, which assisted to confirm that the degree of PPD in the treated group was lower than that in the control group.

3.7. Changes in Microstructure

To investigate whether formic acid effects the cell microstructure, we performed microscopic observations on cassava slices of 6 h after treatment with (0.5%) or without formic acid. The result showed that, compared with the fresh cassava, in the cassava sample at 6 h after treatment without formic acid, the intercellular space was enlarged, indicating that the intercellular layer of the cell wall was degraded. By comparison, 6 h after treatment with 0.5% formic acid, the intercellular space was still as small as that in fresh cassava, indicating that the intercellular layer of the cell wall was not degraded (Figure 7). These results suggested that formic acid suppressed the cell wall degradation, leaving the cells in an intact and stable state and thus not susceptible to deterioration.

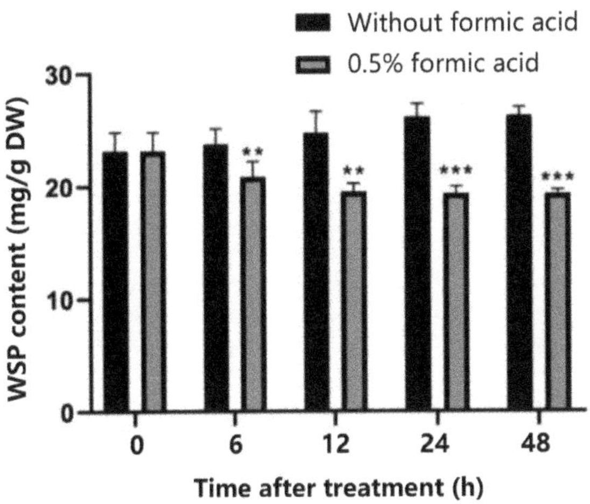

Figure 6. Effect of formic acid on the water-soluble pectin (WSP) content during cassava PPD development. ** indicates the significant difference $p \leq 0.01$, *** indicates the significant difference $p \leq 0.001$.

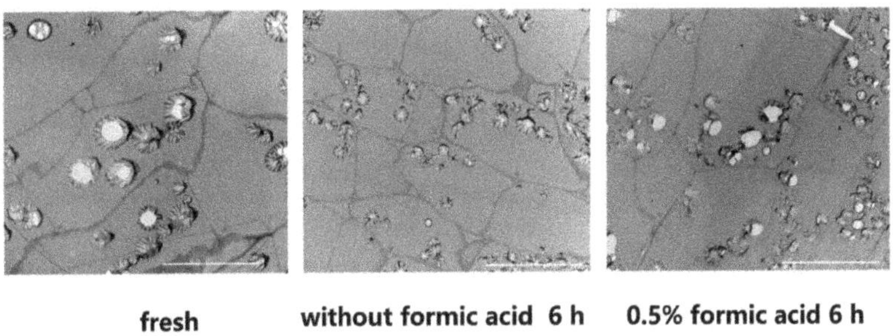

Figure 7. Effect of formic acid on microstructure of cassava slices 6 h after treatment. Control: fresh-cut cassava; without formic acid 6 h: 6 h after treatment without formic acid; 0.5% formic acid 6 h: samples of 6 h after treatment by 0.5% formic acid. The blue arrows point to the intercellular space formed by three cells. The scale bar indicates 50 μm.

4. Discussion

As an important food, feed, and biomass energy raw material, cassava plays an crucial role in human life and production. PPD is the main cause of cassava postharvest loss of quality and yield. Multiple chemicals have been tested to control PPD in cassava, such as ethanol [20], lysozyme [21], ethephon [22], acetic acid [23], and chitosan [24]. Formic acid has been confirmed to control food, fruit, and feed storage quality [7,8,10,11,16,18]. However, the impact of formic acid on the PPD process of cassava remains unexplored. In this study, we used formic acid to treat fresh-cut cassava to verify its effects on controlling PPD. The results showed that formic acid has a significant inhibitory effect on PPD in cassava, and the onset of PPD symptoms was postponed even when treated at a low concentration (0.1% v/v) (Figure 1). PPD often leads to the degradation of starch [20]. In this study, the decreased starch content was observed in control cassava during storage, and 0.5% formic acid prevented the decline in starch content (Figure 2). These results confirm the effectiveness of formic acid in delaying the postharvest deterioration of cassava.

The accumulation of ROS has been revealed to be one of the main reasons for PPD, of which H_2O_2 is a common substance being valuated. In this study, the H_2O_2 content increased during the storage period in control cassava, and formic acid significantly suppressed ROS accumulation (Figure 3). The accumulation of ROS in vivo will lead to membrane lipid peroxidation in plants, and the final product of oxidation is malondialdehyde (MDA). Therefore, the determination of MDA content can reflect the degree of membrane damage by peroxide. Its content has been widely used as a marker to assess the degree of oxidative stress damage to plants [25]. In this study, the MDA content increased gradually in control cassava during storage, while that in the formic acid-treated cassava was significantly reduced (Figure 4). The suppression of both H_2O_2 and MDA content in delaying cassava PPD occurrence has also been observed in a previous study [22].

Ascorbate peroxidase (APX) is the most important peroxidase in H_2O_2 detoxification, it has a greater affinity for H_2O_2 than catalase [26]. Previous studies have found that increased APX activity is positive to preserve fruit quality during postharvest storage [27–29]. The increased APX activity was also associated with delayed PPD occurrence in cassava, which is similar to this study [30]. In this study, formic acid significantly promoted the increase in the APX activity during storage, making cassava free of PPD (Figure 5A). This study further confirmed the relationship between APX and PPD in cassava. Glutathione reductase (GR) is another important enzyme in H_2O_2 detoxification in plants. It belongs to the same ROS scavenging pathway as APX in the ascorbate–glutathione (AsA-GSH) cycle. Numerous studies have found that inducing increased GR activity is consistent with maintaining fruit quality during storage [26,31,32]. In cassava, melatonin delayed cassava PPD by increasing the GR activity, but not the APX activity [33]. In this study, formic acid significantly increased the GR and APX activity, suggesting that formic acid may delay cassava PPD by inducing the AsA-GSH cycle pathway to clear H_2O_2.

Previous studies have found that there is a certain correlation between cell wall structure changes and the PPD of fresh products [34]. In the process of postharvest browning of potato, the intercellular layer of the cell wall was partially degraded and the intercellular space increased [35]. The collapsed cell wall structure was also observed in grapes when panicle browning occurred [36]. In this study, formic acid inhibited the intercellular layer degradation during cassava storage, keeping the intercellular space as small as in fresh cassava. The intercellular layer of the cell wall is mainly composed of complex pectin substances, and the increase in water-soluble pectin (WSP) content is due to the degradation of intercellular layer and the destruction of the cell wall structure [34,37]. During storage, the water-soluble pectin (WSP) increased along with PPD; however, when the occurrence of PPD was delayed by some treatments, the increase in the WSP content was also inhibited in many kinds of fresh agriculture products, such as litchi [38], grape [36], banana [39], *Zizania latifolia* [40], and tomato [41]. In this study, formic acid inhibited the increase in the WSP content, indicating that formic acid can maintain the integrity of the cell wall by inhibiting the increase in the WSP content and reducing the degradation of the intercellular layer, so as to delay the occurrence of PPD.

People may be confused about the use of formic acid because it is toxic in high concentrations. However, formate is a central metabolite, and its addition will as such be unlikely to have major metabolic deleterious consequences [42]. It is found naturally in many foods such as honey, wine, coffee fruit, milk, cheese, and yogurt [7]. Ingesting normal levels of formic acid is easily and quickly metabolized by the human body without causing harm [42]. Formic acid is approved as a component of adjuvants and synthetic flavorings to be directly added into food for human use, as a component of paper materials for food wrapping, as a flavoring, and as a preservative agent [7]. Formic acid can be used in foods and products provided it does not pose significant health concerns. It is also utilized as an additive in animal food and drinking water [7]. The accepted daily intake for human consumption is 0–3 mg/kg [7]. The European Food Safety Authority classifies formic acid as safe, with no restrictions on its use in food [7]. The purpose of the application of formic acid in this study is only to study the internal mechanism controlling the occurrence of PPD in cassava. Its addition and application in food should be carried out in accordance with regulations.

According to a previous study, mechanical injury during cassava harvest does lead to the production of cyanide, which is toxic to human and animal [43]. However, the relationship between cyanide and PPD remains controversial. For instance, cyanide content decreased during cassava postharvest storage [3,44]. The hot water dip delayed cassava PPD occurrence, but increased the cyanide (hydrocyanic acid) content [45]. These studies suggested that high cyanide content can delay cassava PPD. However, in another study, the cassava genotypes containing a low level of cyanide showed less PPD than genotypes containing a high level of cyanide [46]. Moreover, in a study on coating methods to retain the cassava postharvest quality, the cyanide content decreased compared to that in the control [3]. Therefore, we did not detect the cyanide in this study. As a C1 metabolite itself, formic acid is involved in the metabolism of many substances, such as polyamine, sterol, formaldehyde, folate, tryptophan, and purine [42]. Whether formic acid is involved in the synthesis of cyanide and whether the exogenous use of formic acid affects the synthesis of cyanide remains to be carefully investigated.

5. Conclusions

This study provides information on the use of formic acid in cassava slicing to delay PPD occurrence. In fact, the use of formic acid helps to significantly extend shelf life by increasing APX and GR enzyme activity, reducing H_2O_2 and MDA content, and discoloration. In addition, formic acid also inhibited the degradation of intercellular layer by reducing the content of WSP, maintaining the stability of cell wall, inhibiting the occurrence of PPD, and inhibiting the degradation of starch, thus maintaining the quality of cassava during storage. The results of this study provide a method for effectively delaying the occurrence of PPD in cassava, and also provide a direction for future targeted improvement of cassava storage quality. However, whether formic acid affects cassava cyanide content is unclear and needs further investigation in the future.

Author Contributions: Methodology, Y.C. and R.L.; Investigation, Y.C., Z.D., C.S. and X.T.; Writing—original draft, Y.C., A.R.F. and R.L.; Writing—review and editing, A.R.F., Y.Y., J.L., Y.W. and J.G.; Supervision, R.L. and J.G. All authors have read and agreed to the published version of the manuscript.

Funding: This research was funded by the Provincial Natural Fund of Hainan Province: 323RC537; the Earmarked Fund for China Agriculture Research System (CARS-11-HNGJC), and the Chinese Academy of Tropical Agricultural Sciences for Science and Technology Innovation Team of National Tropical Agricultural Science Center (CATASCXTD202301).

Institutional Review Board Statement: Not applicable.

Informed Consent Statement: Not applicable.

Data Availability Statement: The datasets generated during and/or analyzed during the current study are available from the corresponding author on reasonable request.

Conflicts of Interest: The authors declare no conflicts of interest.

References

1. Mbinda, W.; Mukami, A. Breeding for postharvest physiological deterioration in cassava: Problems and strategies. *CABI Agric. Biosci.* **2022**, *3*, 30. [CrossRef]
2. Li, R.; Yuan, S.; Zhou, Y.; Wang, S.; Zhou, Q.; Ding, Z.; Wang, Y.; Yao, Y.; Liu, J.; Guo, J. Comparative transcriptome profiling of cassava tuberous roots in response to postharvest physiological deterioration. *Int. J. Mol. Sci.* **2023**, *24*, 246. [CrossRef] [PubMed]
3. Atieno, L.; Owino, W.; Ateka, E.M.; Ambuko, J. Influence of coating application methods on the postharvest quality of cassava. *Int. J. Food Sci.* **2019**, *2019*, 2148914. [CrossRef] [PubMed]
4. Drapal, M.; Ovalle Rivera, T.M.; Luna Meléndez, J.L.; Perez-Fons, L.; Tran, T.; Dufour, D.; Becerra Lopez-Lavalle, L.A.; Fraser, P.D. Biochemical characterisation of a cassava (*Manihot esculenta* crantz) diversity panel for post-harvest physiological deterioration; metabolite involvement and environmental influence. *J. Plant Physiol.* **2024**, *301*, 154303. [CrossRef]
5. Iyer, S.; Mattinson, D.S.; Fellman, J.K. Study of the early events leading to cassava root postharvest deterioration. *Trop. Plant Biol.* **2010**, *3*, 151–165. [CrossRef]
6. Noon, R.A.; Booth, R.H. Nature of post-harvest deterioration of cassava roots. *Trans. Br. Mycol. Soc.* **1977**, *69*, 287–290. [CrossRef]

7. Ajingi, Y.u.S.; Ruengvisesh, S.; Khunrae, P.; Rattanarojpong, T.; Jongruja, N. The combined effect of formic acid and Nisin on potato spoilage. *Biocatal. Agric. Biotechnol.* **2020**, *24*, 101523. [CrossRef]
8. Lei, Y.; Fan, X.; Li, M.; Chen, Y.; Li, P.; Xie, Y.; Zheng, Y.; Sun, H.; Wang, C.; Dong, R.; et al. Effects of formic acid and lactic acid bacteria on the fermentation products, bacterial community diversity and predictive functional characteristics of perennial ryegrass silage in karst regions. *Fermentation* **2023**, *9*, 675. [CrossRef]
9. Champati, A.; Bhanja, S.K.; Rokade, J.J.; Nayak, N.; Yadav, A.S.; Biswas, A.; Sharma, D.; Chakma, J.; Sky; Mishra, J.; et al. Evaluation of in-feed supplementation of formic acid and thymol as non-antibiotic growth promoters and assessing their effect on antimicrobial resistant *E. coli* isolated in Turkey. *Vet. Res. Commun.* **2024**, *48*, 1741–1754. [CrossRef]
10. Her, J.-Y.; Cho, H.; Kim, M.K.; Lee, K.-G. Organic acids as a freshness indicator for tofu (soybean curd). *J. Food Sci. Technol.* **2017**, *54*, 3443–3450. [CrossRef]
11. Gandra, J.R.; Takiya, C.S.; Del Valle, T.A.; Pedrini, C.D.; Gandra, E.R.; Antônio, G.; de Oliveira, E.R.; Severo, I.K.; Rennó, F.P. Effect of chemical and microbial additives on fermentation profile, chemical composition, and microbial populations of whole-plant soybean silage. *Fermentation* **2024**, *10*, 204. [CrossRef]
12. Koivunen, E.; Jaakkola, S.; Heikkilä, T.; Lampi, A.-M.; Halmemies-Beauchet-Filleau, A.; Lee, M.R.F.; Winters, A.L.; Shingfield, K.J.; Vanhatalo, A. Effects of plant species, stage of maturity, and level of formic acid addition on lipolysis, lipid content, and fatty acid composition during ensiling. *J. Anim. Sci.* **2015**, *93*, 4408–4423. [CrossRef] [PubMed]
13. Wang, X.; Deng, T.; Zhou, X.; Chu, L.; Zeng, X.; Zhang, S.; Guan, W.; Chen, F. A mixture of formic acid, benzoic acid, and essential oils enhanced growth performance via modulating nutrient uptake, mitochondrion metabolism, and immunomodulation in weaned piglets. *Antioxidants* **2024**, *13*, 246. [CrossRef] [PubMed]
14. Torrecillas, S.; Aboelsaadat, E.; Carvalho, M.; Acosta, F.; Monzón-Atienza, L.; Gordillo, Á.; Montero, D. Dietary supplementation of a combination of formic acid and sodium formate in practical diets promotes gilthead sea bream (*Sparus aurata*) gut morphology and disease resistance. *Aquac. Rep.* **2024**, *35*, 101951. [CrossRef]
15. Sholberg, P.L. Fumigation of fruit with short-chain organic acids to reduce the potential of postharvest decay. *Plant Dis.* **1998**, *82*, 689–693. [CrossRef]
16. Romanazzi, G.; Feliziani, E.; Santini, M.; Landi, L. Effectiveness of postharvest treatment with chitosan and other resistance inducers in the control of storage decay of strawberry. *Postharvest Biol. Technol.* **2013**, *75*, 24–27. [CrossRef]
17. Romanazzi, G.; Gabler, F.M.; Margosan, D.; Mackey, B.E.; Smilanick, J.L. Effect of chitosan dissolved in different acids on its ability to control postharvest gray mold of table grape. *Phytopathology* **2009**, *99*, 1028–1036. [CrossRef]
18. Mukhopadhyay, S.; Ukuku, D.O.; Olanya, O.M.; Sokorai, K.; Fan, X. Effects of pulsed light and aerosolized formic acid treatments on inactivation of Salmonella enterica on cherry tomato, reduction of microbial loads, and preservation of fruit quality. *Food Control* **2022**, *136*, 108667. [CrossRef]
19. Bazoti, S.F.; Camargo, A.F.; Bonatto, C.; Kubeneck, S.; Longo, V.D.; Romani, L.C.; Fongaro, G.; de Oliveira, D.; Treichel, H. Levulinic acid as a strategy for control of postharvest citrus blue mold by a newly isolated Penicillium italicum. *J. Chem. Technol. Biotechnol.* **2024**, *99*, 1589–1595. [CrossRef]
20. Liu, G.; Li, B.; Wang, Y.; Wei, B.; He, C.; Liu, D.; Shi, H. Novel role of ethanol in delaying postharvest physiological deterioration and keeping quality in cassava. *Food Bioprocess Technol.* **2019**, *12*, 1756–1765. [CrossRef]
21. Wu, X.; Xu, J.; Ma, Q.; Ahmed, S.; Lu, X.; Ling, E.; Zhang, P. Lysozyme inhibits postharvest physiological deterioration of cassava. *J. Integr. Plant Biol.* **2022**, *64*, 621–624. [CrossRef] [PubMed]
22. Yan, Y.; Li, M.; Ding, Z.; Yang, J.; Xie, Z.; Ye, X.; Tie, W.; Tao, X.; Chen, G.; Huo, K.; et al. The regulation mechanism of ethephon-mediated delaying of postharvest physiological deterioration in cassava storage roots based on quantitative acetylproteomes analysis. *Food Chem.* **2024**, *458*, 140252. [CrossRef] [PubMed]
23. Liu, G.; Zhang, X.; Wei, Y.; Zhao, H.; Wei, J.; Shi, H. Acetic acid delays fresh-cut cassava browning through fine-tunning redox homeostasis. *Sci. Hortic.* **2024**, *326*, 112746. [CrossRef]
24. Wang, C.; Chen, Y.; Chen, S.; Min, Y.; Tang, Y.; Ma, X.; Li, H.; Li, J.; Liu, Z. Spraying chitosan on cassava roots reduces postharvest deterioration by promoting wound healing and inducing disease resistance. *Carbohydr. Polym.* **2023**, *318*, 121133. [CrossRef]
25. Morales, M.; Munné-Bosch, S. Malondialdehyde: Facts and artifacts. *Plant Physiol.* **2019**, *180*, 1246–1250. [CrossRef]
26. Corpas, F.J.; González-Gordo, S.; Palma, J.M. Ascorbate peroxidase in fruits and modulation of its activity by reactive species. *J. Exp. Bot.* **2024**, *75*, 2716–2732. [CrossRef]
27. Fan, S.; Li, Q.; Feng, S.; Lei, Q.; Abbas, F.; Yao, Y.; Chen, W.; Li, X.; Zhu, X. Melatonin maintains fruit quality and reduces anthracnose in postharvest papaya via enhancement of antioxidants and inhibition of pathogen development. *Antioxidants* **2022**, *11*, 804. [CrossRef]
28. González-Gordo, S.; Rodríguez-Ruiz, M.; López-Jaramillo, J.; Muñoz-Vargas, M.A.; Palma, J.M.; Corpas, F.J. Nitric oxide (NO) differentially modulates the ascorbate peroxidase (APX) isozymes of sweet pepper (*Capsicum annuum* L.) fruits. *Antioxidants* **2022**, *11*, 765. [CrossRef]
29. Zhong, T.-Y.; Yao, G.-F.; Wang, S.-S.; Li, T.-T.; Sun, K.-K.; Tang, J.; Huang, Z.-Q.; Yang, F.; Li, Y.-H.; Chen, X.-Y.; et al. Hydrogen sulfide maintains good nutrition and delays postharvest senescence in postharvest tomato fruits by regulating antioxidative metabolism. *J. Plant Growth Regul.* **2021**, *40*, 2548–2559. [CrossRef]

30. Uarrota, V.G.; Moresco, R.; Schmidt, E.C.; Bouzon, Z.L.; da Costa Nunes, E.; de Oliveira Neubert, E.; Peruch, L.A.M.; Rocha, M.; Maraschin, M. The role of ascorbate peroxidase, guaiacol peroxidase, and polysaccharides in cassava (*Manihot esculenta* Crantz) roots under postharvest physiological deterioration. *Food Chem.* **2016**, *197*, 737–746. [CrossRef]
31. Xu, Y.; Cai, Z.; Ba, L.; Qin, Y.; Su, X.; Luo, D.; Shan, W.; Kuang, J.; Lu, W.; Li, L.; et al. Maintenance of postharvest quality and reactive oxygen species homeostasis of pitaya fruit by essential oil p-anisaldehyde treatment. *Foods* **2021**, *10*, 2434. [CrossRef] [PubMed]
32. Huang, R.; Cheng, Y.; Li, C.; Guo, M.; Zhu, J.; Ge, Y.; Sun, T. Postharvest application of acibenzolar- S-methyl delays the senescence of pears by mediating the ascorbate-glutathione cycle. *Sci. Hortic.* **2022**, *293*, 110741. [CrossRef]
33. Ma, Q.; Zhang, T.; Zhang, P.; Wang, Z.-Y. Melatonin attenuates postharvest physiological deterioration of cassava storage roots. *J. Pineal Res.* **2016**, *60*, 424–434. [CrossRef] [PubMed]
34. Li, R.; Rosado-Souza, L.; Sampathkumar, A.; Fernie, A.R. The relationship between cell wall and postharvest physiological deterioration of fresh produce. *Plant Physiol. Biochem.* **2024**, *210*, 108568. [CrossRef] [PubMed]
35. Wang, T.; Yan, T.; Shi, J.; Sun, Y.; Wang, Q.; Li, Q. The stability of cell structure and antioxidant enzymes are essential for fresh-cut potato browning. *Food Res. Int.* **2023**, *164*, 112449. [CrossRef] [PubMed]
36. Li, J.; Wu, Z.; Zhu, Z.; Xu, L.; Wu, B.; Li, J. Botrytis cinerea mediated cell wall degradation accelerates spike stalk browning in Munage grape. *J. Food Biochem.* **2022**, *46*, e14271. [CrossRef]
37. Cui, K.; Yang, L.; Shu, C.; Liu, J.; Zhu, Z.; Yang, Z.; Zhu, X.; Jiang, W. Near freezing temperature storage alleviates cell wall polysaccharide degradation and softening of apricot (*Prunus armeniaca* L.) fruit after simulated transport vibration. *Sci. Hortic.* **2021**, *288*, 110296. [CrossRef]
38. Guo, X.; Li, Q.; Luo, T.; Han, D.; Zhu, D.; Wu, Z. Postharvest calcium chloride treatment strengthens cell wall structure to maintain litchi fruit quality. *Foods* **2023**, *12*, 2478. [CrossRef]
39. Yu, Z.; Yuan, R.; Yang, J.; Pan, Y. Hydrogen sulfide delays softening of banana fruit by inhibiting cell wall polysaccharides disassembly. *Int. J. Food Sci. Technol.* **2023**, *58*, 5236–5246. [CrossRef]
40. Huang, J.; Wu, W.; Fang, X.; Chen, H.; Han, Y.; Niu, B.; Gao, H. Zizania latifolia cell wall polysaccharide metabolism and changes of related enzyme activities during postharvest storage. *Foods* **2022**, *11*, 392. [CrossRef]
41. Fanourakis, D.; Nikoloudakis, N.; Paschalidis, K.; Christopoulos, M.V.; Goumenaki, E.; Tsantili, E.; Delis, C.; Tsaniklidis, G. Gene expression, activity and localization of beta-galactosidases during late ripening and postharvest storage of tomato fruit. *Agriculture* **2022**, *12*, 778. [CrossRef]
42. Pietzke, M.; Meiser, J.; Vazquez, A. Formate metabolism in health and disease. *Mol. Metab.* **2020**, *33*, 23–37. [CrossRef] [PubMed]
43. McMahon, J.; Sayre, R.; Zidenga, T. Cyanogenesis in cassava and its molecular manipulation for crop improvement. *J. Exp. Bot.* **2021**, *73*, 1853–1867. [CrossRef] [PubMed]
44. Akintimehin, E.S.; Adetuyi, F.O.; Karigidi, K.O.; Okonji, R.E.; Akinnubi, C.O. Effects of postharvest storage on cyanide content and activity of partially purified rhodanese from bitter cassava (*Manihot utilissima*) Tubers. *Food Sci. Appl. Biotechnol.* **2020**, *3*, 157–166. [CrossRef]
45. Acedo, J.Z.; Acedo, A.L., Jr. Controlling postharvest physiological deterioration and surface browning in cassava (*Manihot esculenta* crantz) roots with hot water treatment. *Acta Hortic.* **2013**, *989*, 357–361. [CrossRef]
46. Sowmyapriya, S.; Kalarani, M.K.; Manivannan, K.; Mohan, C. Evaluation of postharvest physiological deterioration of cassava (*Manihot esculenta*) genotypes. *Int. J. Ecol. Environ. Sci.* **2020**, *2*, 661–666.

antioxidants

MDPI

Article

Employing Titanium Dioxide Nanoparticles as Biostimulant against Salinity: Improving Antioxidative Defense and Reactive Oxygen Species Balancing in Eggplant Seedlings

Muhammad Fasih Khalid [1], Muhammad Zaid Jawaid [1], Muddasir Nawaz [2], Rana Abdul Shakoor [2,3] and Talaat Ahmed [1,*]

[1] Environmental Science Center, Qatar University, Doha 2713, Qatar; mfasih.khalid@qu.edu.qa (M.F.K.); mk1208853@student.qu.edu.qa (M.Z.J.)
[2] Center for Advanced Materials, Qatar University, Doha 2713, Qatar; m.nawaz@qu.edu.qa (M.N.); shakoor@qu.edu.qa (R.A.S.)
[3] Department of Mechanical and Industrial Engineering, Qatar University, Doha 2713, Qatar
* Correspondence: t.alfattah@qu.edu.qa

Abstract: Salinity is a major abiotic stress that affects the agricultural sector and poses a significant threat to sustainable crop production. Nanoparticles (NPs) act as biostimulants and significantly mitigate abiotic stress. In this context, this experiment was designed to assess the effects of foliar application of titanium dioxide (TiO_2) nanoparticles at 200 and 400 ppm on the growth of eggplant (*Solanum melongena*) seedlings under moderate (75 mM) and high (150 mM) salinity stress. The TiO_2-NPs employed were characterized by X-ray diffraction (XRD), Fourier transform infrared spectroscopy (FTIR), thermal gravimetric analysis (TGA), and scanning electron microscopy (SEM) analysis. The seedlings were assessed physiologically, growth-wise, and biochemically. The seedlings were significantly affected by their physiological attributes (Fv′/Fm′, Fv/Fm, NPQ), growth (root length, shoot length, number of leaves, fresh biomass, dry biomass, leaf greenness), antioxidative enzymes (SOD, POD, CAT, APx, GR), stress indicators (H_2O_2, MDA), and toxic ion (Na^+) concentrations. The maximum decrease in physiological and growth attributes in eggplant seedling leaves was observed with no TiO_2-NP application at 150 mM NaCl. Applying TiO_2-NPs at 200 ppm showed significantly less decrease in Fv′/Fm′, root length, shoot length, number of leaves, fresh biomass, dry biomass, and leaf greenness. In contrast, there were larger increases in SOD, POD, CAT, APx, GR, and TSP. This led to less accumulation of H_2O_2, MDA, and Na^+. No significant difference was observed in higher concentrations of TiO_2-NPs compared to the control. Therefore, TiO_2-NPs at 200 ppm might be used to grow eggplant seedlings at moderate and high salinity.

Keywords: antioxidative enzymes; reactive oxygen species; nanotechnology; sodium toxicity; vegetables

1. Introduction

Salinity is a significant abiotic stress that constantly threatens crop production and global food security. Currently, about 23% of farmlands and 25–30% of irrigated areas are affected by salt stress [1]. Due to urbanization and increased food demand, the move to drier farmland reduces land and water resources. Salinity stress affects plants by decreasing the soil's osmotic ability, causing nutritional imbalances, and increasing ionic toxicity [2]. The high salinity in the soil is causing a significant decrease in crop growth, yield, and overall agricultural production worldwide [3]. In high-salinity conditions, cells expel Na^+ ions and produce osmoprotectants, changing their photosynthetic and antioxidant levels [4]. High salt levels in the soil can disturb the balance of ions in plants, hindering water uptake and causing oxidative damage, leading to lower crop yields and reduced crop quality [5]. When plants are under salinity stress, their cells gather reactive oxygen species (ROS), leading to ionic, osmotic, and nutritional stress. The most common ROS produced by plant cells are superoxide anion radicals ($O_2^{\bullet-}$), hydrogen peroxide (H_2O_2), and hydroxyl radicals ($OH^{\bullet}$). This problem is expected to worsen due to high-salinity irrigation water due to low rainfall and inadequate agricultural practices, especially in dry regions where the evapotranspiration exceeds the precipitation [6]. However, adopting sustainable agricultural practices offers hope in mitigating this issue. Developing and implementing these sustainable agricultural practices to mitigate salinity stress's harmful impact on agricultural production is crucial and necessary. It is the key to meeting the growing food demand and feeding the increasing global population [5]. Plants have developed defense mechanisms to cope with harmful ROS from toxic metals by activating the antioxidant defense system. Salt stress increases antioxidant enzymes in plants, such as superoxide dismutase (SOD), peroxidase (POD), catalase (CAT), glutathione reductase (GR), and ascorbate peroxidase (APx). Long-term exposure to salt stress leads to enhanced antioxidant activity that indicates the scavenging of H_2O_2 [7]. There have been reports of the applications of different phytohormones used to reduce salt stress, such as strigolactones [8–10], gibberellin [11], salicylic acid [12], and ascorbic acid [13,14]. The harmful effects of saline soils on crops have led to the use of expensive technologies, which can also harm the ecosystem. Therefore, there is a need to develop cost-effective, eco-friendly, and practical solutions. Currently, nanoparticles (NPs) are widely used and are significant contributors across the globe as plant growth regulators [15]. As tiny, unique entities, their properties improve plants' growth and performance when faced with different stress factors, such as salt stress [16]. Lashkary et al. [17] reported the application of NPs for inducing oxidative stress in numerous plants.

Titanium dioxide (TiO_2) NPs are some of the most commonly used nanoparticles to address environmental challenges, improving numerous plants' biochemical and physiological properties [18]. Lei et al. [19] reported improvements in antioxidant enzyme activity and photosynthetic rate with the application of TiO_2, leading to enhanced crop growth. The enhanced antioxidant enzyme activity reduces plants' H_2O_2 and MDA production, thus improving their proline content and plant growth [20]. Another study found that exogenously applied TiO_2-NPs improved salt stress in maize crops by enhancing their chlorophyll contents, phenolics, and yield [21]. Khan [22] discovered that applying TiO_2-NPs can mitigate tomato salt stress, improving the chlorophyll content, enzyme activities, and yield. TiO_2 acts as a plant stimulant, activating defense mechanisms against different stress factors [23]. However, several reports have indicated that excess nanoparticle application can cause toxic effects when dosed in high concentrations [24]. These effects have been reported to vary with the environment and plant species and the applied concentration. As per Gohari et al. [24], TiO_2-NPs at the rate of 200 ppm should lead to significant improvements in Moldavian plants against salinity. Similarly, Amooghaie et al. [25] studied the application of 80 ppm TiO_2 NPs against *Carum copticum*. High concentrations of TiO_2 in plants have been reported to cause toxic effects, resulting in stunted growth. Rico et al. [26] reported that TiO_2 concentration leads to higher ROS production, chlorophyll degradation, and increased cellular toxicity. Therefore, reducing the concentration of TiO_2 is essential [27].

One of these vegetables is eggplant (*Solanum melongena* L.), which belongs to the Solanaceous family and is cultivated worldwide [28]. Eggplant is a plant that is sensitive to salt stress. When exposed to high salt levels, its growth and yield decrease [29]. However, different varieties of eggplant show varying responses to different types of stress. This variation is due to differences in their physical, biochemical, and physiological traits [30]. Eggplants contain high levels of phenolic acids, which have numerous health benefits. This has led to research on the salt tolerance of eggplants growing in harsh, high-salinity conditions [31]. TiO_2-NP foliar application has been used on eggplant against fungus (*Phomopsis vexans*), bacteria (*Ralstonia solanacearum*), and root-rot nematode (*Meloidogyne incognita*) at 200 ppm, and significant changes were observed [32]. So, we aim to explore the effects TiO_2-NPs on the growth, physiological responses, and overall health of eggplant (*Solanum melongena*) seedlings when subjected to moderate and high salinity stress by assessing various growth parameters, photosynthetic efficiency, toxic ion content, stress indicators, and antioxidative enzymatic activity. This research seeks to explain the possible influence of TiO_2 nanoparticles in lessening or intensifying the negative consequences of salinity stress on eggplant seedlings.

2. Materials and Methods

2.1. Characterization of TiO_2-NPs

The solid TiO_2-NPs were purchased from Sigma-Aldrich (St. Louis, MO, USA). Field emission scanning electron microscopy (FE-SEM-Nova Nano-450, FEI, USA) was used for morphological analysis of the TiO_2 nanoparticles. X-ray diffraction (XRD-PANalytical Empyrean, Malvern, Almelo, The Netherlands) was used for analysis with a scanning angle in the range of $0° \leq 2\theta \leq 90°$ at a scanning rate of 2°/min to study the structure and phase purity of the TiO_2 nanoparticles. Fourier transform infrared spectroscopy (FTIR-PerkinElmer, Waltham, MA, USA) was utilized to analyze the TiO_2 particles in the 5000–500 cm^{-1} range. Thermogravimetric analysis (TGA-pyris 4000 PerkinElmer, Waltham, MA, USA) was used to study the thermal stability of the TiO_2 nanoparticles.

2.2. Plant Growth and Experimental Conditions

Plants were obtained from Agrico-Qatar when they were two weeks old. The seedlings were placed in a greenhouse facility at Qatar University. After two weeks of acclimatization, the uniformly sized seedlings were transplanted into peat moss pots (Ø 20 cm) (one plant in each pot). Before the experiment, plants were only irrigated with a Hoagland nutrient solution [33] for one week to maintain growth and prevent stress. To obtain precise results from our experiments, healthy plants were selected after one week to perform the experiments. The experiment was performed in the greenhouse facility at Qatar University with a mean daytime temperature of 28 °C and nighttime temperature of 14 °C. The relative humidity varied between 60% and 80%, and the amount of light (PAR) ranged from 300 to 500 $\mu mols/m^2/s$. A completely randomized design was used with two factors, i.e., nanoparticles and salt stress. The seedlings were subjected to three levels of salinity (0 mM, 75 mM, and 150 mM) through irrigation with the Hoagland solution by using sodium chloride (NaCl) powder (99% Sigma Aldrich). Furthermore, the exogenous foliar application of titanium dioxide (TiO_2) nanoparticles (NPs) was applied at the rates of 0 ppm, 200 ppm, and 400 ppm twice a week. The titanium (IV) oxide (TiO_2, 99.7%) trace nanoparticles were purchased from Sigma-Aldrich, and later, dispersed solutions of 200 and 400 ppm were prepared by mixing them into deionized water followed by ultra-sonication treatment to prepare a homogenous and well-dispersed TiO_2 solution. The experiment had three replications and five pots in each replication (Figure 1).

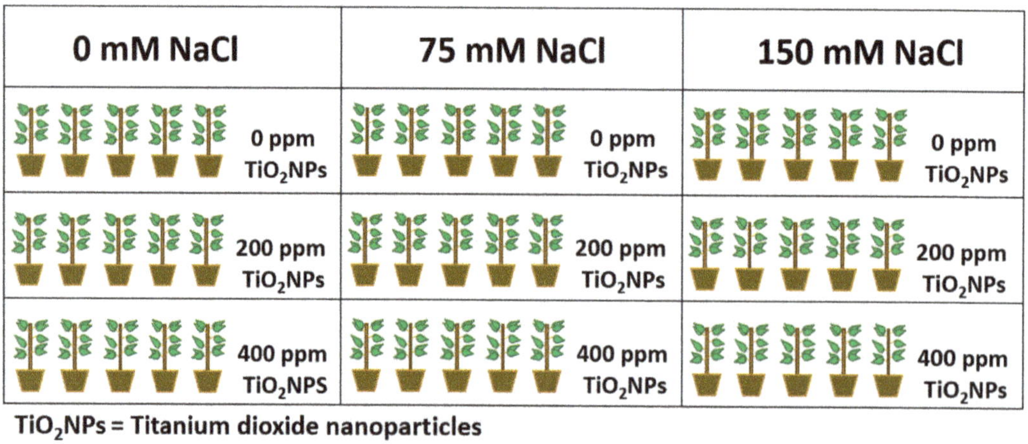

Figure 1. Experimental layout of one replication for eggplant seedlings.

2.3. Chlorophyll Fluorescence

The chlorophyll fluorescence and nonphotochemical quenching (NPQ) of eggplant leaves were measured at the end of the trial using the FluorPen (FP-100, Photon Systems Instruments, Drásov, Czech Republic). Fully expanded leaves of eggplant plants were used for Fv′/Fm′ in the daytime. For Fv/Fm and NPQ, leaves were exposed to dark conditions for 1 h using light-retaining clips, and then, readings were taken [34].

2.4. Growth Attributes

The relative chlorophyll content was measured using a SPAD meter (SPAD-502 plus Konica Minolta, Tokyo, Japan). After harvesting the eggplant seedlings, the number of leaves, the length of shoots, the length of roots, and the fresh biomass of the plants were measured. Dry biomass was then generated by placing the harvested plants in a hot air oven at 65 °C for five days.

2.5. Enzymatic Activity and TSP

The freshly harvested leaves were crushed in liquid nitrogen and stored at −80 °C to measure the enzymatic activity and for further analysis. To measure SOD (1.15.1.1), POD (1.11.1.7), and CAT (1.11.1.6) activity, 300 mg of powdered leaves was homogenized in chilled conditions with 3 mL sodium phosphate buffer (pH 7.8) and centrifuged at 15,000 rpm at 4 °C for 5 min [34]. For SOD activity, 1 mL reaction solution contained 75 mM ethylenediamine tetra-acetic acid, 50 mM nitroblue tetrazolium, 50 mM sodium phosphate buffer, 1.3 μM riboflavin, and 13 mM methionine, and the absorbance was measured at 560 nm using a spectrophotometer [35]. Change and Maehly's [36] method was used to estimate the POD and CAT activity. The reaction solution of CAT contained 5.9 mM hydrogen peroxide and 50 mM sodium phosphate buffer. In comparison, the POD activity reaction solution consisted of 20 mM guaiacol, 50 mM sodium phosphate buffer, and 40 mM hydrogen peroxide. The absorbance was read at 240 and 470 nm.

The APx (1.11.1.11) activity was measured using [37]. The reaction solution for APx contained 50 mM sodium phosphate buffer, 0.5 mM ascorbate, 0.1 mM ethylenediamine tetra-acetic acid, and 1.2 mM hydrogen peroxide, and absorbance was measured at 290 nm. For GR (1.6.4.2) activity, the reaction solution contained 100 mM potassium phosphate buffer, 0.5 mM glutathione oxidase, 2 mM ethylenediamine tetra-acetic acid, and 0.2 mM nicotinamide adenine dinucleotide phosphate hydrogen, and the absorbance was measured at 340 nm [38]. To measure the total soluble protein (TSP) content, 0.5 g of fresh leaves was homogenized with 1 mL phosphate saline buffer (pH 7.2) at 10,000 rpm for 5 min. The extracted solution was then added to deionized water and Coomassie blue dye. The absorbance was measured at 595 nm [39] using a UV/Vis spectrophotometer, Lambda-25 (Perkins Elmer, Waltham, MA, USA).

2.6. Stress Indicators and Toxic Ion

For stress indicators, H_2O_2 and MDA were measured. For H_2O_2 and MDA estimation, 300 mg of crushed leaves was taken and homogenized in 3 mL of 0.1% trichloroacetic acid and centrifuged at 12,000 rpm for 15 min. For H_2O_2, the reaction solution contained 1 M potassium iodide and 10 mM potassium phosphate buffer, and the absorbance was read at 390 nm [40]. The MDA was carried out by following [41]. The absorbance was measured at 532 and 600 nm using a UV/Vis spectrophotometer, Lambda-25 (Perkins Elmer, Waltham, MA, USA). The reaction solution contained 0.5% thiobarbituric acid and 20% trichloroacetic acid.

To measure the Na^+ ions in eggplant plants, 0.1 g dried leaves were wet-digested at 350 °C by adding 2 mL sulfuric acid and 0.2 mL hydrogen peroxide to stabilize the phosphorous to obtain a colorless solution [42]. The inductively coupled plasma optical emission spectrometer, (ICP-OES) Optima 5300 DV (Perkins Elmer, Waltham, MA, USA) was used to measure the Na concentration.

2.7. Statistical Analysis

The data were analyzed using the Stastix 8.1 (Tallahassee, FL, USA) software package, and an analysis of variance (ANOVA) is shown in Supplementary Table S1. The least significant difference test (LSD) was conducted to compare the means at a probability level of 5% [43]. Pearson's correlation and principal component analysis (PCA) were constructed using Rstat.

3. Results

Under salt stress conditions, the eggplant seedling leaves exhibited necrotic symptoms, alleviated by applying TiO_2-NPs to the foliage (Supplementary Figure S1).

3.1. Characterization of Nanoparticles

The characterization of TiO_2-NPs was confirmed before their application by XRD, FTIR, TGA, and SEM.

3.1.1. XRD and FTIR Analysis

Structural analysis of the TiO_2 was carried out using an XRD analysis (Figure 2A), where the pure crystalline behavior of the TiO_2 nanoparticles can be observed, with sharp peaks present. The XRD pattern shows the characteristic peaks at 25.2°, 37.6°, 48°, 53.8°, 55°, 62.4°, 68.8°, and 74.8°, which are in agreement with the (JPCD:01-078-2486), which confirms the anatase pattern of TiO_2-NPs [44]. The FTIR analysis was conducted and is depicted in Figure 2B. The broad peak at 3400–3500 cm^{-1} in the FTIR spectra of TiO_2-NPs indicates the hydroxyl group O-H due to the moisture presence of TiO_2 [45]. The Ti-OH stretching vibration can also be observed at 1633 cm^{-1} of the TiO_2 spectra. Furthermore, the distinct peaks of Ti-O-Ti vibrations were noted at wavenumbers below 1000 cm^{-1} [46].

3.1.2. SEM and TGA

Scanning electron microscopy (SEM) images of the TiO_2 were taken at 100,000× magnification and are depicted in Figure 2C, which shows the spherical morphology of the TiO_2 nanoparticles. TiO_2 was distributed homogeneously without agglomeration, and the particle size range was observed to be from 30 to 50 nm. A few nanoparticles above 50 nm in size were also observed. However, most of the nanoparticles were smaller than 50 nm in size.

The results depicted in Figure 2D demonstrate that the TiO_2 sample maintained its stability without any notable weight loss, confirming its pure composition. A minor total weight loss of around 1.4% was observed in the temperature range of 600 °C due to some moisture removal. So, the TGA of titania (TiO_2) shows that it is a very thermally and chemically stable compound [47].

3.2. Chlorophyll Fluorescence

The chlorophyll attributes of eggplant seedlings were significantly affected by varying levels of salt stress when TiO_2-NPs were applied (Supplementary Table S1, Figure 3). The eggplant seedling leaves exhibited significantly reduced Fv'/Fm' and Fv/Fm when exposed to salt stress conditions. The decrease was more significant at 150 mM than at 0 and 75 mM NaCl. The seedlings treated with different concentrations of TiO_2-NPs displayed smaller reductions in Fv'/Fm' and Fv/Fm. At 75 mM NaCl, the seedlings treated with 200 ppm of TiO_2-NPs showed no significant difference in Fv'/Fm' value compared with 0 mM (Figure 3A); instead, the higher concentration (400 ppm) of TiO_2-NPs showed a larger decrease in Fv'/Fm' at 75 mM. However, at 150 mM, both TiO_2-NP concentrations showed higher Fv'/Fm' values than 0 mM. Moreover, the TiO_2-NPs at 200 ppm showed less decline than 400 ppm under 150 mM salt stress condition (Figure 3A).

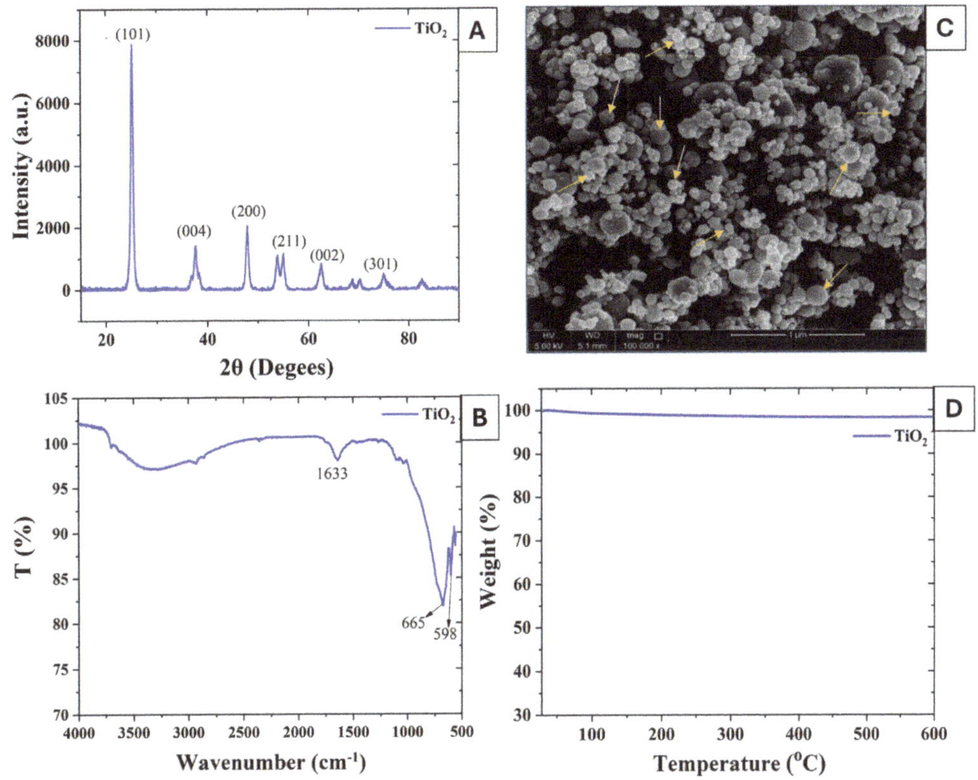

Figure 2. (**A**) XRD; (**B**) FTIR; (**C**) SEM; (**D**) TGA of TiO_2-NPs. Arrows in (**C**) indicate the size distribution of TiO_2-NPs.

Similarly, 200 and 400 ppm TiO_2-NPs exhibited smaller decreases in Fv/Fm than 75 mM of NaCl compared to 0 mM (Figure 3B). At 150 mM NaCl stress, 400 ppm showed no significant difference compared to 0 mM. The NPQ increased with increasing salt levels. The maximum NPQ was observed in eggplant seedling leaves at 150 mM with no TiO_2-NP application (Figure 3C). The application of TiO_2-NPs at 200 and 400 ppm showed smaller increases at 75 and 150 mM of NaCl, while 200 ppm of TiO_2-NPs showed the minimum increase in NPQ (Figure 3C).

3.3. Growth Attributes

The eggplant seedlings' growth attributes were significantly affected by salinity with TiO_2-NPs (Figure 4). Leaf greenness declined by increasing the salinity level (Figure 4A). The minimum decline was observed at 75 mM NaCl with 200 ppm TiO_2-NPs. Similarly, at 150 mM of NaCl, 200 ppm TiO_2-NPs showed a smaller decline in leaf greenness than 0 and 150 mM of NaCl (Figure 4A). The maximum decline was observed at 75 mM of NaCl without TiO_2-NP application, while at 150 mM NaCl with 0, and 400 ppm TiO_2-NPs, there was a maximum decrease. The growth attributes of eggplant seedlings were decreased by increasing the salinity levels (Figure 4B–D). A higher decrease was observed at 150 mM than at 75 mM NaCl. TiO_2-NP application at 200 ppm showed a smaller decrease than 0 and 400 ppm, but the ANOVA showed that the difference was statistically non-significant (Supplementary Table S1). Under salt stress, fresh biomass and dry biomass also showed similar results in eggplant seedlings (Figure 4E,F). The decrease in fresh biomass was greater at 150 mM NaCl with 0 ppm TiO_2-NP application. The minimum dry biomass was measured at 0 and 400 ppm under 150 mM NaCl. Meanwhile, 200 ppm TiO_2-NPs exhibited a smaller decrement than 400 ppm TiO_2-NPs at 75 and 150 mM NaCl.

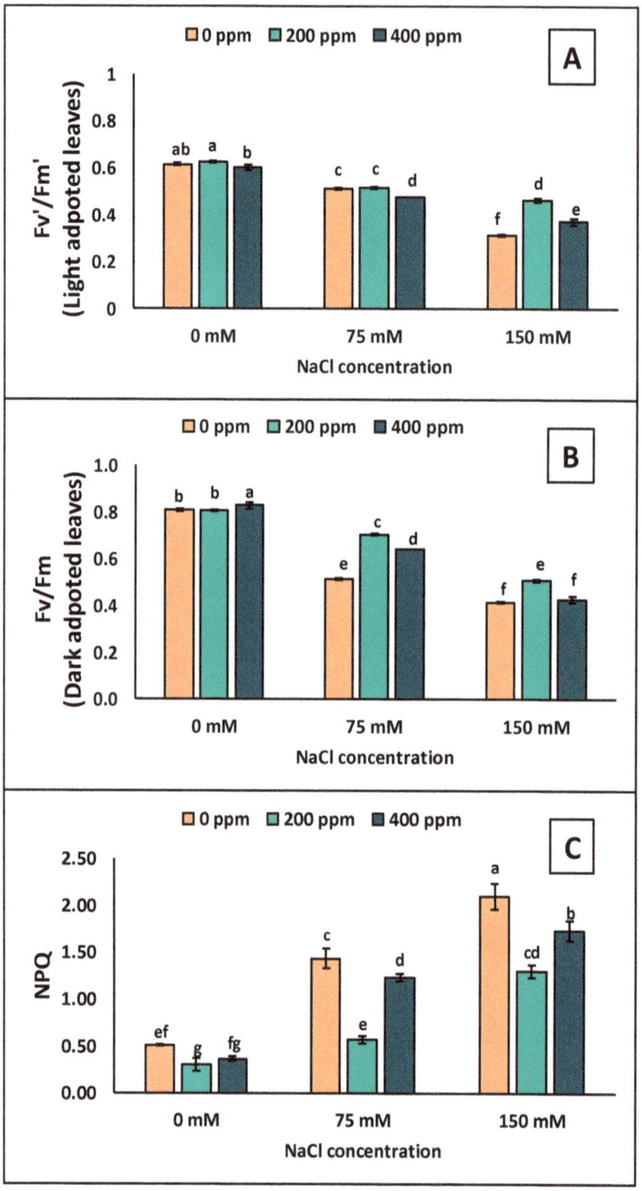

Figure 3. Measurements of the chlorophyll fluorescence in the leaves of eggplant seedlings grown under different levels of salinity stress (0, 75, 150 mM) with TiO_2-NP application (0, 200, 400 ppm). (**A**) Fv′/Fm′; (**B**) Fv/Fm; (**C**) NPQ. Lettering showing the statistical differences according to LSD test. Values are mean ±S.E. at $p < 0.05$.

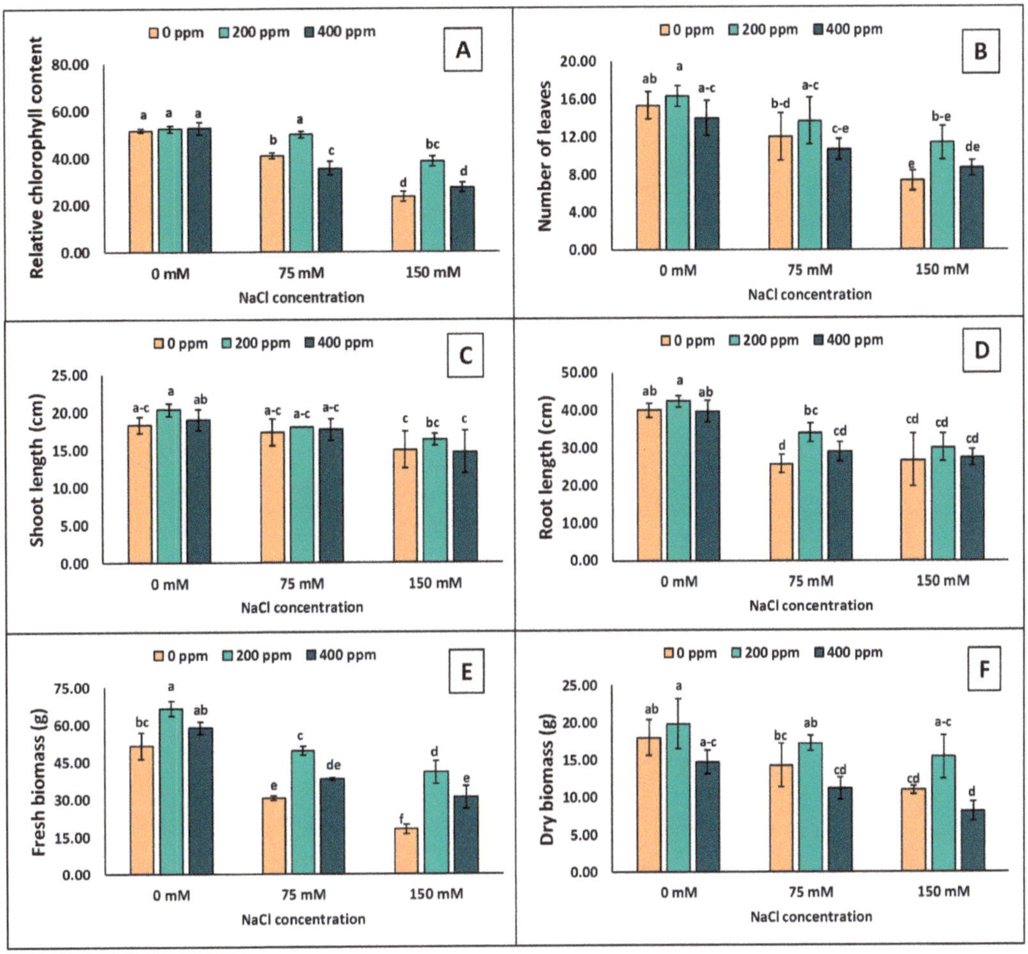

Figure 4. Measurements of the growth attributes in the leaves of eggplant seedlings grown under different levels of salinity stress (0, 75, 150 mM) with TiO_2-NP application (0, 200, 400 ppm). (**A**) Relative chlorophyll content; (**B**) number of leaves; (**C**) shoot length; (**D**) root length; (**E**) fresh biomass; (**F**) dry biomass. Lettering showing the statistical differences according to LSD test. Values are mean ±S.E. at $p < 0.05$.

3.4. Enzymatic Activity and TSP

The antioxidative enzyme activity and TSP were significantly increased by increasing salt stress by spraying with TiO_2-NPs (Figure 5). APx, CAT, SOD, POD, and GR showed higher activity when exposed to salt stress conditions. The maximum SOD, POD, CAT, APx, and GR activity was detected at 150 mM NaCl with 200 ppm foliar TiO_2-NP application (Figure 5). Meanwhile, the minimum activity was observed with 0 ppm and 400 ppm TiO_2-NPs at 75 and 150 mM NaCl. The 200 ppm TiO_2-NPs significantly increased the antioxidative enzyme activity at 0, 75, and 150 mM NaCl (Figure 5). However, the TSP content increased at 150 mM, while no significant difference was observed with TiO_2-NPs (Figure 5F).

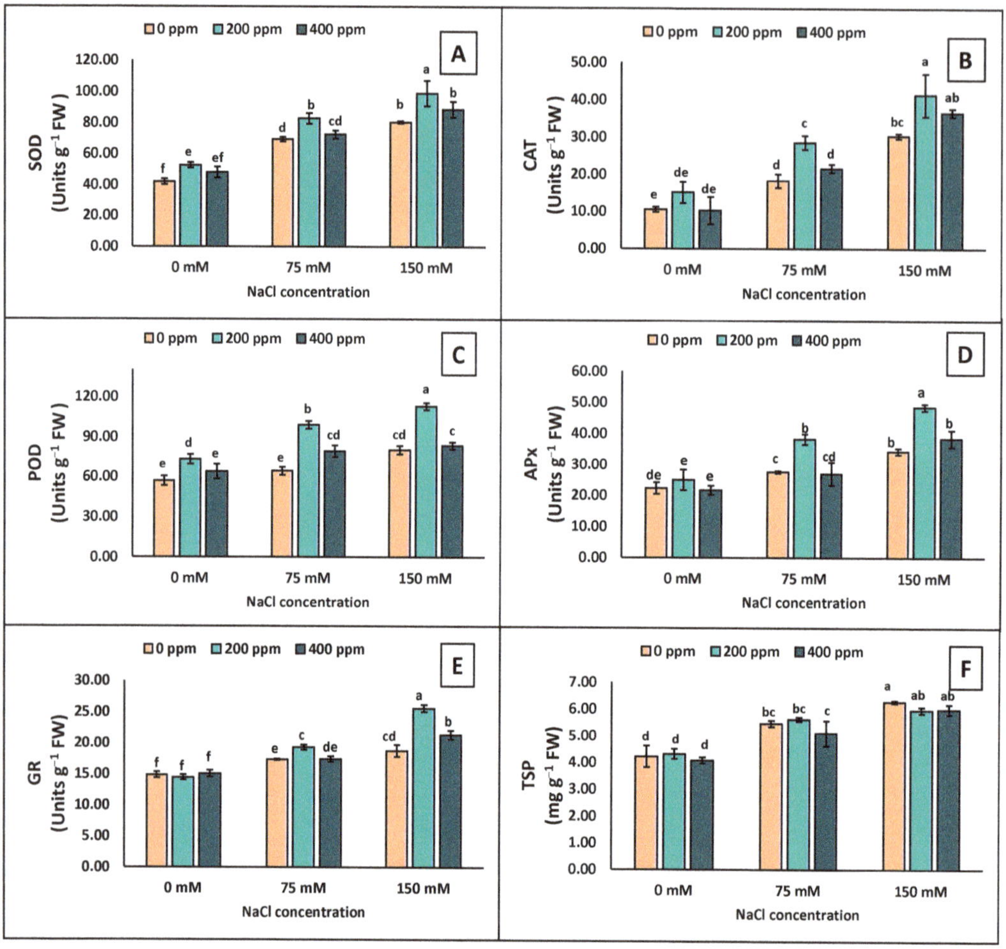

Figure 5. Measurements of the antioxidative enzymes in the leaves of eggplant seedlings grown under different levels of salinity stress (0, 75, 150 mM) with TiO_2-NP application (0, 200, 400 ppm). (**A**) SOD; (**B**) CAT; (**C**) POD; (**D**) APx; (**E**) GR; (**F**) TSP. Lettering showing the statistical differences according to LSD test. Values are mean ±S.E. at $p < 0.05$.

3.5. Stress Indicators and Toxic Ions

The stress indicators in the leaves of eggplant seedlings were observed to be higher by increasing the salinity stress (Figure 6A,B). The maximum H_2O_2 and MDA contents were observed at 150 mM NaCl with no application of TiO_2-NPs. The spraying of TiO_2-NPs on leaves at 200 ppm showed significantly lower H_2O_2 and MDA contents than 0 and 400 ppm TiO_2-NPs (Figure 6A,B). The only significant difference was at 400 ppm TiO_2-NPs and 150 mM H_2O_2 content.

Under salinity stress, the Na ion showed an increment in the leaves of eggplant seedlings by increasing its NaCl concentration (Figure 6C). The maximum Na^+ was observed in the leaves exposed to 150 mM NaCl. No statistically significant difference was observed at 75 mM NaCl with foliar application of TiO_2-NPs. Meanwhile, at 150 mM, 200 ppm showed less Na^+ accumulation in the leaves of eggplant seedlings (Figure 6C).

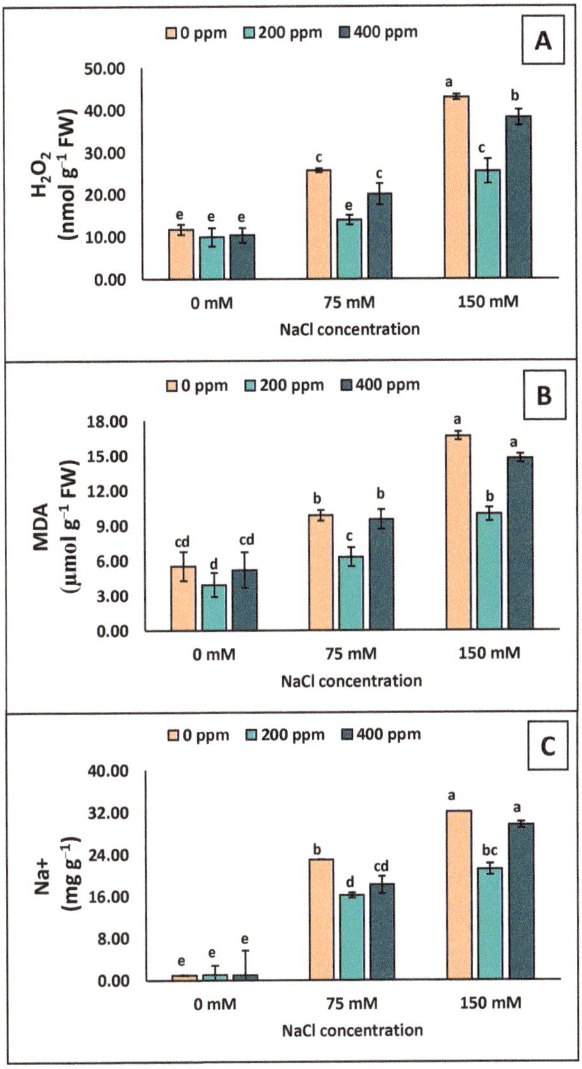

Figure 6. Measurements of the stress indicators and toxic ions in the leaves of eggplant seedlings grown under different levels of salinity stress (0, 75, 150 mM) with TiO_2-NP application (0, 200, 400 ppm). (**A**) H_2O_2; (**B**) MDA; (**C**) Na^+. Lettering showing the statistical differences according to LSD test. Values are mean ±S.E. at $p < 0.05$.

3.6. Pearson's Correlation

Leaf greenness was significantly and positively correlated with the number of leaves, Fv/Fm, Fv′/Fm′, and fresh biomass and was negatively correlated with NPQ and H_2O_2. MDA and H_2O_2 were negatively correlated with the shoot length, number of leaves, fresh biomass, Fv/Fm, and Fv′/Fm′, but positively correlated with NPQ (Figure 7). SOD was positively correlated with CAT, APx, GR, and TSP but negative with physiological attributes. Na^+ ions showed a significant positive correlation with MDA, TSP, H_2O_2, and NPQ. They also showed a negative correlation with the root length, shoot length, number of leaves, fresh biomass, leaf greenness, Fv/Fm, and Fv′/Fm′ (Figure 7).

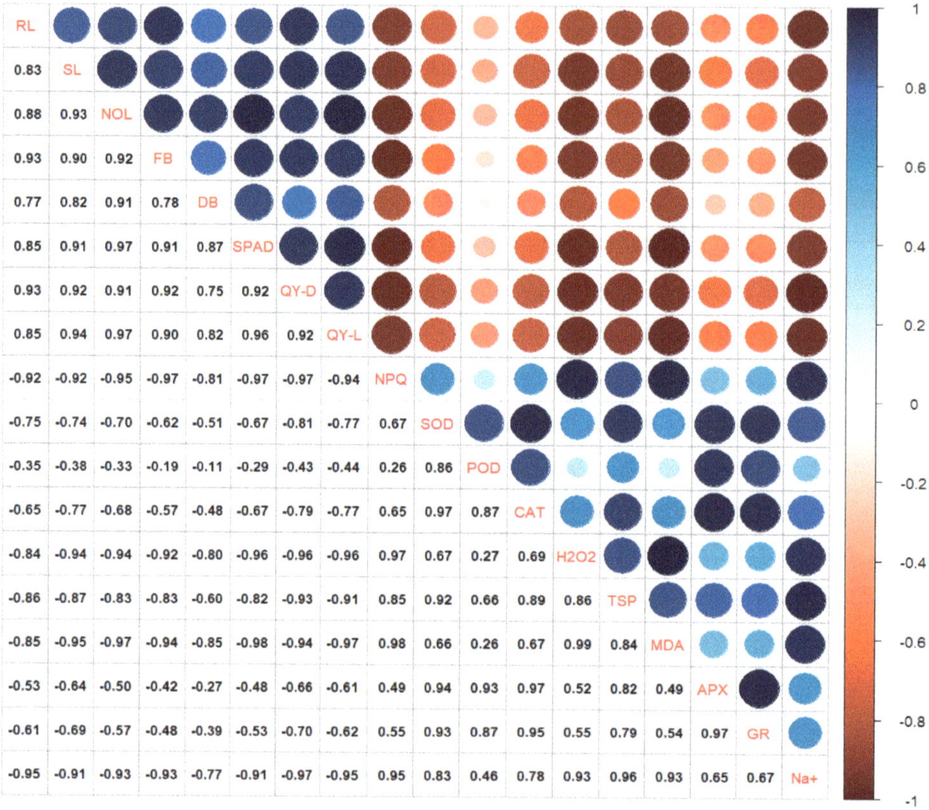

Figure 7. The Pearson's correlation matrix of exogenous application of TiO_2-NPs on eggplant seedlings grown under salt toxicity. RL = root length; SL = shoot length; NOL = no. of leaves; FB = fresh biomass; DB = dry Biomass; SPAD = relative chlorophyll content; QY-D = Fv/Fm; QY-L; Fv′/Fm′; NPQ = non photochemical quenching; SOD = superoxide dismutase; POD = peroxidase; CAT = catalase; TSP = total soluble proteins; MDA = malondialdehyde; APX = ascorbate peroxidase; GR = glutathione reductase.

3.7. PCA

A PCA was conducted for different investigated variables of eggplant seedlings under salt toxicity by applying TiO_2-NPs. The analysis depicted the separation due to foliar application of TiO_2-NPs and salinity stress. PC1 separated the salinity levels, while PC2 separated TiO_2-NPs. A variation of 58.3% was observed, while PC2 showed 13.9% variability (Figure 8A). A variation of 58.3% was observed in PC1, and NPQ, SOD, CAT, H_2O_2, MDA, TSP, GR, and Na^+ contributed positively, while all growth attributes contributed negatively (Supplementary Table S2). Eighteen variable groups over four sections were explored, which showed negative and positive factor spaces in 2D (Figure 8B). POD, CAT, APx, GR, SOD, and TSP are positioned on the positive section of both axes (X and Y), showing positive associations. The remaining twelve attributes were negatively associated with at least one parameter, as they were positioned on the negative axis.

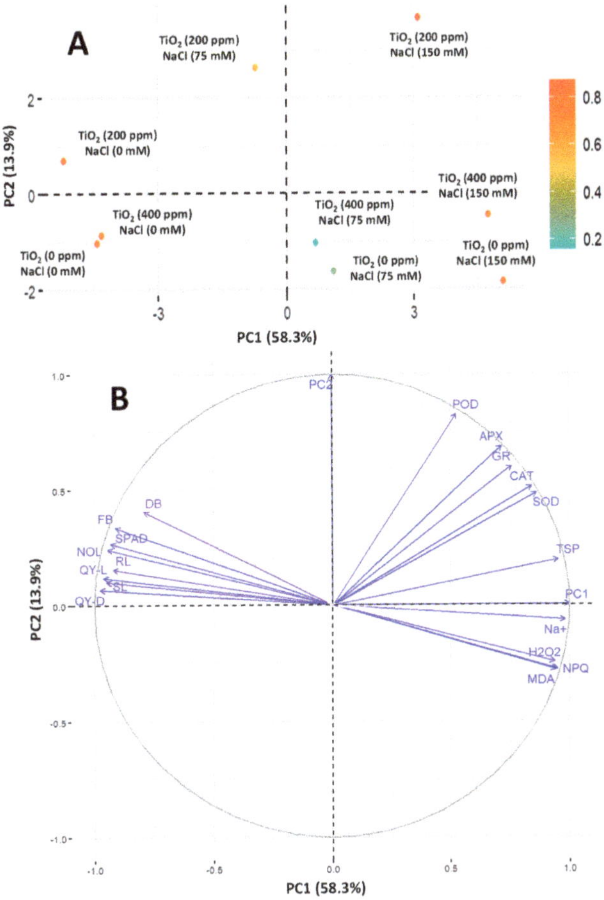

Figure 8. PCA representing eggplant seedlings grown under different levels of salinity stress (0, 75, 150 mM) with TiO_2-NP application (0, 200, 400 ppm). Contribution of each attribute on axes (**A**) and studied attributes (physiological and biochemical) on axes (**B**).

4. Discussion

Agricultural industries are plagued by salinity, one of the oldest and most concerning challenges. Salty conditions significantly impact plant productivity and growth, disrupting various physiological and biochemical processes. Physiological and biochemical processes in a saline environment are affected by the accumulation of Na^+, which affects nutrient and ion balances [48,49]. Salinity stress affects photosynthesis. NaCl stress disrupts photosystem II (PSII) due to its toxicity to Na^+, the most sensitive photochemical process [15,50]. Our results indicated that Fv′/Fm′ and Fv/Fm showed a decline under salt stress conditions. Plants face environmental stress when they encounter challenging conditions, as the Fv′/Fm′ and Fv/Fm of PSII decline. This implies that when exposed to environmental stress, the plant suffers from photoinhibition due to excess photon flux density inside the plant [51]. An increase in NPQ is mainly caused by releasing extra energy that can cause damage. This increase is linked to a negative relationship with PSII (Figure 7). It is possible that this could rival photochemical quenching and decrease the number of electrons engaged in photosynthesis, as demonstrated by a decrease in Fv′/Fm′ [15,34]. PSII and NPQ mainly affected plants under stress conditions due to electron acceptor concentrations ($NADP^+$ is on the acceptor side of PSI) [52]. Wu et al. [53] and AlKhatib et al. [54] reported results that are in alignment with our results, showing a decrease in Fv′/Fm′ and Fv/Fm and NPQ of eggplant seedlings under salinity stress. At the same time, the TiO_2-NPs limit the decrease in PSII, which makes the plant tolerate salinity (Figure 3). The foliar application of NPs on plants can synthesize more complexes for light harvesting, increasing photosynthesis and light absorption and leading to tolerance against salinity [16,55,56].

The decrease in photosynthetic attributes was also led by the plants' chlorophyll content, estimated by leaf greenness. The decline in leaf greenness also decreases plant growth, as illustrated in Figures 7 and 8. The leaf greenness is significant and positively correlated with plant growth attributes. Our results indicated that leaf greenness was positively correlated with PSII and negatively correlated with Na^+. Increasing the salinity level in plants decreased their growth and leaf greenness (Figure 4). Our results also depicted that applying TiO_2-NPs under normal conditions enhanced the growth and development of the plants, maintained their growth, and showed smaller decreases under salinity. Previous studies also showed increased the growth and agronomic traits of different plant species exposed to NPs under salinity [24,57,58]. Due to their small size and large surface area, NPs penetrate plant cells easily and alleviate the adverse effects of salinity relatively quickly by improving plant growth. The enhanced growth characteristics of plants might be explained by increased water absorption and less leaf transpiration, which results in improved water relations [59–61].

When plants undergo salinity stress, they enhance the production of ROS, i.e., $O_2^{\bullet-}$, H_2O_2, and $OH^{\bullet}$ [16]. Plants that are better at disposing of reactive oxygen species (ROS) are more resistant to oxidative damage than those with weaker ROS removal abilities [56,58]. Exposure to salinity leads to higher ROS production in plants, causing damage to nucleic acids, membrane integrity, cellular metabolism, and various metabolic processes in chloroplasts and mitochondria, inevitably leading to cell death [62]. Our results suggested that eggplant seedlings exposed to salinity showed more accumulation of H_2O_2 and MDA (Figure 6). Our correlation and PCA also showed a significant and positive relation with Na^+ ions, which demonstrated that by increasing the Na^+ concentration in plants, H_2O_2 and MDA also increased (Figures 7 and 8). However, applying TiO_2-NPs significantly decreases the production of H_2O_2 and MDA. Similar findings that NPs decreased the accumulation of H_2O_2 and MDA in Zea mays [63], grapes [64], and Moldavin Balm [58] were observed. Plants have a mechanism to maintain the equilibrium in the production and expulsion of ROS. To keep the balance in response to the overproduction of ROS in saline conditions, plants enhance the production of several scavenging antioxidative enzymes. It was convincingly reported that SOD is the primary enzyme that activates and converts the $O_2^{\bullet-}$ to H_2O_2, which CAT and APx further detoxify. The scavenging of $O_2^{\bullet-}$ by SOD reduced the lipid peroxidation [62]. H_2O_2 is also scavenged by POD in the chloroplast [65]. Similarly, the GR enzyme plays a vital role in the ascorbate–glutathione cycle. GR catalyzes the NADPH-dependent reduction in oxidized glutathione, helping protect many plants from oxidative damage induced by salt stress [66]. Our study also observed increased antioxidative enzymes when exposed to salinity stress (Figure 5). However, an increase was observed in the seedlings treated with foliar application of TiO_2-NPs. Similar findings that TiO_2-NPs enhanced the production of antioxidative enzymes under salinity stress were observed by Shah et al. [63] and Gohari et al. [24]. Omar et al. [67] reported that applying TiO_2-NPs under salinity stress on Vicia faba seedlings increased antioxidative enzymes and heat shock proteins (HSPs). TiO_2-NPs, which are up-regulated HSPs (HSP17.9 and HSP70), act as molecular chaperones and prevent misfolded proteins from aggregating, and their application can therefore protect cells and support the proper folding of newly synthesized proteins [68]. However, the increase in TiO_2-NPs also caused plant toxicity [24]. The higher concentration of NPs interacts with DNA and damages the root meristem [69]. A higher concentration of NPs increases the ROS, which damage lipids and proteins [70]. In our findings, we also observed that a higher concentration of TiO_2-NPs (400 ppm) led to lower antioxidative enzymes and higher stress indicators (H_2O_2, MDA) than 200 ppm of TiO_2-NPs. The Na^+ accumulation in the leaves was also limited following foliar application of TiO_2-NPs in eggplant under 75 and 150 mM NaCl (Figure 6C). Less accumulation of Na^+ was shown with higher concentrations of TiO_2-NPs (400 ppm) at 75 and 150 mM of NaCl, but it did not significantly differ from the control. The Na^+ accumulation was also found to be positive and significantly correlated with H_2O_2 and MDA, which indicated that by increasing the Na^+ accumulation in leaves, the plant stress indicators also increased (Figure 7). However, Na^+ is negatively correlated with PSII variables, leaf greenness, the no. of leaves, etc., which demonstrated that all growth and physiological attributes showed decrements by increasing the toxic ion concentration in the plant.

5. Conclusions

Moderate (75 mM) and high (150 mM) salt stress caused a significant decrease in the growth and development of the eggplant seedlings (Figure 9). The damage due to salinity was reduced by the application of TiO_2-NPs in eggplant seedlings by activating antioxidative enzymes and restricting the toxic ion accumulation in leaves. The concentration of 200 ppm of TiO_2-NPs significantly reduced the salinity effect in plants and improved their growth (shoot and root) and fresh and dry biomass. PSII was maintained in the plants that were treated with 200 ppm TiO_2-NPs. The application increased the oxidative enzyme activity (SOD, POD, and CAT) and simultaneously limited H_2O_2 and MDA accumulation. At a concentration of 400 ppm, high levels of TiO_2-NPs did not result in any notable variances in the enzymatic activity and growth of the plants compared to the control.

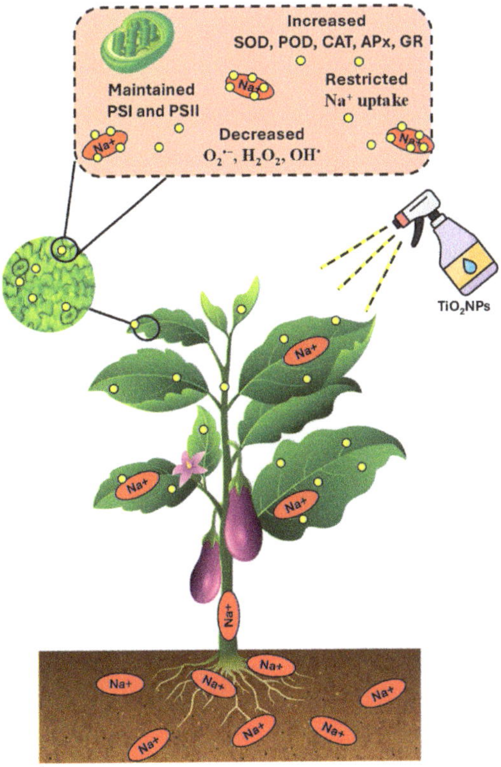

Figure 9. Schematic diagram showing how TiO_2-NPs mitigate the salt toxicity in eggplant plants.

Supplementary Materials: The following supporting information can be downloaded at https://www.mdpi.com/article/10.3390/antiox13101209/s1: Figure S1: Effect of TiO_2NPs on eggplant seedlings at 150 mM NaCl; Table S1: Analysis of Variance for eggplants seedlings under salinity stress with foliar application of TiO_2-NPs; Table S2: Principal Components for physiological and biochemical responses in the leaves of eggplant seedlings under salinity with application of TiO_2-NPs seedlings under salinity.

Author Contributions: M.F.K.: Conceptualization, Methodology, Investigation, Writing—original draft, and Visualization. M.Z.J.: Formal analysis and Writing—review and editing. M.N.: NP analysis—investigation and Writing—review and editing. R.A.S.: Investigation, Formal analysis, and Writing—review and editing. T.A.: Supervision, Resources, Funding acquisition, and Writing—review and editing. All authors have read and agreed to the published version of the manuscript.

Funding: This work was funded by Qatar University, Collaborative grant QUCG-ESC-24/25-507.

Institutional Review Board Statement: Not applicable.

Informed Consent Statement: Not applicable.

Data Availability Statement: All data are available within the article.

Conflicts of Interest: The authors declare no conflicts of interest.

References

1. Karaca, C.; Aslan, G.E.; Buyuktas, D.; Kurunc, A.; Bastug, R.; Navarro, A. Effects of salinity stress on drip-irrigated tomatoes grown under mediterranean-type greenhouse conditions. *Agronomy* **2022**, *13*, 36. [CrossRef]
2. Ahmed, M.Z.; Gul, B.; Khan, M.A.; Watanabe, K.N. Characterization and function of sodium exchanger genes in Aeluropus lagopoides under NaCl stress. In *Halophytes for Food Security in Dry Lands*; Academic Press: Cambridge, MA, USA, 2016; pp. 1–16.
3. Daba, A.W.; Qureshi, A.S. Review of soil salinity and sodicity challenges to crop production in the lowland irrigated areas of Ethiopia and its management strategies. *Land* **2021**, *10*, 1377. [CrossRef]
4. Balasubramaniam, T.; Shen, G.; Esmaeili, N.; Zhang, H. Plants' response mechanisms to salinity stress. *Plants* **2023**, *12*, 2253. [CrossRef] [PubMed]
5. Junedi, M.A.; Mukhopadhyay, R.; Manjari, K.S. Alleviating salinity stress in crop plants using new engineered nanoparticles (ENPs). *Plant Stress* **2023**, *9*, 100184. [CrossRef]
6. Amer, R. Spatial relationship between irrigation water salinity, waterlogging, and cropland degradation in the arid and semi-arid environments. *Remote Sens.* **2021**, *13*, 1047. [CrossRef]
7. Khalid, M.F.; Abou Elezz, A.; Jawaid, M.Z.; Ahmed, T. Salicylic acid restricts mercury translocation by activating strong antioxidant defense mechanisms in sweet pepper (*Capsicum annum* L.). *Environ. Technol. Innov.* **2023**, *32*, 103283. [CrossRef]
8. Khalid, M.F.; Shafqat, W.; Khan, R.I.; Jawaid, M.Z.; Hussain, S.; Saqib, M.; Rizwan, M.; Ahmed, T. Unveiling the resilience mechanism: Strigolactones as master regulators of plant responses to abiotic stresses. *Plant Stress* **2024**, *12*, 100490. [CrossRef]
9. Zhang, X.; Zhang, L.; Ma, C.; Su, M.; Wang, J.; Zheng, S.; Zhang, T. Exogenous strigolactones alleviate the photosynthetic inhibition and oxidative damage of cucumber seedlings under salt stress. *Sci. Hortic.* **2022**, *297*, 110962. [CrossRef]
10. Ling, F.; Su, Q.; Jiang, H.; Cui, J.; He, X.; Wu, Z.; Zhang, Z.; Liu, J.; Zhao, Y. Effects of strigolactone on photosynthetic and physiological characteristics in salt-stressed rice seedlings. *Sci. Rep.* **2020**, *10*, 6183. [CrossRef]
11. Liu, J.; Wu, Y.; Dong, G.; Zhu, G.; Zhou, G. Progress of research on the physiology and molecular regulation of sorghum growth under salt stress by gibberellin. *Int. J. Mol. Sci.* **2023**, *24*, 6777. [CrossRef]
12. Mir, R.A.; Somasundaram, R. Salicylic acid and salt stress tolerance in plants: A review. *J. Stress Physiol. Biochem.* **2021**, *17*, 32–50.
13. Khan, A.; Ashraf, M. Exogenously applied ascorbic acid alleviates salt-induced oxidative stress in wheat. *Environ. Exp. Bot.* **2008**, *63*, 224–231.
14. El-Beltagi, H.S.; Ahmad, I.; Basit, A.; Shehata, W.F.; Hassan, U.; Shah, S.T.; Haleema, B.; Jalal, A.; Amin, R.; Khalid, M.A.; et al. Ascorbic acid enhances growth and yield of sweet peppers (*Capsicum annum*) by mitigating salinity stress. *Gesunde Pflanz.* **2022**, *74*, 423–433. [CrossRef]
15. Mahawar, L.; Živčák, M.; Barboricova, M.; Kovár, M.; Filaček, A.; Ferencova, J.; Vysoká, D.M.; Brestič, M. Effect of copper oxide and zinc oxide nanoparticles on photosynthesis and physiology of *Raphanus sativus* L. under salinity stress. *Plant Physiol. Biochem.* **2024**, *206*, 108281. [CrossRef]
16. Khalid, M.F.; Iqbal Khan, R.; Jawaid, M.Z.; Shafqat, W.; Hussain, S.; Ahmed, T.; Alina Marc, R. Nanoparticles: The plant saviour under abiotic stresses. *Nanomaterials* **2022**, *12*, 3915. [CrossRef]
17. Lashkary, M.; Moghaddam, M.; Asgharzade, A.; Tatari, M. Titanium dioxide nanoparticle is involved in mitigating NaCl-induced *Calendula officinalis* L. by activation of antioxidant defense system and accumulation of osmolytes. *Plant Physiol. Biochem.* **2021**, *166*, 31–40. [CrossRef]

18. Mustafa, N.; Raja, N.I.; Ilyas, N.; Ikram, M.; Mashwani, Z.U.R.; Ehsan, M. Foliar applications of plant-based titanium dioxide nanoparticles to improve agronomic and physiological attributes of wheat (*Triticum aestivum* L.) plants under salinity stress. *Green Process. Synth.* **2021**, *10*, 246–257. [CrossRef]
19. Lei, Z.; Mingyu, S.; Xiao, W.; Chao, L.; Chunxiang, Q.; Liang, C.; Fashui, H. Antioxidant stress is promoted by nano-anatase in spinach chloroplasts under UV-B radiation. *Biol. Trace Elem. Res.* **2008**, *121*, 69–79. [CrossRef]
20. Abdel Latef, A.A.H.; Srivastava, A.K.; El-sadek, M.S.A.; Kordrostami, M.; Tran, L.S.P. Titanium dioxide nanoparticles improve growth and enhance tolerance of broad bean plants under saline soil conditions. *Land Degrad. Dev.* **2018**, *29*, 1065–1073. [CrossRef]
21. Hu, J.; Wu, X.; Wu, F.; Chen, W.; White, J.C.; Yang, Y.; Wang, X. Potential application of titanium dioxide nanoparticles to improve the nutritional quality of coriander (*Coriandrum sativum* L.). *J. Hazard. Mater.* **2020**, *389*, 121837. [CrossRef]
22. Khan, M.N. *Nano-Titanium Dioxide (nano-TiO_2) Mitigates NaCl Stress by Enhancing Antioxidative Enzymes and Accumulation of Compatible Solutes in Tomato (Lycopersicon esculentum Mill.)*; CABI Digital Library: Wallingford, UK, 2016.
23. Mustafa, N.; Raja, N.I.; Ilyas, N.; Abasi, F.; Ahmad, M.S.; Ehsan, M.; Prońków, J. Exogenous application of green titanium dioxide nanoparticles (TiO_2 NPs) to improve the germination, physiochemical, and yield parameters of wheat plants under salinity stress. *Molecules* **2022**, *27*, 4884. [CrossRef] [PubMed]
24. Gohari, G.; Mohammadi, A.; Akbari, A.; Panahirad, S.; Dadpour, M.R.; Fotopoulos, V.; Kimura, S. Titanium dioxide nanoparticles (TiO_2 NPs) promote growth and ameliorate salinity stress effects on the essential oil profile and biochemical attributes of Dracocephalum Moldavia. *Sci. Rep.* **2020**, *10*, 912. [CrossRef] [PubMed]
25. Amooaghaie, R.; Majidi, M.; Farhadian, S. Impact of nano-TiO_2 on salt stress tolerance of Carum copticum. *J. Plant Process Funct.* **2022**, *11*, 19–34.
26. Rico, C.M.; Peralta-Videa, J.R.; Gardea-Torresdey, J.L. Chemistry, biochemistry of nanoparticles, and their role in antioxidant defense system in plants. In *Nanotechnology and Plant Sciences: Nanoparticles and Their Impact on Plants*; Springer: Berlin/Heidelberg, Germany, 2015; pp. 1–17.
27. Tan, W.; Peralta-Videa, J.R.; Gardea-Torresdey, J.L. Interaction of titanium dioxide nanoparticles with soil components and plants: Current knowledge and future research needs–a critical review. *Environ. Sci. Nano* **2018**, *5*, 257–278. [CrossRef]
28. Jiang, Z.; Shen, L.; He, J.; Du, L.; Xia, X.; Zhang, L.; Yang, X. Functional analysis of SmMYB39 in salt stress tolerance of eggplant (*Solanum melongena* L.). *Horticulturae* **2023**, *9*, 848. [CrossRef]
29. Hanachi, S.; Van Labeke, M.C.; Mehouachi, T. Application of chlorophyll fluorescence to screen eggplant (*Solanum melongena* L.) cultivars for salt tolerance. *Photosynthetica* **2014**, *52*, 57–62. [CrossRef]
30. Mustafa, Z.; Ayyub, C.M.; Amjad, M.; Ahmad, R. Assessment of biochemical and ionic attributes against salt stress in eggplant (*Solanum melongena* L.) genotypes. *J. Anim. Plant Sci.* **2017**, *27*, 503–509.
31. Brenes, M.; Solana, A.; Boscaiu, M.; Fita, A.; Vicente, O.; Calatayud, Á.; Plazas, M. Physiological and biochemical responses to salt stress in cultivated eggplant (*Solanum melongena* L.) and in *S. insanum* L., a close wild relative. *Agronomy* **2020**, *10*, 651. [CrossRef]
32. Khan, M.; Siddiqui, Z.A.; Parveen, A.; Khan, A.A.; Moon, I.S.; Alam, M. Elucidating the role of silicon dioxide and titanium dioxide nanoparticles in mitigating the disease of the eggplant caused by Phomopsis vexans, Ralstonia solanacearum, and root-knot nematode Meloidogyne incognita. *Nanotechnol. Rev.* **2022**, *11*, 1606–1619. [CrossRef]
33. Hoagland, D.R.; Arnon, D.I. *The Water-Culture Method for Growing Plants without Soil*; CABI Digital Library: Wallingford, UK, 1938; Volume 347.
34. Khalid, M.F.; Vincent, C.; Morillon, R.; Anjum, M.A.; Ahmad, S.; Hussain, S. Different strategies lead to a common outcome: Different water-deficit scenarios highlight physiological and biochemical strategies of water-deficit tolerance in diploid versus tetraploid Volkamer lemon. *Tree Physiol.* **2021**, *41*, 2359–2374. [CrossRef]
35. Giannopolitis, C.N.; Ries, S.K. Superoxide dismutases: I. Occurrence in higher plants. *Plant Physiol.* **1977**, *59*, 309–314. [CrossRef] [PubMed]
36. Change, B.; Maehly, A.C. Assay of catalases and peroxidase. *Methods Enzym.* **1955**, *2*, 764–775.
37. Nakano, Y.; Asada, K. Hydrogen peroxide is scavenged by ascorbate-specific peroxidase in spinach chloroplasts. *Plant Cell Physiol.* **1981**, *22*, 867–880.
38. Foyer, C.H.; Halliwell, B. The presence of glutathione and glutathione reductase in chloroplasts: A proposed role in ascorbic acid metabolism. *Planta* **1976**, *133*, 21–25. [CrossRef] [PubMed]
39. Bradford, M.M. A rapid and sensitive method for the quantitation of microgram quantities of protein utilizing the principle of protein-dye binding. *Anal. Biochem.* **1976**, *72*, 248–254. [CrossRef] [PubMed]
40. Velikova, V.; Yordanov, I.; Edreva, A.J.P.S. Oxidative stress and some antioxidant systems in acid rain-treated bean plants: Protective role of exogenous polyamines. *Plant Sci.* **2000**, *151*, 59–66. [CrossRef]
41. Heath, R.L.; Packer, L. Photoperoxidation in isolated chloroplasts: I. Kinetics and stoichiometry of fatty acid peroxidation. *Arch. Biochem. Biophys.* **1968**, *125*, 189–198. [CrossRef]
42. Ryan, J.; Estefan, G.; Rashid, A. *Soil and Plant Analysis Laboratory Manual*; ICARDA: Beirut, Lebanon, 2001.
43. Gomez, K.A.; Gomez, A.A. *Statistical Procedures for Agricultural Research*; John Wiley & Sons: Hoboken, NJ, USA, 1984.
44. Matouke, M.M. FTIR study of the binary effect of titanium dioxide nanoparticles ($nTiO_2$) and copper (Cu^{2+}) on the biochemical constituents of liver tissues of catfish (*Clarias gariepinus*). *Toxicol. Rep.* **2019**, *6*, 1061–1070. [CrossRef]
45. Hamdan, S.A.; Ibrahim, I.M.; Ali, I.M. Comparison of anatase and rutile TiO_2 nanostructure for gas sensing application. *Dig. J. Nanomater. Biostructures* **2020**, *15*, 1001–1008. [CrossRef]

46. Sacco, A.; Mandrile, L.; Tay, L.L.; Itoh, N.; Raj, A.; Moure, A.; Del Campo, A.; Fernandez, J.F.; Paton, K.R.; Wood, S.; et al. Quantification of titanium dioxide (TiO_2) anatase and rutile polymorphs in binary mixtures by Raman spectroscopy: An interlaboratory comparison. *Metrologia* **2023**, *60*, 055011. [CrossRef]
47. Alsaiari, M.A.; Alhemiary, N.A.; Umar, A.; Hayden, B.E. Growth of amorphous, anatase and rutile phase TiO_2 thin films on Pt/TiO_2/SiO_2/Si (SSTOP) substrate for resistive random access memory (ReRAM) device application. *Ceram. Int.* **2020**, *46*, 16310–16320. [CrossRef]
48. Munns, R.; Tester, M. Mechanisms of salinity tolerance. *Annu. Rev. Plant Biol.* **2008**, *59*, 651–681. [CrossRef] [PubMed]
49. Khalid, M.F.; Hussain, S.; Anjum, M.A.; Ali, M.A.; Ahmad, S.; Ejaz, S.; Ali, S.; Usman, M.; Haque, E.U.; Morillon, R. Efficient compartmentalization and translocation of toxic minerals lead tolerance in volkamer lemon tetraploids more than diploids under moderate and high salt stress. *Fruits* **2020**, *75*, 204–215. [CrossRef]
50. Jajoo, A. Changes in photosystem II in response to salt stress. In *Ecophysiology and Responses of Plants under Salt Stress*; Springer New York: New York, NY, USA, 2012; pp. 149–168.
51. Bjorkman, O. Photon yield of O_2 evolution and chlorophyll fluorescence characteristics at 77 K among vascular plants of diverse origins. *Planta* **1987**, *170*, 489–504. [CrossRef] [PubMed]
52. Maxwell, K.; Johnson, G.N. Chlorophyll fluorescence—A practical guide. *J. Exp. Bot.* **2000**, *51*, 659–668. [CrossRef]
53. Wu, X.; Zhu, Z.; Li, X.; Zha, D. Effects of cytokinin on photosynthetic gas exchange, chlorophyll fluorescence parameters and antioxidative system in seedlings of eggplant (*Solanum melongena* L.) under salinity stress. *Acta Physiol. Plant.* **2012**, *34*, 2105–2114. [CrossRef]
54. Alkhatib, R.; Abdo, N.; Mheidat, M. Photosynthetic and ultrastructural properties of eggplant (*Solanum melongena*) under salinity stress. *Horticulturae* **2021**, *7*, 181. [CrossRef]
55. Ali, S.; Mehmood, A.; Khan, N. Uptake, translocation, and consequences of nanomaterials on plant growth and stress adaptation. *J. Nanomater.* **2021**, *2021*, 6677616. [CrossRef]
56. Li, Z.; Juneau, P.; Lian, Y.; Zhang, W.; Wang, S.; Wang, C.; Shu, L.; Yan, Q.; He, Z.; Xu, K. Effects of titanium dioxide nanoparticles on photosynthetic and antioxidative processes of Scenedesmus obliquus. *Plants* **2020**, *9*, 1748. [CrossRef]
57. Singh, S.; Husen, A. Role of nanomaterials in the mitigation of abiotic stress in plants. In *Nanomaterials and Plant Potential;* Springer: Berlin/Heidelberg, Germany, 2019; pp. 441–471.
58. Mohammadi, M.H.Z.; Panahirad, S.; Navai, A.; Bahrami, M.K.; Kulak, M.; Gohari, G. Cerium oxide nanoparticles (CeO_2-NPs) improve growth parameters and antioxidant defense system in Moldavian Balm (*Dracocephalum moldavica* L.) under salinity stress. *Plant Stress* **2021**, *1*, 100006. [CrossRef]
59. Martínez-Ballesta, M.C.; Zapata, L.; Chalbi, N.; Carvajal, M. Multiwalled carbon nanotubes enter broccoli cells enhancing growth and water uptake of plants exposed to salinity. *J. Nanobiotechnol.* **2016**, *14*, 42. [CrossRef] [PubMed]
60. Alsaeedi, A.; El-Ramady, H.; Alshaal, T.; El-Garawany, M.; Elhawat, N.; Al-Otaibi, A. Silica nanoparticles boost growth and productivity of cucumber under water deficit and salinity stresses by balancing nutrients uptake. *Plant Physiol. Biochem.* **2019**, *139*, 1–10. [CrossRef] [PubMed]
61. Chen, D.; Wang, S.; Yin, L.; Deng, X. How does silicon mediate plant water uptake and loss under water deficiency? *Front. Plant Sci.* **2018**, *9*, 281. [CrossRef] [PubMed]
62. Gill, S.S.; Tuteja, N. Reactive oxygen species and antioxidant machinery in abiotic stress tolerance in crop plants. *Plant Physiol. Biochem.* **2010**, *48*, 909–930. [CrossRef]
63. Shah, T.; Latif, S.; Saeed, F.; Ali, I.; Ullah, S.; Alsahli, A.A.; Jan, S.; Ahmad, P. Seed priming with titanium dioxide nanoparticles enhances seed vigor, leaf water status, and antioxidant enzyme activities in maize (*Zea mays* L.) under salinity stress. *J. King Saud Univ. -Sci.* **2021**, *33*, 101207. [CrossRef]
64. Mozafari, A.A.; Ghadakchi asl, A.; Ghaderi, N. Grape response to salinity stress and role of iron nanoparticle and potassium silicate to mitigate salt induced damage under in vitro conditions. *Physiol. Mol. Biol. Plants* **2018**, *24*, 25–35. [CrossRef]
65. Dionisio-Sese, M.L.; Tobita, S. Antioxidant responses of rice seedlings to salinity stress. *Plant Sci.* **1998**, *135*, 1–9. [CrossRef]
66. Foyer, C.; Lelandais, M.; Galap, C.; Kunert, K.J. Effects of elevated cytosolic glutathione reductase activity on the cellular glutathione pool and photosynthesis in leaves under normal and stress conditions. *Plant Physiol.* **1991**, *97*, 863–872. [CrossRef]
67. Omar, S.A.; Elsheery, N.I.; Pashkovskiy, P.; Kuznetsov, V.; Allakhverdiev, S.I.; Zedan, A.M. Impact of titanium oxide nanoparticles on growth, pigment content, membrane stability, DNA damage, and stress-related gene expression in Vicia faba under saline conditions. *Horticulturae* **2023**, *9*, 1030. [CrossRef]
68. Tichá, T.; Samakovli, D.; Kuchařová, A.; Vavrdová, T.; Šamaj, J. Multifaceted roles of HEAT SHOCK PROTEIN 90 molecular chaperones in plant development. *J. Exp. Bot.* **2020**, *71*, 3966–3985. [CrossRef]
69. Cox, A.; Venkatachalam, P.; Sahi, S.; Sharma, N. Silver and titanium dioxide nanoparticle toxicity in plants: A review of current research. *Plant Physiol. Biochem.* **2016**, *107*, 147–163. [CrossRef]
70. Demir, E.; Kaya, N.; Kaya, B. Genotoxic effects of zinc oxide and titanium dioxide nanoparticles on root meristem cells of Allium cepa by comet assay. *Turk. J. Biol.* **2014**, *38*, 31–39. [CrossRef]

antioxidants

MDPI

Article

The Abscisic Acid Receptor Gene *StPYL8-like* from *Solanum tuberosum* Confers Tolerance to Drought Stress in Transgenic Plants

Panfeng Yao [1], Chunli Zhang [1,2], Chao Sun [1,2], Yuhui Liu [1], Zhen Liu [1], Jia Wei [1,2], Xinglong Su [1,2], Jiangping Bai [1,2], Junmei Cui [1] and Zhenzhen Bi [1,2,*]

1 State Key Laboratory of Aridland Crop Science, Gansu Agricultural University, Lanzhou 730070, China; yaopf@gsau.edu.cn (P.Y.); zhangchunl@st.gsau.edu.cn (C.Z.); sunc@gsau.edu.cn (C.S.); lyhui@gsau.edu.cn (Y.L.); liuzhen@gsau.edu.cn (Z.L.); wej@st.gsau.edu.cn (J.W.); suxl@st.gsau.edu.cn (X.S.); baijp@gsau.edu.cn (J.B.); cuijm@gsau.edu.cn (J.C.)

2 College of Agronomy, Gansu Agricultural University, Lanzhou 730070, China

* Correspondence: bizz@gsau.edu.cn

Abstract: Pyrabactin resistance 1-like (PYL) proteins are abscisic acid (ABA) receptors that play a crucial role in the plant's response to adverse environmental conditions. However, as of yet, there is limited research on the role of PYL proteins in potato. In this study, a potato *PYL* gene, *StPYL8-like*, was identified through transcriptome analysis under drought stress. Molecular characterization revealed that the StPYL8-like protein possesses a highly conserved PYL family domain. Evolutionary analysis demonstrated that StPYL8-like protein clusters with various PYL proteins are involved in stress responses across different species. Functional assays showed that *StPYL8-like* robustly responds to different abiotic stresses, including drought and ABA treatment. Furthermore, the transient and stable expressions of *StPYL8-like* in tobacco enhanced their drought resistance, leading to increased plant height, leaf number, and fresh weight, as well as an improved root system. Transgenic tobacco carrying the *StPYL8-like* gene exhibited lower malondialdehyde (MDA) levels and higher proline accumulation and antioxidant enzyme activity compared to wild-type plants under drought conditions. Moreover, *StPYL8-like* upregulated the expression of stress-responsive genes (*NtRD29A*, *NtLEA5*, *NtP5CS*, *NtPOD*, *NtSOD*, and *NtCAT*) in transgenic plants subjected to drought stress. Collectively, these findings highlight the positive regulatory role of the *StPYL8-like* gene in enhancing potato plants' response to drought stress.

Keywords: potato; drought tolerance; *StPYL8-like*; expression analysis

1. Introduction

The growth and development of plants are frequently impacted by a variety of abiotic stresses, including drought, low temperature, high salinity and so on. Abiotic stress poses a significant latent threat to agricultural production, often resulting in pervasive reductions in crop yield. In response to abiotic stress, plants must orchestrate a range of physiological and biochemical responses, in addition to gene regulation, to effectively acclimate to adverse environmental conditions [1,2]. Gene regulation is intricately linked to stress-induced hormonal signaling pathways. Studies have demonstrated the critical involvement of ABA as a stress hormone in mediating plant responses to abiotic stressors [3,4]. It is crucial to conduct in-depth research on the ABA signaling pathway to improve plant adaptability to stressful environments.

Previous studies have identified the three fundamental components of the ABA signaling pathway: the ABA receptor PYL protein, the negative regulatory factor 2C protein phosphatase (PP2C), and the positive regulatory factor SNF1-related protein kinase 2 (SnRK2). These components collectively establish a dual-negative regulatory system, known as PYL-PP2C-SnRK2, which governs ABA signaling and subsequent responses [5,6]. Under normal circumstances, plants maintain low ABA levels, leading to the dimeric form of PYL that is unable to interact with PP2C. Consequently, PP2C exhibits high phosphatase activity, binding to SnRK2 to suppress its kinase function, thereby impeding downstream transcription-factor phosphorylation and gene expression within the ABA signaling pathway. However, in response to stressors like drought, ABA levels surge, facilitating the binding of ABA to and inducing a conformational change in PYL from its dimeric to its monomeric form. The monomeric PYL-ABA complex then associates with PP2C, forming a ternary complex that obstructs the active site of PP2C's phosphatase activity. This interaction releases SnRK2 activity, enabling the phosphorylation of downstream targets and the subsequent regulation of stomatal closure and resistance mechanisms [7,8]. In essence, the ABA receptor protein PYL functions upstream of the PYL-PP2C-SnRK2 regulatory system within the ABA signaling pathway, crucially involved in ABA signal recognition, PP2C inhibition, and the initiation of ABA signaling transduction.

As a stress hormone, ABA content in plants rapidly elevates under abiotic-stress conditions, particularly drought and salt stress. In recent years, the role of ABA receptors in plant responses to abiotic stress has been gradually elucidated. To date, researchers have conducted functional analyses on 35 PYL proteins across 13 species, revealing the pivotal role of PYL in plant responses to abiotic stress [9], root development [10], fruit ripening [11], and various physiological processes. Primarily, PYL's function lies in mediating signal transduction in plant responses to abiotic stress, with divergent mechanisms of action observed. For instance, Arabidopsis *PYL5* bolsters drought resistance by directly inhibiting type A PP2Cs [12]; wheat *PYL4* enhances water-use efficiency and drought tolerance by modulating stomatal conductance and elevating photosynthetic activity in transgenic plants [13]; rice *PYL3/10* heightens seed sensitivity to ABA during germination and early seedling stages, thereby enhancing adaptability to stress factors [14]; and maize *PYL8/9/12* enhances drought tolerance in transgenic plants through increased proline accumulation [15]. Furthermore, investigations on *PYL* gene function in various plant species such as grapes [16], tomatoes [17], rapeseed [18], strawberries [19], and apples [20] have demonstrated that modulating the expression of this gene family can enhance drought resistance in plants. In essence, the *PYL* gene significantly contributes to plant responses to drought stress, with its regulatory potential offering a means to enhance plant resilience, albeit through diverse mechanisms of action.

Potato (*Solanum tuberosum* L.) plays a critical role in ensuring global food security as the fourth-largest staple crop globally. It is primarily grown in northwestern China, where water scarcity frequently induces drought stress in potatoes, consequently exerting a significant influence on both local and national commercial potato production [21]. Consequently, there is a growing emphasis on exploring and characterizing drought-resistant genes in potato to enhance stress tolerance. Some progress has also been made in the functional study of key genes in the ABA signaling pathway for potato drought tolerance. For instance, Bai et al. [22] identified the *SnRK* family genes in potatoes, and Yao et al. [23] further validated the drought-resistance function of this gene family. Wang et al. [24] conducted a comprehensive screening of the potato *PP2C* gene family at the genome level and observed the response of each gene to abiotic stress. Garia et al. [25] discovered that *StPP2Ac2b* plays a positive regulatory role in tuber formation in creeping stems. Moreover, Gui et al. [26] recently identified the *PYL* family genes in potatoes and assessed the response of each gene to various abiotic stresses. In the initial phase of this study, comprehensive whole-genome profiling of the potato *PYL* gene family was carried out, alongside the assessment of the expression profiles of individual *PYL* genes under drought conditions. Subsequently, *StPYL8-like*, exhibiting a pronounced response to drought stress, was chosen as the target for further exploration, and its role in drought resistance mechanisms was investigated. Then, the drought-resistance function of *StPYL8-like* was verified through transient and stable transformation methods. The outcomes of this investigation advance and diversify research on the ABA signaling pathway in potatoes, offering insights for exploring the molecular regulatory network of potato drought tolerance and enhancing the molecular breeding of drought-resistant potato cultivars.

2. Materials and Methods

2.1. Plant Material and Growth Conditions

The potato cultivars "Qingshu 9" (QS9) and the tobacco variety NT12 were conserved at the State Key Laboratory of Aridland Crop Science of Gansu Agricultural University. Potato plants were cultivated in an artificial growth chamber under a light regimen of 16 h of illumination and 8 h of darkness, maintained at a temperature of 22 ± 2 °C, with a relative humidity of 60%. Tobacco seedlings were cultured in a greenhouse with a temperature of 25 ± 2 °C, a light cycle of 12 h of light and 12 h of darkness, and a relative humidity of 60%. The plant expression vector pCAMBIA1304 and *Agrobacterium* tumefaciens GV3101 were housed within our laboratory.

2.2. Isolation and Characterization of StPYL8-like

The potato reference-genome sequence and GFF annotation file were acquired from the potato genome website (http://spuddb.uga.edu/, accessed on 5 March 2022), while all AtPYL protein sequences were sourced from TAIR (https://www.arabidopsis.org/, accessed on 5 March 2022). The StPYL protein sequences were derived through sequence alignment and screening using TBtools. Subsequently, the expression profile of *StPYLs* was analyzed through transcriptome data of potatoes under drought stress. Finally, the *StPYL8-like* gene (Gene ID: PGSC0003DMG400005016) that significantly responds to drought stress was cloned. Following the sequencing outcomes, specific primers were designed for the amplification of the *StPYL8-like* gene (534 bp, GenBank ID: XP_006360709), which was subsequently cloned using potato young-leaf cDNA as a template. The protein sequence encoded by *StPYL8-like* was then subjected to multiple-sequence alignment and phylogenetic analysis using the Cluster X and MEGA software tools (https://app.clusterx.com/, https://www.megasoftware.net/, accessed on 8 April 2022).

2.3. Expression Analysis of StPYL8-like

The response of *StPYL8-like* to diverse abiotic stresses was detected using qRT-PCR. Three-week-old potato plants were subjected to stress treatments, including 200 mM mannitol, 100 mM NaCl, 100 μM ABA, 100 μM SA, and 100 μM MeJA. Control conditions consisted of a normal MS medium for drought and salt treatments, while alcohol spraying served as the control for hormone treatments. Samples were collected 0, 1, 3, 6, and 12 h post stress induction, followed by RNA extraction post liquid-nitrogen freezing. Each stress treatment was replicated thrice. The qRT-PCR primers are detailed in Supplementary Table S1.

2.4. Promoter cis-Element Analysis

The 2000 bp promoter sequence upstream of the start codon of the *StPYL8-like* gene was extracted using TBtools. Subsequently, the cis-acting elements of the *StPYL8-like* promoter sequence were analyzed utilizing the PlantCARE online tool (http://bioinformatics.psb.ugent.be/webtools/plantcare/html/, accessed on 15 April 2022). Following this, *StPYL8-like* promoter was incorporated into the plant expression vector pBI101-GUS through homologous recombination. The transient transformation of 4-week-old tender tobacco leaves was carried out using *Agrobacterium* tumefaciens GV3101, with samples collected 2 days later for GUS histological staining at 37 °C to assess the activity of the *StPYL8-like* promoter.

2.5. Evaluation of Drought Resistance after Transient Transformation of Tobacco

The coding sequence (CDS) of the *StPYL8-like* gene was amplified by PCR using gene-specific primers. The PCR product was subsequently digested with *Bgl* II and *Spe* I restriction enzymes and ligated into the pCAMBIA1304-GUS vector to generate the recombinant plasmid pC1304-StPYL8-like-GUS. The construct was then transformed into

Agrobacterium tumefaciens GV3101 cells using the heat-shock method. To assess the biological function of the *StPYL8-like* gene, transient transformation was performed in tobacco plants. Tobacco plants grown under normal conditions for approximately 4 weeks were transferred to Hoglund nutrient solution for a 2-day culture period. Following this, GV3101 harboring pC1304-StPYL8-like-GUS was infiltrated into tobacco leaves, with the empty vector pC1304 serving as a negative control. After 2 days of recovery, all the plants were subjected to stress treatment with 20% PEG600 in the Hoglund nutrient solution. GUS staining was carried out after 6 h to confirm successful transformation. Subsequently, various physiological parameters associated with stress responses were measured to evaluate the impact of the *StPYL8-like* gene on drought tolerance in transgenic tobacco plants.

2.6. Evaluation of Drought Resistance after Stable Transformation of Tobacco

The recombinant plasmid pC1304-StPYL8-like-GUS was introduced into tobacco T12 through the *Agrobacterium*-mediated leaf-disk transformation method. Subsequently, transgenic plants were cultivated on MS medium and screened for resistant seedlings using 1/2 MS medium supplemented with hygromycin (50 mg L^{-1}, *w/v*). Positive lines were then confirmed by qRT-PCR and GUS staining. Based on the qRT-PCR results, three transgenic lines exhibiting the highest *StPYL8-like* expression levels were selected. These lines were subsequently subjected to drought stress by cutting them into uniform stem segments and transferring them to MS medium supplemented with 100 mM, 200 mM, and 300 mM mannitol. A control group was maintained under normal MS medium conditions. Following a 30-day drought-stress treatment period, various indicators related to stress response were evaluated.

2.7. β-Glucuronidase (GUS) Staining

GUS histological staining was performed as detailed in a previous publication. Transgenic plant leaves were collected and immersed in a reaction solution containing 2 mM 5-bromo-4-chloro-3-indolyl-β-D-glucuronic acid, 50 mM sodium phosphate, 10 mM EDTA, 2 mM ferrocyanide, and 0.1% Triton X-100 at pH 7.0. The samples were then incubated at 37 °C in darkness for 12 h. Subsequently, the stained leaves were subjected to a gradient elution process using 30%, 60%, and 100% ethanol in an 80 °C water bath. Once chlorophyll was completely removed, micrographs were captured using an optical microscope.

2.8. Determination of Phenotypic and Physiological Indicators

The proline and malondialdehyde (MDA) contents were quantified following the methodologies outlined in previous studies. Enzyme activities of superoxide dismutase (SOD), peroxidase (POD), and catalase (CAT) were assessed in both transgenic and control plants utilizing commercially available assay kits (BC0170, BC0220, BC0200, and BC0200 from Beijing Solaibao Technology Co., Ltd., Beijing, China). Phenotypic characteristics primarily encompass plant height, stem diameter, and root length. Furthermore, a root scanner was employed for the evaluation of root parameters including total root length, number of root tips, root surface area, and root volume.

2.9. Expression Measurement of Stress-Responsive Genes

Total RNA was extracted from tobacco samples subjected to various stress conditions and reverse-transcribed to cDNA. Subsequently, specific primers (Table S1) were employed to quantify the expression levels of stress-responsive genes via qRT-PCR.

2.10. Statistical Analysis

The expression analysis and physiological index determination under abiotic stress were both performed for three independent biological replicates with three technical repeats. The mean and standard deviation (SD) were calculated for each treatment based on the collected data from these replicates. A one-way analysis of variance (ANOVA) was conducted using SPSS v19.0 (SPSS Inc., Chicago, IL, USA) to assess the variability among the three overexpression lines (OE) and control plants, with statistical significance defined as $p < 0.05$ or $p < 0.01$.

3. Results

3.1. Molecular Characterization Analysis of *StPYL8-like*

The whole-genome identification results revealed the presence of 20 *PYL* genes in potatoes, among which three genes exhibited two different transcripts (Table S2). All PYL proteins contain the distinctive conserved domains characteristic of this protein family (Figure S1). Subsequently, the response of all *StPYL* genes to drought stress was evaluated through the transcriptome analysis of 'QS9' potato under drought conditions, revealing distinct expression patterns among *PYL* family members under such stress. Notably, *StPYL16*, displaying a significant response to drought stress, was singled out for further investigation (Figure S2). Nevertheless, the sequence was submitted to the NCBI database a few years ago and named *StPYL8-like* (GenBank ID: XP_006360709). To ensure coherence, we ultimately labeled the *PYL* sequence under investigation in this study *StPYL8-like*.

The *StPYL8-like* gene comprises a 534 bp open reading frame that encodes a protein consisting of 177 amino acids. Blastx analysis revealed that StPYL8-like shares sequence identities with AtPYL9 and AtPYL8 of 78% and 77% in Arabidopsis, respectively. Subsequent sequence alignment demonstrated a high similarity between the StPYL8-like protein sequence and the conserved domains of PYL proteins found in Arabidopsis, apple, cotton, and rice. Moreover, typical Gate and Latch motifs, which are key features of the PYL family proteins, were identified (Figure 1a). Additionally, phylogenetic analysis placed StPYL8-like in subgroup III, where it clustered with most PYL proteins involved in abiotic stress regulation, such as MdPYL9, GhPYL26, and OsPYL7/10/11 (Figure 1b). This suggested that StPYL8-like may possess similar biological functions to those proteins.

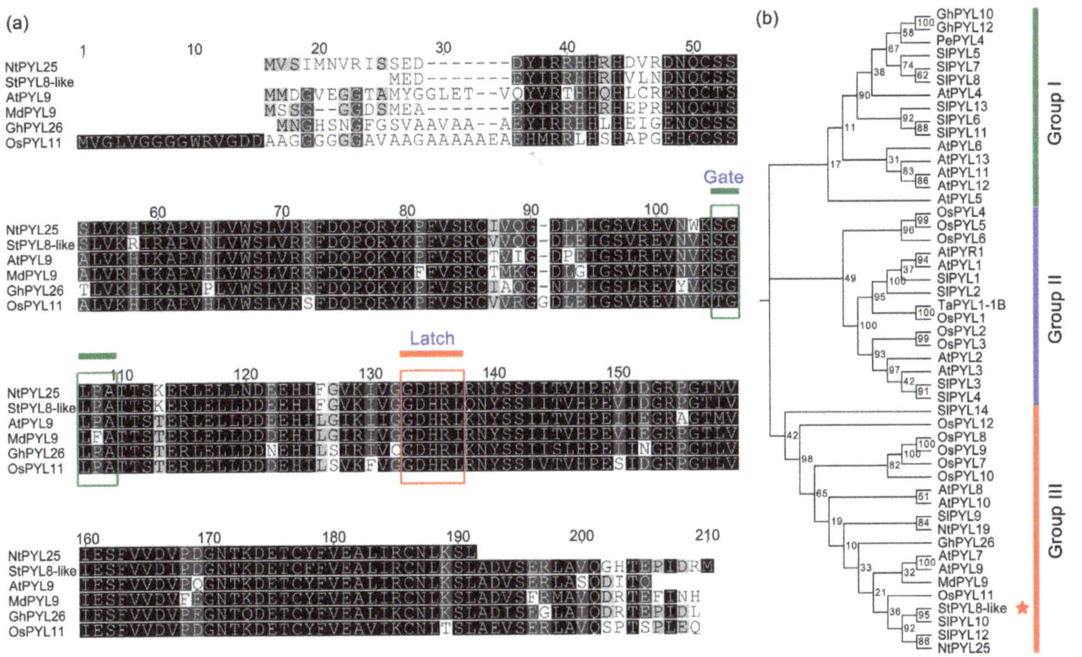

Figure 1. Analysis of the structure of *StPYL8-like* protein. (**a**) Sequence alignment of *StPYL8-like* with other PYL proteins. Gate and Latch conserved domains are indicated; (**b**) Phylogenetic relationship of *StPYL8-like* with other PYL proteins. The red asterisk represents potato *StPYL8-like* protein.

3.2. *StPYL8-like* Is Strongly Induced by Drought Treatment

To explore the reaction of *StPYL8-like* to a range of abiotic stresses, such as drought, salt, 4 °C, and ABA and SA treatments, experiments were conducted (Figure 2). In general, the *StPYL8-like* gene displays diverse expression patterns in response to the different abiotic-stress treatments. The findings revealed that in QS9, *StPYL8-like* demonstrated a pattern of initial increase followed by decrease across multiple stressors. Specifically, under drought, salt, 4 °C, and ABA stimuli, *StPYL8-like* exhibited a notable rise to its peak value after 1 h of stress, followed by a gradual decline after 3 h. Conversely, under SA and MeJA stimuli, a continuous upward trajectory was observed.

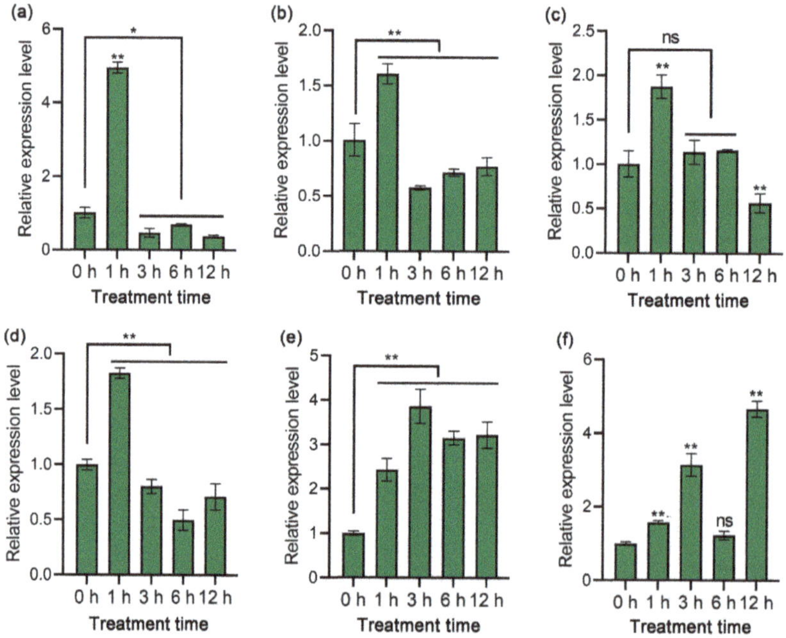

Figure 2. The relative expression level of the *StPYL8-like* gene under abiotic-stress conditions. (**a**–**f**) represent 200 mM mannitol, 100 mM NaCl, 100 μM ABA, 100 μM SA, and 100 μM MeJA stress, respectively. Data represent the means ± SD of three replicates. * and ** indicate significant difference at $p < 0.05$ and $p < 0.01$ levels, respectively. ns indicates that the difference is not significant. The horizontal line above the column indicates that the data below the line have the same level of significance difference compared to the data at 0 h.

3.3. Transient Transformation of *StPYL8-like* Improves Drought Tolerance of Transgenic Tobacco

To preliminarily investigate the function of *StPYL8-like*, we transiently expressed it in tobacco and exposed the plants to drought stress (Figure 3a). Initially, GUS staining was conducted on the transgenic plants, revealing a distinct blue color in *StPYL8-like*-transformed leaves as opposed to slightly yellow control leaves, indicating the successful integration of the target gene (Figure 3b). Furthermore, to elucidate the effects of *StPYL8-like* on transgenic plants, we evaluated the activities of antioxidant enzymes (SOD, POD, and CAT) closely associated with plant stress responses, along with the accumulation levels of MDA and Pro (Figure 3c). Our findings demonstrated that the overexpression of *StPYL8-like* resulted in a substantial increase in Pro accumulation in transgenic plants under drought-stress conditions while decreasing MDA levels. Conversely, no significant differences were observed between the two indicators under normal conditions. The activity of antioxidant enzymes followed a similar trend. Under normal circumstances, there was a certain variance in antioxidant enzyme activity between control and transgenic plants, albeit not statistically significantly; however, under drought-stress conditions, the activity of all three antioxidant enzymes in transgenic plants exhibited a noteworthy increase compared to the control.

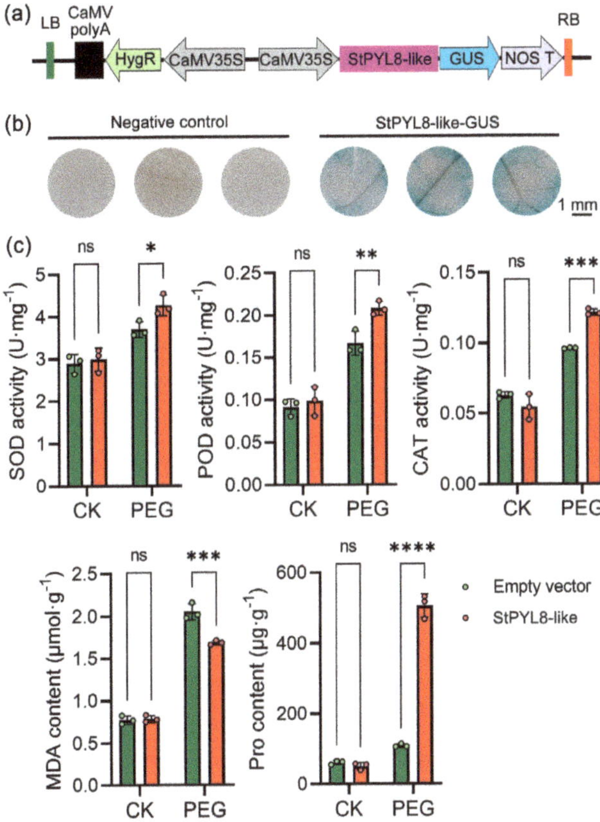

Figure 3. Identification of drought resistance in tobacco after transient transformation of the *StPYL8-like* gene. (**a**) Schematic diagram of the *StPYL8-like* gene plant-overexpression vector; (**b**) GUS histochemical staining of *StPYL8-like* transgenic plants; (**c**) Determination of physiological indexes related to stress. CK and PEG represent normal Hogland nutrient solution and Hogland nutrient solution containing 20% PEG6000, respectively. *, **, ***, and **** indicate significant difference at $p < 0.05$, $p < 0.01$, $p < 0.001$, and $p < 0.0001$ levels, respectively. ns indicates that the difference is not significant.

3.4. Generation of Stable Transgenic Tobacco

To further explore the role of *StPYL8-like* in response to drought stress, we established an overexpression of *StPYL8-like* in tobacco via an *Agrobacterium*-mediated leaf-disc transformation technique (Figure 4a). Subsequent hygromycin selection yielded 20 transgenic plants exhibiting resistance. Following RT-PCR analysis, the presence of the vector tag gene was confirmed in 13 out of the 20 resistant plants (Figure 4b). This finding was consistent with positive GUS histological staining in seedlings (Figure 4c). Subsequently, the quantitative assessment of *StPYL8-like* expression levels in each positive plant was performed using qRT-PCR, leading to the selection of three transgenic lines (OE-6, OE-8, and OE-9), characterized by high *StPYL8-like* expression levels, for further investigation (Figure 4d).

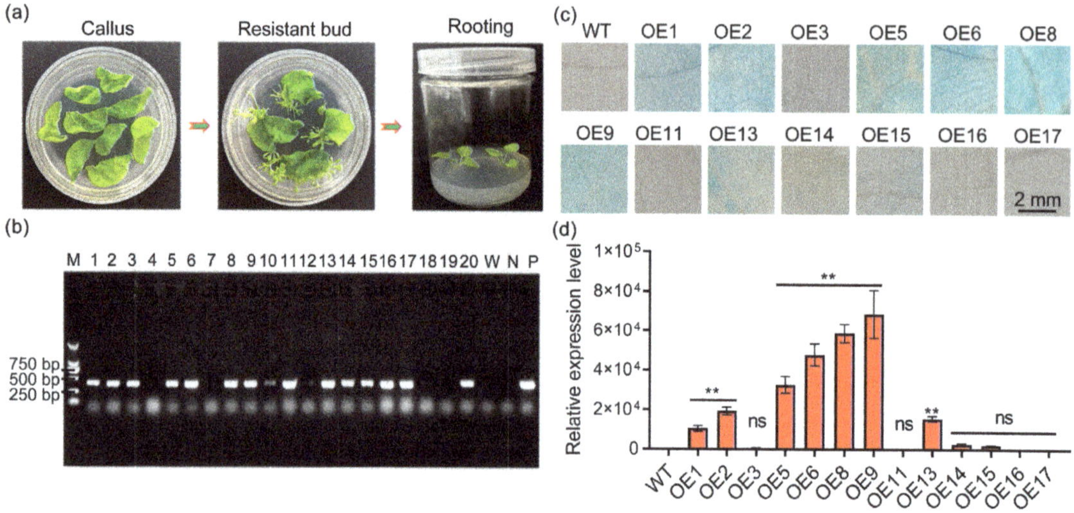

Figure 4. Positive identification of *StPYL8-like* transgenic tobacco. (**a**) The process of transforming *StPYL8-like* into tobacco; (**b**) PCR molecular identification of positive *StPYL8-like* transgenic lines; (**c**) GUS histochemical staining of *StPYL8-like* transgenic lines and control plants; (**d**) Analysis of expression level of *StPYL8-like* in transgenic lines. ** represents significant differences between transgenic lines and WT at $p < 0.01$. ns indicates that the difference is not significant. The horizontal line above the column indicates that the data below the line have the same level of significance difference compared to the data at WT.

3.5. Overexpression of StPYL8-like Increases Tobacco Drought Tolerance

To determine the drought-resistance function of *StPYL8-like*, transgenic tobacco and control plants were subjected to drought-stress treatments. Initially, images of each plant genotype were captured under stress conditions for visual assessment. The results indicated discernible disparities in the growth patterns between transgenic and control strains under normal conditions, with transgenic plants displaying superior growth performance. Upon exposure to stress, the growth of all plants was impeded, with control plants exhibiting a more pronounced inhibition compared to transgenic plants. Furthermore, this discrepancy magnified with escalating stress severity. A notable observation was that transgenic plants maintained normal root growth at a concentration of 200 mM mannitol, whereas the root growth of control plants was completely suppressed (Figure 5a).

To more specifically illustrate the impact of *StPYL8-like* overexpression on the growth of transgenic plants, detailed measurements were conducted on parameters including leaf number, plant height and fresh weight (Figure 5b). The results indicated that under normal conditions, the *StPYL8-like* transgenic plants exhibited significantly higher leaf numbers, plant heights, and fresh weights compared to control plants, suggesting that the overexpression of *StPYL8-like* may contribute to both plant development and enhanced drought resistance in transgenic plants. This difference was even more pronounced under 100 mM mannitol stress conditions. When the root growth of wild-type plants was inhibited (200 mM mannitol), no significant differences were observed in leaf number and fresh weight compared to transgenic plants; however, the plant height remained notably lower in wild-type plants. Under 300 mM mannitol conditions, no significant variations were detected in the aforementioned parameters.

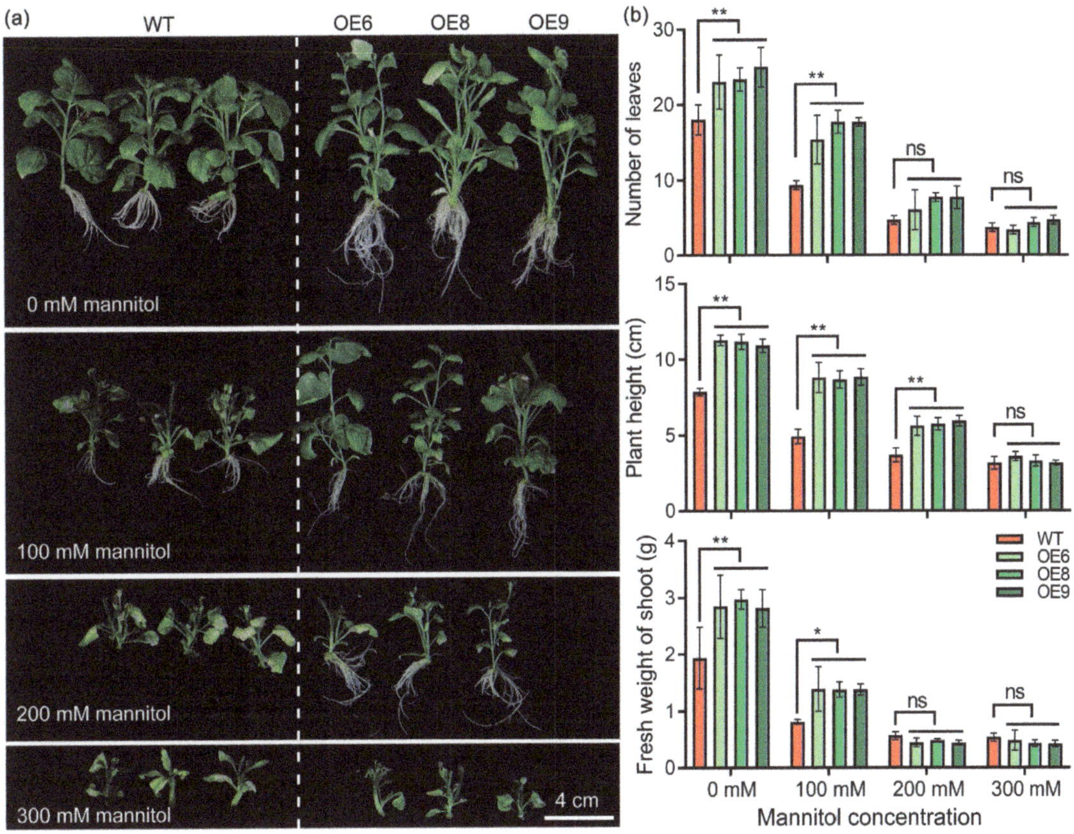

Figure 5. Drought-resistance function of *StPYL8-like* after stable transformation of tobacco. Transgenic tobacco-stem segments with consistent growth were transferred to MS medium as well as MS medium supplemented with 100 mM, 200 mM, and 300 mM mannitol for 30 d to stress, followed by the measurement of various phenotypic traits. (**a**) Phenotypic identification of plants under drought stress; (**b**) Statistics of leaf number, plant height and fresh weight of plants under normal and drought stress. * and ** indicate significant difference at $p < 0.05$ and $p < 0.01$ levels, respectively. ns indicates that the difference is not significant. The horizontal line above the column indicates that the data below the line have the same level of significance difference compared to the data at WT.

The aforementioned findings demonstrated that the overexpression of *StPYL8-like* has a substantial impact on the growth of transgenic plant roots. Subsequently, we conducted a comprehensive assessment of root phenotypes and associated metrics across different plant strains subjected to varying stress conditions. Morphologically, the root system of *StPYL8-like* transgenic plants exhibited superior development compared to control plants under both normal and stress conditions (Figure 6a). These observations were substantiated by the evaluation of pertinent metrics. Specifically, parameters such as primary root length, total root length, root area, number of root tips, and total root volume displayed a consistent pattern. Under normal circumstances, the metrics for transgenic plants were markedly higher than those for control plants, whereas under stress conditions, all line indicators experienced inhibition; however, the discrepancy persisted, peaking at 200 mM mannitol (Figure 6b). Notably, there was no significant alteration in root diameter pre- and post-stress during standard root growth.

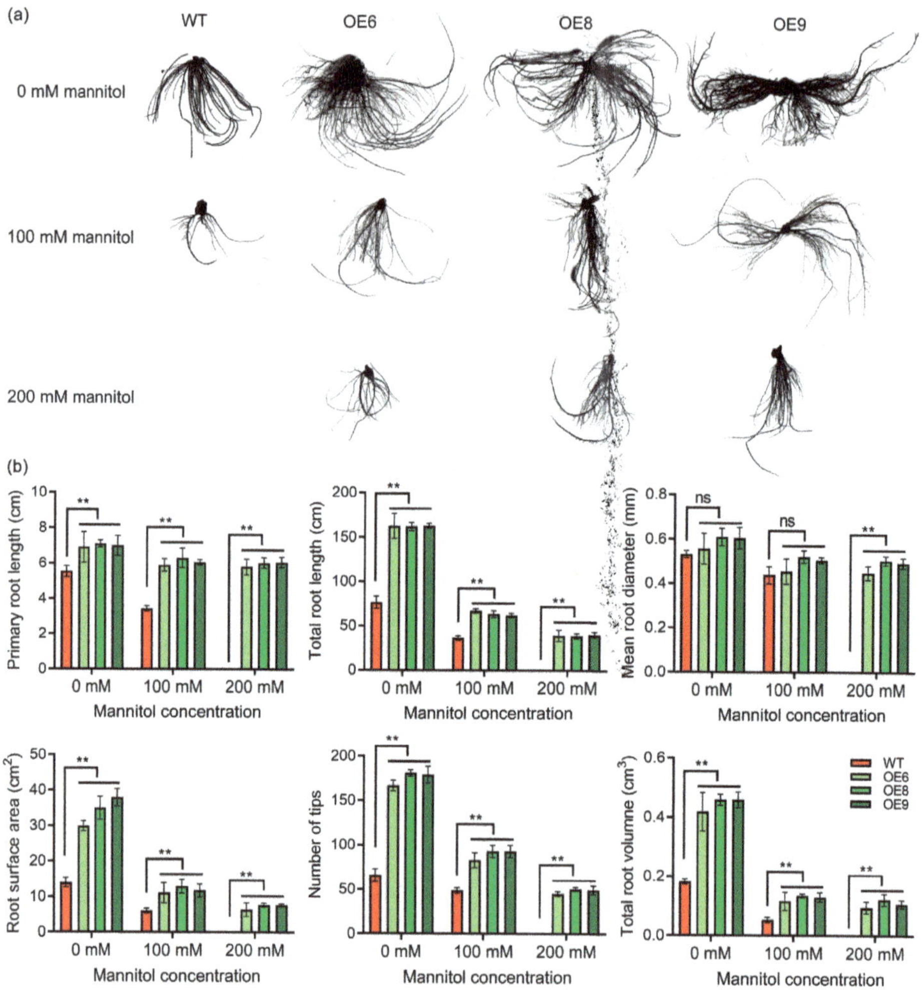

Figure 6. Identification of root phenotype of *StPYL8-like* transgenic plants under stress conditions (**a**) Root-scanning diagram of each plant under different treatment conditions; (**b**) Determination of root system-related indexes of various plants under different treatment conditions. ** represents significant differences between transgenic lines and WT at $p < 0.01$. ns indicates that the difference is not significant. The horizontal line above the column indicates that the data below the line have the same level of significance difference compared to the data at WT.

Finally, the activity of three antioxidant enzymes associated with anti-stress was further measured. As showed in the figure, under normal conditions, the POD activity of transgenic plants was significantly higher than that of the control, whereas the SOD and CAT activities did not display significant deviations from the control. Conversely, when subjected to stress conditions, the activities of all three enzymes followed a similar trend of alteration, with enzyme levels in transgenic plants markedly surpassing those in control plants. Notably, this disparity widened concomitantly with escalating stress intensity. Moreover, two key physiological indicators closely associated with drought stress, MDA and Pro, were assessed (Figure 7). While the overexpression of *StPYL8-like* led to alterations in the growth phenotype of transgenic plants under normal conditions, there was no discernible impact on the accumulation of MDA and Pro compared to the control group. Specifically, no significant differences were observed between the transgenic plants and the control group under normal conditions. However, when exposed to stress conditions of 200 mM and 300 mM mannitol, the MDA levels in each transgenic strain were markedly lower than those in the control plants, with a noticeable concentration-dependent effect. Similarly, under stress conditions, the Pro levels in transgenic plants were considerably higher than those in the control plants, and this disparity intensified with the severity of the stress.

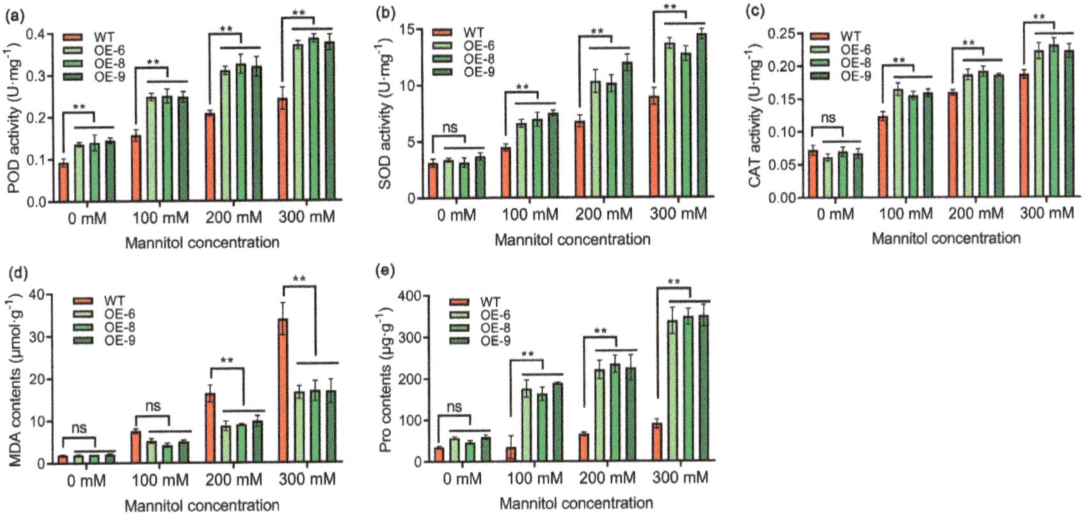

Figure 7. Determination of stress-related physiological index contents in different plants under different treatment conditions. (**a**) POD activity; (**b**) SOD activity; (**c**) CAT activity; (**d**) MDA content; (**e**) Pro content. WT represents wild tobacco. OE-6, OE-8 and OE-9 represent three transgenic lines. ** indicates significant difference at $p < 0.01$ level. ns indicates that the difference is not significant. The horizontal line above the column indicates that the data below the line have the same level of significance difference compared to the data at WT.

3.6. Stress-Related Gene Expression in Transgenic Tobacco Plants under Drought Stress Mediated by StPYL8-like

To investigate the potential mechanism underlying the enhancement of drought tolerance in transgenic plants expressing *StPYL8-like*, we analyzed the expression patterns of stress-related genes in both *StPYL8-like* transgenic lines and control plants under normal and drought conditions. Our findings revealed that upon exposure to drought stress, the transcript levels of various stress-related genes, such as *NtRD29A*, *NtLEA5*, *NtP5CS*, *NtPOD*, *NtSOD*, and *NtCAT*, were markedly elevated in *StPYL8-like* transgenic plants compared to the control plants (Figure 7). Interestingly, no significant differences were observed in gene expression levels between *StPYL8-like* transgenic plants and control plants under normal growth conditions, except for the *NtPOD* gene. Importantly, under drought-stress conditions, the expression levels of all identified genes in both wild-type and transgenic plants exhibited a substantial increase relative to normal conditions (Figure 8). These results suggest that the overexpression of *StPYL8-like* could potentially modulate the expression of these stress-related genes either directly or indirectly, thereby bolstering the drought resistance of transgenic plants.

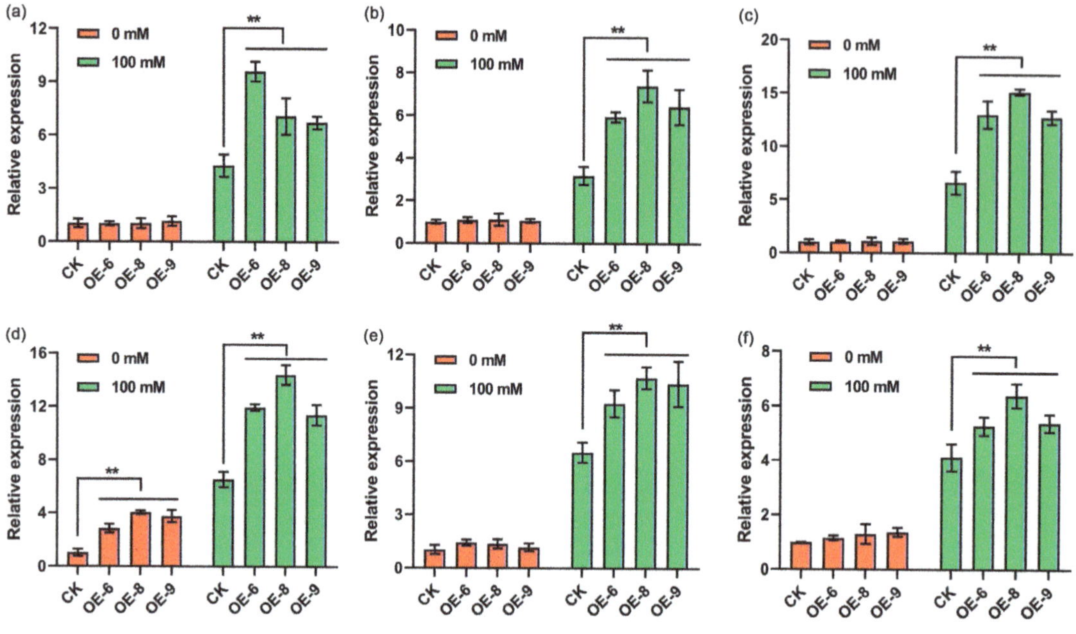

Figure 8. Expression analysis of stress-related genes in transgenic plants and wild-type plants. (**a**–**f**) represents *NtRD29A*, *NtLEA5*, and *NtP5CS*, *NtPOD*, *NtSOD*, *NtCAT* genes, respectively. ** represents significant differences between transgenic lines and WT at $p < 0.01$. CK represents wild tobacco. OE-6, OE-8 and OE-9 represent three transgenic lines. 0 mM and 100 mM represent normal MS medium and MS medium containing 100 mM mannitol, respectively. The horizontal line above the column indicates that the data below the line have the same level of significance difference compared to the data at CK.

4. Discussion

The plant hormone ABA plays a pivotal role in regulating plant growth, development, and responses to stress [27]. Drought stress typically results in heightened intracellular ABA synthesis, with the ABA receptor protein PYL playing a central role in the ABA-mediated signaling pathway that governs plant drought-resistance mechanisms [28]. This current study successfully cloned the *StPYL8-like*, a member of the *PYL* gene family, from potato. Subsequently, following a comprehensive molecular characterization, we confirmed its regulatory impact on drought resistance by assessing transient and stable expression in transgenic tobacco plants.

While ABA plays a pivotal role in this process, it was not until 2009 that researchers unearthed PYL in an investigation on a synthetic ABA agonist known as pyranose, designating PYL as an ABA receptor [29]. Afterwards, extensive research endeavors have been directed towards the screening and identification of PYL family proteins across diverse plant species. In Arabidopsis, 14 PYL proteins have been categorized into three subfamilies based on phylogenetic analysis and structural and functional characterization [30]. Discrepancies in the number of *PYL* family genes have been unveiled among different plant species through whole-genome identification. For instance, there are 15 in tomato [31], 40 in cotton [32], 14 in poplar [33], 8 in grapes [16], 38 in wheat [34], and 13 in rice [35]. Despite variations in the quantity of *PLY* family genes across species, the clustering patterns of these genes exhibit substantial consistency among plants. The classification of PYL family members into three subfamilies in various species indicates a relatively conserved function of PYL proteins throughout the evolutionary process. Recent investigations, utilizing transcriptome data from drought-stressed potato, have pinpointed the *StPYL8-like* gene as significantly responsive to drought stress. Further phylogenetic analysis has revealed that StPYL8-like protein clusters with PYLs are implicated in the abiotic stress response in organisms such as Arabidopsis, rice, and cotton (Figure 1). This finding underscores the potential significance of *StPYL8-like* in mediating the potato's response to drought stress.

Numerous studies have validated its function in enhancing plant drought tolerance. For instance, the overexpression of *AtPYR1* and *AtPYL1/2/3/8/9* notably boosted the drought tolerance of transgenic Arabidopsis [36]. Comparable outcomes have likewise been observed in other plant species like maize [15], rice [14], poplar [37], wheat [13], and cotton [38]. Wheat *PYL4* demonstrates the ability to enhance water-use efficiency and drought resistance by reducing stomatal aperture and increasing photosynthetic activity in transgenic plants [13]. The overexpression of rice *PYL3/10* [14,39] and cotton *PYL10/12/26* [38] heightens seed sensitivity to ABA during the germination and seedling stages, thereby enhancing plant adaptability to abiotic stress. Maize *PYL3/9/10/13* enhances the drought-stress tolerance of transgenic plants by elevating proline accumulation [15]. Research on *PYL* in crops such as grapes [16], tomato [17], rapeseed [18], strawberries [19], and apple [20] has also exhibited that upregulating this gene family can enhance plant resistance to abiotic stress. This investigation identifies *StPYL8-like*, a member of the *PYL* gene family, through transcriptome analysis under early drought stress, potentially participating in the potato response to drought stress. Similar to previously studied *PYL* genes, the overexpression of *StPYL8-like* mitigated damage to drought-stressed transgenic plants and improved their growth status under stress conditions. Moreover, the response to ABA results showed that the expression level of the *StPYL8-like* gene was significantly affected after 1 h of stress, and the expression trend was almost identical to that under drought, salt, and low-temperature stress. These findings suggested that the *StPYL8-like* gene may be involved in modulating abiotic stress responses through the ABA signaling pathway, not only in drought stress but also potentially in salt and low-temperature stress. There have been similar reports before, such as the *OsPYL10* gene in rice, which simultaneously enhances the survival rate of transgenic plants under drought- and low-temperature stress conditions [39]; the grape *VaPYL4* gene simultaneously enhances the tolerance of transgenic plants to three abiotic stresses: drought, salt, and low temperature [40]; and the *TaPYL1-1B* gene in wheat not only enhances the drought resistance of transgenic plants but also increases field yield [41]. These collective results highlight the multi-faceted role of ABA receptor PYL in plants, and whether the *StPYL8-like* gene also has the function of regulating multiple signal pathways needs further study and confirmation.

When plants are exposed to drought stress, the equilibrium between reactive-oxygen-species production and clearance in cells is disrupted, leading to an accumulation of reactive oxygen species, resulting in oxidative damage and eventual cell death [42]. This phenomenon has been extensively researched and confirmed in various plant species. In response to drought stress, plants typically upregulate the transcription levels of antioxidant enzyme genes such as *SOD*, *POD*, and *CAT*. Consequently, the activity of SOD/CAT/POD enzymes increases, ultimately facilitating the elimination of reactive oxygen species [43,44]. These response mechanisms are pivotal in enhancing plant resilience to drought stress. This study found that under drought-stress conditions, the expression levels of *SOD/CAT/POD* genes and the activities of SOD/CAT/POD enzymes were significantly elevated in *StPYL8-like* transgenic plants as compared to the control plants (Figures 3 and 8). Notably, under normal growth conditions, the expression of the *NtPOD* gene in *StPYL8-like*-overexpressing plants exceeded that in control plants. The reduced MDA content in transgenic plants lends further support to this observation (Figure 7), indicating a decrease in oxidative damage. The scavenging system for reactive oxygen species is crucial for plants to combat drought stress, involving a variety of enzymatic and non-enzymatic scavengers. These enzymes work collectively to mitigate the deleterious effects of reactive oxygen species [45,46]. MDA, a product of lipid oxidation, serves as a marker of oxidative damage in plants [47]. Proline, an osmolyte present in plant cytoplasm, plays a critical role in maintaining membrane and protein structure, eliminating reactive oxygen species, and minimizing photodamage to chloroplast thylakoid membranes [48]. In this study, the overexpression of *StPYL8-like* led to an elevation in free proline content in transgenic plants under drought stress, accompanied by the upregulation of the key gene *NtP5CS* involved in proline synthesis (Figure 8). Plants possess two pathways for proline synthesis: the glutamate pathway and the ornithine pathway, with the former predominantly active under osmotic stress. In the glutamate pathway, glutamate (Glu) is converted by pyrroline-5-carboxylic acid synthase (P5CS) into glutamyl semialdehyde (GSA), which cyclizes spontaneously to form pyrroline-5-carboxylic acid (P5C). P5C is then converted to proline through the action of pyrroline-5-carboxylic acid reductase (P5CR) [49,50]. The *P5CS* gene, serving as the primary rate-limiting enzyme in this synthesis pathway, is crucial for proline biosynthesis [51,52]. Therefore, the findings of this study suggest that *StPYL8-like* may enhance the drought tolerance of transgenic plants by influencing proline synthesis, necessitating further experimental investigations to elucidate its molecular mechanisms. Furthermore, prior investigations have indicated that *PYL* genes may be involved in plant growth and development or contribute to plant drought resistance through the modulation of growth and development processes. For instance, the overexpression of *GhPYL10/12/26* genes in cotton has been shown to improve drought resistance in transgenic plants, as well as promote enhanced root-system development under normal growth conditions [38]. Arabidopsis *PYL9* not only enhances drought resistance by reducing water evaporation but also by promoting leaf senescence in old leaves and suppressing growth in young tissues under severe drought stress [10]. Intriguingly, parallel findings were observed in our study. Specifically, under standard growth conditions, transgenic plants with *StPYL8-like* characteristics exhibited superior growth compared to control plants, notably reflected in their more robust root systems. Thus, it is hypothesized that *StPYL8-like* genes could not only boost plant drought resistance through facilitating proline synthesis but could also be involved in modulating plant growth and development. The upregulation of the *StPYL8-like* gene results in transgenic plants with well-developed root systems, enabling them to absorb more water during drought stress, thereby improving drought resistance. Hence, our hypothesis is that the *StPYL8-like* gene could modulate transgenic plant-root development via ABA signaling, concurrently facilitating proline biosynthesis, thereby augmenting the drought-stress tolerance of transgenic plants. In light of this hypothesis, we propose to extend our investigation by screening the interacting proteins of StPYL8-like under drought-stress conditions and evaluating the transcriptional impact of gene overexpression on transgenic plants. Through a holistic approach encompassing physiological, biochemical, transcriptional, and proteomic analyses, we could comprehensively elucidate the molecular mechanisms underlying the involvement of *StPYL8-like* genes in plant drought resistance. Furthermore, our upcoming research endeavors will involve an investigation into the presence of single-nucleotide polymorphisms within the *StPYL8-like* gene across various potato genotypes. Subsequently, we aim to elucidate the potential association between such polymorphisms and drought-resistance levels in different potato genotypes. This comprehensive analysis is poised to enhance our understanding of the biological role of the *StPYL8-like* gene while serving as a valuable resource for the establishment of molecular markers linked to drought resistance in potatoes. Ultimately, these findings are anticipated to significantly advance the molecular breeding efforts targeted at developing drought-tolerant potato cultivars.

5. Conclusions

In conclusion, in this research, we successfully cloned a member of the potato *PYL* gene family: *StPYL8-like*. Stress-response assays revealed that *StPYL8-like* exhibited significant responses to a range of stress stimuli, such as drought and ABA. Subsequent functional characterization demonstrated that the overexpression of *StPYL8-like* led to elevated levels of proline accumulation and the enhanced activity of antioxidant enzymes in transgenic plants when exposed to drought stress, ultimately resulting in increased drought-stress tolerance.

Supplementary Materials: The following supporting information can be downloaded at: https://www.mdpi.com/article/10.3390/antiox13091088/s1, Figure S1: Analysis of conserved domains of PYL proteins in Arabidopsis (A) and potatoes (B); Figure S2: Expression level analysis of *StPYL* genes under drought stress; Table S1: The sequences of primers were used in the study; Table S2: Molecular characteristics of potato PYL family genes.

Author Contributions: Conceptualization, P.Y. and Z.B.; methodology, C.Z., J.C. and C.S.; software, P.Y., Y.L. and Z.L.; validation, C.Z., J.W. and X.S.; formal analysis, C.Z., J.W., X.S. and Y.L.; investigation, P.Y. and Z.B.; data curation, J.B. and J.C.; writing—original draft preparation, P.Y. and Z.B.; writing—review and editing, P.Y., J.B. and Z.B. All authors have read and agreed to the published version of the manuscript.

Funding: The Gansu Provincial Science and Technology Department (Grant No. 23JRRA1414); the Gansu Province Education Science and Technology Innovation Project (Grant No. 2024A-052); the Gansu Provincial Natural Science Foundation (Grant No. 22JR5RA833); the University Industry Support Program (Grant No. 2024CYZC-29); the China Agricultural University Corresponding Support Research Joint Fund (Grant No. GSAU-DKZY-2024-005); the National Natural Science Foundation of China (32060502), and the Research Program Sponsored by State Key Laboratory of Aridland Crop Science of China (Grant No. GSCS-2022-Z01).

Institutional Review Board Statement: Not applicable.

Informed Consent Statement: Not applicable.

Data Availability Statement: All data are available within the manuscript.

Conflicts of Interest: The authors declare no conflicts of interest.

References

1. Arbona, V.; Manzi, M.; Zandalinas, S.I.; Vives-Peris, V.; Pérez-Clemente, R.M.; Gómez-Cadenas, A. Physiological, metabolic, and molecular responses of plants to abiotic stress. In *Stress Signaling in Plants: Genomics and Proteomics Perspective*; Springer: Berlin, Germany, 2017; Volume 2, pp. 20–35.
2. Ben Rejeb, I.; Pastor, V.; Mauch-Mani, B. Plant responses to simultaneous biotic and abiotic stress: Molecular mechanisms. *Plants* **2014**, *3*, 458–475. [CrossRef] [PubMed]
3. Zhang, J.; Jia, W.; Yang, J.; Ismail, A.M. Role of ABA in integrating plant responses to drought and salt stresses. *Field Crop. Res.* **2006**, *97*, 111–119. [CrossRef]
4. Sakata, Y.; Komatsu, K.; Takezawa, D. ABA as a universal plant hormone. In *Progress in Botany*; Springer: Berlin, Germany, 2013; Volume 75, pp. 57–96.
5. Miyazono, K.I.; Miyakawa, T.; Sawano, Y.; Kubota, K.; Kang, H.J.; Asano, A.; Miyauchi, Y.; Takahashi, M.; Zhi, Y.; Fujita, Y. Structural basis of abscisic acid signalling. *Nature* **2009**, *462*, 609–614. [CrossRef]
6. Melcher, K.; Ng, L.M.; Zhou, X.E.; Soon, F.F.; Xu, Y.; Suino-Powell, K.M.; Park, S.Y.; Weiner, J.J.; Fujii, H.; Chinnusamy, V. A gate–latch–lock mechanism for hormone signalling by abscisic acid receptors. *Nature* **2009**, *462*, 602–608. [CrossRef] [PubMed]
7. De Zélicourt, A.; Colcombet, J.; Hirt, H. The role of MAPK modules and ABA during abiotic stress signaling. *Trends Plant Sci.* **2016**, *21*, 677–685. [CrossRef]
8. Saddhe, A.A.; Kundan, K.; Padmanabh, D. Mechanism of ABA signaling in response to abiotic stress in plants. In *Mechanism of Plant Hormone Signaling under Stress*; Wiley: Hoboken, NJ, USA, 2017; Volume 1, pp. 173–195.
9. Zhang, H.; Zhu, J.; Gong, Z.; Zhu, J.K. Abiotic stress responses in plants. *Nat. Rev. Genet.* **2022**, *23*, 104–119. [CrossRef]
10. Zhao, Y.; Chan, Z.; Gao, J.; Xing, L.; Cao, M.; Yu, C.; Hu, Y.; You, J.; Shi, H.; Zhu, Y. ABA receptor PYL9 promotes drought resistance and leaf senescence. *Proc. Natl. Acad. Sci. USA* **2016**, *113*, 1949–1954. [CrossRef] [PubMed]
11. Gao, Z.; Li, Q.; Li, J.; Chen, Y.; Luo, M.; Li, H.; Wang, J.; Wu, Y.; Duan, S.; Wang, L. Characterization of the ABA receptor VlPYL1 that regulates anthocyanin accumulation in grape berry skin. *Front. Plant Sci.* **2018**, *9*, 592. [CrossRef]
12. Quan, W.; Hu, Y.; Mu, Z.; Shi, H.; Chan, Z. Overexpression of *AtPYL5* under the control of guard cell specific promoter improves drought stress tolerance in Arabidopsis. *Plant Physiol. Biochem.* **2018**, *129*, 150–157. [CrossRef]
13. Mega, R.; Abe, F.; Kim, J.S.; Tsuboi, Y.; Tanaka, K.; Kobayashi, H.; Sakata, Y.; Hanada, K.; Tsujimoto, H.; Kikuchi, J. Tuning water-use efficiency and drought tolerance in wheat using abscisic acid receptors. *Nat. Plants* **2019**, *5*, 153–159. [CrossRef] [PubMed]
14. Lenka, S.K.; Muthusamy, S.K.; Chinnusamy, V.; Bansal, K.C. Ectopic expression of rice *PYL3* enhances cold and drought tolerance in *Arabidopsis thaliana*. *Mol. Biotechnol.* **2018**, *60*, 350–361. [CrossRef] [PubMed]
15. He, Z.; Zhong, J.; Sun, X.; Wang, B.; Terzaghi, W.; Dai, M. The maize ABA receptors ZmPYL8, 9, and 12 facilitate plant drought resistance. *Front. Plant Sci.* **2018**, *9*, 422. [CrossRef] [PubMed]

16. Zhao, Y.; Qi, G.; Ren, F.; Wang, Y.; Wang, P.; Wu, X. Analysis of *PYL* genes and their potential relevance to stress tolerance and berry ripening in grape. *J. Am. Soc. Hortic. Sci.* **2020**, *145*, 308–317. [CrossRef]
17. Li, Y.; Zhang, X.; Jiang, J.; Zhao, T.; Xu, X.; Yang, H.; Li, J. Virus-induced gene silencing of *SlPYL4* decreases the drought tolerance of tomato. *Hortic. Plant J.* **2022**, *8*, 361–368. [CrossRef]
18. Li, Y.; Wang, D.; Sun, C.; Hu, X.; Mu, X.; Hu, J.; Yang, Y.; Zhang, Y.; Xie, C.G.; Zhou, X. Molecular characterization of an AtPYL1-like protein, BrPYL1, as a putative ABA receptor in *Brassica rapa*. *Biochem. Biophys. Res. Commun.* **2017**, *487*, 684–689. [CrossRef]
19. Hou, B.Z.; Chen, X.H.; Shen, Y.Y. Interactions between strawberry ABA receptor PYR/PYLs and protein phosphatase PP2Cs on basis of transcriptome and yeast two-hybrid analyses. *J. Plant Growth Regul.* **2021**, *40*, 594–602. [CrossRef]
20. Yang, J.; Wang, M.; Zhou, S.; Xu, B.; Chen, P.; Ma, F.; Mao, K. The ABA receptor gene *MdPYL9* confers tolerance to drought stress in transgenic apple (*Malus domestica*). *Environ. Exp. Bot.* **2022**, *194*, 104695. [CrossRef]
21. Qin, R.; Zhang, F.; Yu, C.; Zhang, Q.; Qi, J.; Li, F.M. Contributions made by rain-fed potato with mulching to food security in China. *Eur. J. Agron.* **2022**, *133*, 126435. [CrossRef]
22. Bai, J.; Mao, J.; Yang, H.; Khan, A.; Fan, A.; Liu, S.; Zhang, J.; Wang, D.; Gao, H.; Zhang, J. Sucrose non-ferment 1 related protein kinase 2 (*SnRK 2*) genes could mediate the stress responses in potato (*Solanum tuberosum* L.). *BMC Genet.* **2017**, *18*, 41. [CrossRef]
23. Yao, P.; Sun, L.; Dekomah, S.; Bi, Z.; Sun, C.; Mao, J.; Zhang, C.; Qin, T.; Wang, Y.; Liu, Y. Evolutionary analysis of *StSnRK2* family genes and their overexpression in transgenic tobacco improve drought tolerance. *Int. J. Mol. Sci.* **2023**, *24*, 1000. [CrossRef]
24. Wang, Y.F.; Liao, Y.Q.; Wang, Y.P.; Yang, J.W.; Zhang, N.; Si, H.J. Genome-wide identification and expression analysis of *StPP2C* gene family in response to multiple stresses in potato (*Solanum tuberosum* L.). *J. Integr. Agric.* **2020**, *19*, 1609–1624. [CrossRef]
25. Muñiz García, M.N.; Muro, M.C.; Mazzocchi, L.C.; País, S.M.; Stritzler, M.; Schlesinger, M.; Capiati, D.A. The protein phosphatase 2A catalytic subunit *StPP2Ac2b* acts as a positive regulator of tuberization induction in *Solanum tuberosum* L. *Plant Mol. Biol.* **2017**, *93*, 227–245. [CrossRef] [PubMed]
26. Gul, S.; Gul, H.; Shahzad, M.; Ullah, I.; Shahzad, A.; Khan, S.U. Comprehensive analysis of potato (*Solanum tuberosum*) *PYL* genes highlights their role in stress responses. *Funct. Plant Biol.* **2024**, *51*, FP24094. [CrossRef] [PubMed]
27. Bulgakov, V.P.; Wu, H.C.; Jinn, T.L. Coordination of ABA and chaperone signaling in plant stress responses. *Trends Plant Sci.* **2019**, *24*, 636–651. [CrossRef] [PubMed]
28. Saradadevi, R.; Palta, J.A.; Siddique, K.H. ABA-mediated stomatal response in regulating water use during the development of terminal drought in wheat. *Front. Plant Sci.* **2017**, *8*, 1251. [CrossRef] [PubMed]
29. Park, S.Y.; Fung, P.; Nishimura, N.; Jensen, D.R.; Fujii, H.; Zhao, Y.; Lumba, S.; Santiago, J.; Rodrigues, A.; Chow, T.F.F. Abscisic acid inhibits type 2C protein phosphatases via the PYR/PYL family of START proteins. *Science* **2009**, *324*, 1068–1071. [CrossRef] [PubMed]
30. Dittrich, M.; Mueller, H.M.; Bauer, H.; Peirats-Llobet, M.; Rodriguez, P.L.; Geilfus, C.M.; Carpentier, S.C.; Al Rasheid, K.A.; Kollist, H.; Merilo, E. The role of Arabidopsis ABA receptors from the PYR/PYL/RCAR family in stomatal acclimation and closure signal integration. *Nat. Plants* **2019**, *5*, 1002–1011. [CrossRef]
31. González-Guzmán, M.; Rodríguez, L.; Lorenzo-Orts, L.; Pons, C.; Sarrión-Perdigones, A.; Fernández, M.A.; Peirats-Llobet, M.; Forment, J.; Moreno-Alvero, M.; Cutler, S.R. Tomato PYR/PYL/RCAR abscisic acid receptors show high expression in root, differential sensitivity to the abscisic acid agonist quinabactin, and the capability to enhance plant drought resistance. *J. Exp. Bot.* **2014**, *65*, 4451–4464. [CrossRef]
32. Zhang, G.; Lu, T.; Miao, W.; Sun, L.; Tian, M.; Wang, J.; Hao, F. Genome-wide identification of ABA receptor PYL family and expression analysis of PYLs in response to ABA and osmotic stress in Gossypium. *PeerJ* **2017**, *5*, e4126. [CrossRef]
33. Bai, G.; Xie, H.; Yao, H.; Li, F.; Chen, X.; Zhang, Y.; Xiao, B.; Yang, J.; Li, Y.; Yang, D.H. Genome-wide identification and characterization of ABA receptor PYL/RCAR gene family reveals evolution and roles in drought stress in *Nicotiana tabacum*. *BMC Genom.* **2019**, *20*, 575. [CrossRef]
34. Lei, P.; Wei, X.; Gao, R.; Huo, F.; Nie, X.; Tong, W.; Song, W. Genome-wide identification of *PYL* gene family in wheat: Evolution, expression and 3D structure analysis. *Genomics* **2021**, *113*, 854–866. [CrossRef] [PubMed]
35. Yadav, S.K.; Santosh Kumar, V.V.; Verma, R.K.; Yadav, P.; Saroha, A.; Wankhede, D.P.; Chaudhary, B.; Chinnusamy, V. Genome-wide identification and characterization of ABA receptor *PYL* gene family in rice. *BMC Genomics* **2020**, *21*, 676. [CrossRef] [PubMed]
36. Li, X.; Li, G.; Li, Y.; Kong, X.; Zhang, L.; Wang, J.; Li, X.; Yang, Y. ABA receptor subfamily III enhances abscisic acid sensitivity and improves the drought tolerance of Arabidopsis. *Int. J. Mol. Sci.* **2018**, *19*, 1938. [CrossRef] [PubMed]
37. Yu, J.; Yang, L.; Liu, X.; Tang, R.; Wang, Y.; Ge, H.; Wu, M.; Zhang, J.; Zhao, F.; Luan, S. Overexpression of poplar pyrabactin resistance-like abscisic acid receptors promotes abscisic acid sensitivity and drought resistance in transgenic Arabidopsis. *PLoS ONE* **2016**, *11*, e0168040. [CrossRef]
38. Chen, Y.; Feng, L.; Wei, N.; Liu, Z.H.; Hu, S.; Li, X.B. Overexpression of cotton *PYL* genes in Arabidopsis enhances the transgenic plant tolerance to drought stress. *Plant Physiol. Biochem.* **2017**, *115*, 229–238. [CrossRef]
39. Verma, R.K.; Santosh Kumar, V.V.; Yadav, S.K.; Pushkar, S.; Rao, M.V.; Chinnusamy, V. Overexpression of ABA receptor *PYL10* gene confers drought and cold tolerance to indica rice. *Front Plant Sci.* **2019**, *10*, 1488. [CrossRef]
40. Ren, C.; Kuang, Y.; Lin, Y.; Guo, Y.; Li, H.; Fan, P.; Li, S.; Liang, Z. Overexpression of grape ABA receptor gene *VaPYL4* enhances tolerance to multiple abiotic stresses in Arabidopsis. *BMC Plant Biol.* **2022**, *22*, 271. [CrossRef]

41. Mao, H.; Jian, C.; Cheng, X.; Chen, B.; Mei, F.; Li, F.; Zhang, Y.; Li, S.; Du, L.; Li, T. The wheat ABA receptor gene *TaPYL1-1B* contributes to drought tolerance and grain yield by increasing water-use efficiency. *Plant Biotechnol. J.* **2022**, *20*, 846–861. [CrossRef]
42. Sena, L.A.; Chandel, N.S. Physiological roles of mitochondrial reactive oxygen species. *Mol. Cell* **2012**, *48*, 158–167. [CrossRef]
43. Sheng, L.; Sun, X.; Mo, C.; Hao, M.; Wei, X.; Ma, A. Relationship between antioxidant enzymes and sclerotial formation of Pleurotus tuber-regium under abiotic stress. *Appl. Microbiol. Biotechnol.* **2023**, *107*, 1391–1404. [CrossRef]
44. Hou, L.; Zhao, M.; Huang, C.; He, Q.; Zhang, L.; Zhang, J. Alternative oxidase gene induced by nitric oxide is involved in the regulation of ROS and enhances the resistance of *Pleurotus ostreatus* to heat stress. *Microb. Cell Factories* **2021**, *20*, 137. [CrossRef] [PubMed]
45. Sachdev, S.; Ansari, S.A.; Ansari, M.I.; Fujita, M.; Hasanuzzaman, M. Abiotic stress and reactive oxygen species: Generation, signaling, and defense mechanisms. *Antioxidants* **2021**, *10*, 277. [CrossRef] [PubMed]
46. Gill, S.S.; Tuteja, N. Reactive oxygen species and antioxidant machinery in abiotic stress tolerance in crop plants. *Plant Physiol. Biochem.* **2010**, *48*, 909–930. [CrossRef]
47. Zaid, A.; Wani, S.H. Reactive oxygen species generation, scavenging and signaling in plant defense responses. In *Bioactive Molecules in Plant Defense*; Springer: Berlin, Germany, 2019; Volume 23, pp. 111–132.
48. Jaleel, C.A.; Gopi, R.; Sankar, B.; Manivannan, P.; Kishorekumar, A.; Sridharan, R.; Panneerselvam, R. Studies on germination, seedling vigour, lipid peroxidation and proline metabolism in *Catharanthus roseus* seedlings under salt stress. *S. Afr. J. Bot.* **2007**, *73*, 190–195. [CrossRef]
49. Ding, Z.; Ericksen, R.E.; Escande-Beillard, N.; Lee, Q.Y.; Loh, A.; Denil, S.; Steckel, M.; Haegebarth, A.; Ho, T.S.W.; Chow, P. Metabolic pathway analyses identify proline biosynthesis pathway as a promoter of liver tumorigenesis. *J. Hepatol.* **2020**, *72*, 725–735. [CrossRef] [PubMed]
50. Phang, J.M.; Liu, W.; Zabirnyk, O. Proline metabolism and microenvironmental stress. *Annu. Rev. Nutr.* **2010**, *30*, 441–463. [CrossRef]
51. Amini, S.; Ghobadi, C.; Yamchi, A. Proline accumulation and osmotic stress: An overview of *P5CS* gene in plants. *J. Plant Mol. Breed.* **2015**, *3*, 44–55.
52. Rai, A.N.; Penna, S. Molecular evolution of plant P5CS gene involved in proline biosynthesis. *Mol. Biol. Rep.* **2013**, *40*, 6429–6435. [CrossRef]

antioxidants

MDPI

Article

Physiological and Transcriptome Analysis of the Effects of Exogenous Strigolactones on Drought Responses of Pepper Seedlings

Huangying Shu [1,2,3], Muhammad Ahsan Altaf [1,2], Naveed Mushtaq [1,2], Huizhen Fu [1,2,3], Xu Lu [1,2,3], Guopeng Zhu [1,2,3], Shanhan Cheng [1,2,3] and Zhiwei Wang [1,2,3,*]

1 Key Laboratory for Quality Regulation of Tropical Horticultural Crops of Hainan Province, School of Breeding and Multiplication (Sanya Institute of Breeding and Multiplication), Center of Nanfan and High-Efficiency Tropical Agriculture, Hainan University, Sanya 572025, China; shuhuangying@hainanu.edu.cn (H.S.); ahsanaltaf@hainanu.edu.cn (M.A.A.); 2015204041@njau.edu.cn (N.M.); fhzhenz@hainau.edu.cn (H.F.); luxu@hainanu.edu.cn (X.L.); zhuguopeng@hainanu.edu.cn (G.Z.); 990865@hainanu.edu.cn (S.C.)
2 Key Laboratory for Quality Regulation of Tropical Horticultural Crops of Hainan Province, School of Tropical Agriculture and Forestry, Hainan University, Haikou 570228, China
3 Hainan Yazhou Bay Seed Laboratory, Sanya 572025, China
* Correspondence: wangzhiwei@hainanu.edu.cn

Abstract: Drought stress significantly restricts the growth, yield, and quality of peppers. Strigolactone (SL), a relatively new plant hormone, has shown promise in alleviating drought-related symptoms in pepper plants. However, there is limited knowledge on how SL affects the gene expression in peppers when exposed to drought stress (DS) after the foliar application of SL. To explore this, we conducted a thorough physiological and transcriptome analysis investigation to uncover the mechanisms through which SL mitigates the effects of DS on pepper seedlings. DS inhibited the growth of pepper seedlings, altered antioxidant enzyme activity, reduced relative water content (RWC), and caused oxidative damage. On the contrary, the application of SL significantly enhanced RWC, promoted root morphology, and increased leaf pigment content. SL also protected pepper seedlings from drought-induced oxidative damage by reducing MDA and H_2O_2 levels and maintaining POD, CAT, and SOD activity. Moreover, transcriptomic analysis revealed that differentially expressed genes were enriched in ribosomes, ABC transporters, phenylpropanoid biosynthesis, and Auxin/MAPK signaling pathways in DS and DS + SL treatment. Furthermore, the results of qRT-PCR showed the up-regulation of *AGR7*, *ABI5*, *BRI1*, and *PDR4* and down-regulation of *SAPK6*, *NTF4*, *PYL6*, and *GPX4* in SL treatment compared with drought-only treatment. In particular, the key gene for SL signal transduction, *SMXL6*, was down-regulated under drought. These results elucidate the molecular aspects underlying SL-mediated plant DS tolerance, and provide pivotal strategies for effectively achieving pepper drought resilience.

Keywords: pepper; drought stress; strigolactone; transcriptome; molecular

1. Introduction

Plants are subjected to various stresses from their surrounding environment throughout their entire life. Drought stress (DS) is becoming more prominent as an abiotic stress factor that affects the yield of horticultural crops. In recent decades, drought intensity, frequency, and severity have posed a severe threat to worldwide agricultural production [1]. However, drought tolerance in plants is a multifaceted characteristic that is significantly influenced by environmental factors. DS significantly altered the morphological, physiological, and metabolic functions in plants [2]. Plant tolerance mechanisms for DS may give genetic resources to generate drought-resistant crops.

Pepper (*Capsicum* spp.) is one of the world's major vegetable crops, and is widely cultivated due to its exceptional adaptability [3]. Additionally, peppers are rich source of vitamins, pigments, carbohydrates, and antioxidants [4]. The continuing deterioration of the climate, the frequent occurrence of high temperatures, and the shortage of water during the pepper growth and development period have a major impact on pepper quality and output. In light of this, it is crucial to understand the physiological and molecular responses of chili peppers to DS and enhance their capacity for resilience. Strigolactone (SL) is a novel class of plant hormone, characterized by a diverse array of chemical structures. It is a sesquiterpenoid lactone that is generated from a tetracyclic carotenoid. Its primary function is to inhibit plant branching [5,6]. The use of phytohormones in the agricultural sector suggests its potential for enhancing drought tolerance, increasing crop production, and regulating plant growth. SL is a promising tool for improving both crop productivity and resilience in plants [7]. Exogenous SL can alleviate the adverse effects of DS by regulating stomatal closure, chlorophyll synthesis, and photosynthesis, as well as activating the antioxidant defense protection mechanism in grapevines [8]. Importantly, exogenous SL efficiently improved the sensitivity of stomata to abscisic acid (ABA) in a DS environment [9]. In *Arabidopsis*, SL acts as a positive regulator in the regulation of the plant's response to DS. Comparative transcriptome investigations have demonstrated that plants effectively integrate several hormone-signaling pathways, such as SL, ABA, and cytokinin, to cope with environmental stresses [10]. *SsMAX2* has been reported to reduce chlorophyll degradation and water loss, thereby decreasing H_2O_2 levels in *Arabidopsis* [11]. Although SL's regulation of plant stress responses is well documented, its specific impact on enhancing DS resistance in pepper has received comparatively limited attention. Consequently, there is an urgent need to comprehensively explore SLs and their uses, and investigate their potential applications in pepper seedlings under DS conditions.

SL has the potential to serve as an effective approach for mitigating the adverse effects of drought stress on plants. The molecular pathways involved in regulating drought tolerance via exogenous SL, especially when pepper seedlings are exposed to drought conditions, have not yet been determined. The investigation of the role of SL in the mitigation of DS in pepper is an intriguing avenue for research. In order to address these concerns, we conducted an analysis of the phenotypic, physiological, biochemical, and molecular pathways of pepper seedlings under DS. Furthermore, we explored the potential mechanisms of SL in alleviating DS in pepper seedlings and identify genes associated with DS responses.

2. Materials and Methods

2.1. Plant Materials and Treatments

Capsicum chinense (HNUCC16) was cultivated in growth chambers under control conditions (16/8 h day/night photoperiod, 26/22 °C day/night temperature, and relative humidity of 75%). At the 6–8 leaf stage, identical seedlings were transferred into plastic containers containing Hoagland's nutrient solution (HNS). HNS was replaced every three days. After a 7-day adaptation period, the plants were separated into different treatment groups, as follows: (1) control; (2) drought stress (DS), in which plants were placed in 10% (*w*/*v*) polyethylene glycol (PEG-6000) to simulate drought conditions; and (3) 3 μM GR24 pretreatment, in which plants were placed in 10% (*w*/*v*) PEG-6000 (SL + DS). Each treatment group included 15 plants. GR24 (Coolaber, Beijing, China), an SL analog, was dissolved with acetone and 0.1% Tween 20, and the storage concentration was 10 mM. Control and DS were then treated with the equivalent amount of distilled water with the same amount of acetone and Tween 20. SL was applied to the leaves on peppers for three consecutive days (twice a day). After SL application, seedlings were exposed to drought stress treatment for three days. After stress treatment, leaf samples were collected for morphology, physiological, and biochemical assays and RNA sequencing analysis. All samples were immediately frozen in liquid nitrogen and stored at −80 °C. Each treatment was repeated three times. Pepper seeds were supplied by the School of Tropical Agriculture and Forestry, Hainan University (Hainan, China). Three biological replicates were used in each treatment.

2.2. Measurement of Relative Water Content and Root Morphology

The relative water content (RWC) was measured using a modified protocol described previously [8]. The pepper leaves were rinsed with distilled water, and leaves were wiped with absorbent paper and the fresh weight (FW) recorded. The samples were soaked in distilled water for 24 h at 4 °C in darkness, and the turgid weight (TW) was recorded. The dry weight (DW) was then determined after drying at 85 °C for 2 min and 24 h at 60 °C. RWC formula:

$$RWC\ (\%) = (FW - DW)/(TW - DW) \times 100\%$$

Roots' phenotype was recorded after seven days of DS. To examine the root morphology, four plants of similar size were chosen from each duplicate, and their roots were taken. Subsequently, the roots were completely cleaned with running tap water. An Image Scanning Screen (Epson Expression 110000XL, Regent Instruments, Québec, QC, Canada) was used for root scanning, and WinRHIZO 2003a software was used for root data analysis.

2.3. Scanning Electron Microscopy

The leaf tissue blocks were left in an electron microscopy fixative for 2 h at room temperature, then transferred to 4 °C for preservation and transport. Then, leaf tissue blocks were treated with 0.1 M PB (pH 7.4). Then, tissue blocks were transferred to 1% OsO_4 in 0.1 M PB (pH 7.4). After that, leaf tissue blocks were infused with alcohol and isoamyl acetate. Dry samples were dried with a dryer, and specimens were attached to metallic stubs using carbon stickers and sputter coated with gold. Finally, we observed and captured images with a scanning electron microscope (SU8100, Hitachi, Tokyo, Japan).

2.4. Chlorophyll Content

For the determination of pigment molecules, 0.1 g fresh leaves ground, and placed in a 10 mL of acetone extracting solution. The sample was centrifuged at 10,000× *g* for 10 min. The chlorophyll was spectrophotometrically determined in supernatant at 663 and 645 nm, respectively. Determination of chlorophyll content was realized using the formulae reported by [12].

2.5. Measurement of Antioxidant Enzyme Activity and Related Metabolites

The activities of superoxide dismutase (SOD), peroxidase (POD), and catalase (CAT) were analyzed based on the modified protocols reported by [12]. Fresh samples (0.1 g) were ground in 900 μL phosphate buffer (pH 7.8), and the homogenate was centrifuged at 10,000× *g* for 10 min. The supernatant was used to determine the activity of antioxidant enzymes. The MDA level, H_2O_2 contents, the Pro concentration, and the soluble sugar content of the leaves were determined using an assay kit (Jiancheng Bioengineering Institute, Nanjing, China), following the detailed instructions [12]. The colorimetric measurements of MDA, H_2O_2, Pro, and soluble sugar were measured at 530, 405, 520, and 620 nm, respectively.

2.6. Transcriptome Sequencing, Differential Gene Expression, and Enrichment Analysis

We selected explants with normal development for sampling in different treatments, including control, DS, and SL+DS. Total RNA was isolated with Trizol Reagents (Thermo Fisher Scientific, Shanghai, China). Nine nondirectional libraries were produced using the NEBNext® UltraTM RNA Library Prep Kit for Illumina® (NEB, Boston, Massachusetts, USA) and sequenced on the Illumina Novaseq platform. Clean reads were obtained by filtering low-quality reads and aligning them with the *C. chinense* reference genome using HISAT2 (v.2.0.5) [13]. The methods of Benjamini and Hochberg were used to adjust the obtained *p*-value to control the error detection rate. Genes with an adjusted p-value < 0.05 discovered through DESeq2 were assigned as differentially expressed (DEG) [14]. Each subset of the clustering line graph, where the value of differential genes is the union of all comparative combinations of differential genes, and the expression level of differential genes in the FPKM expression matrix of each sample were taken as log2 (fpkm + 1) and centralized for correction. The Gene Ontology (GO) of differentially expressed genes and the statistical enrichment of DEGs in the KEGG pathway were analyzed using the clusterProfiler R package [15,16].

2.7. Gene Expression by qRT-PCR

Total RNA was reversed transcriptionally using Hiscript Q RT SuperMix for qPCR (Vazyme Biotech, Nanjing, China), and qPCRs were analyzed using ChamQ SYBR qPCR Master Mix (Vazyme Biotech, Nanjing, China). The amplification program was based on the QuantStudio™ 5 Real-Time PCR System (Thermo Fisher Scientific, Waltham, Massachusetts, USA) [17]. The specific primers of the selected genes are listed in Table S1.

2.8. Statistical Analysis

Statistical analysis was performed using Microsoft Excel (v16.72) and SPSS (v26). The significance of the difference between the mean values was tested using Duncan's multiple range test ($p < 0.05$). Figures were generated using GraphPad Prism software (v9.5.0), representing mean values with a standard deviation of three replicates per treatment.

3. Results

3.1. SL Alleviated the Negative Effects of Drought Stress on the Morphology of Pepper

To confirm the possible effects of SL on pepper growth in a DS environment, we examined the pepper morphology in the control, DS, and DS + SL groups. DS significantly inhibited the root growth of pepper seedlings. However, pretreatment with SL can effectively alleviate the inhibitory effect of DS. Under DS conditions, the root length, surface area, volume, tips, crossing, and forks of the plant decreased by 49.56%, 59.45%, 65.27%, 32.31%, 55.7%, and 56.62%, respectively, compared with the control treatments. Furthermore, after SL pretreatment, the reduction was only 20.13%, 35.48%, 35.71%, 15.4%, 29.11%, and 22.63%, respectively, compared to the control group. Importantly, foliar application of SL significantly promoted the root architecture system of pepper seedlings in a DS environment (Figure 1). These findings suggested that SL could maintain normal root growth and balanced water uptake.

Figure 1. *Cont.*

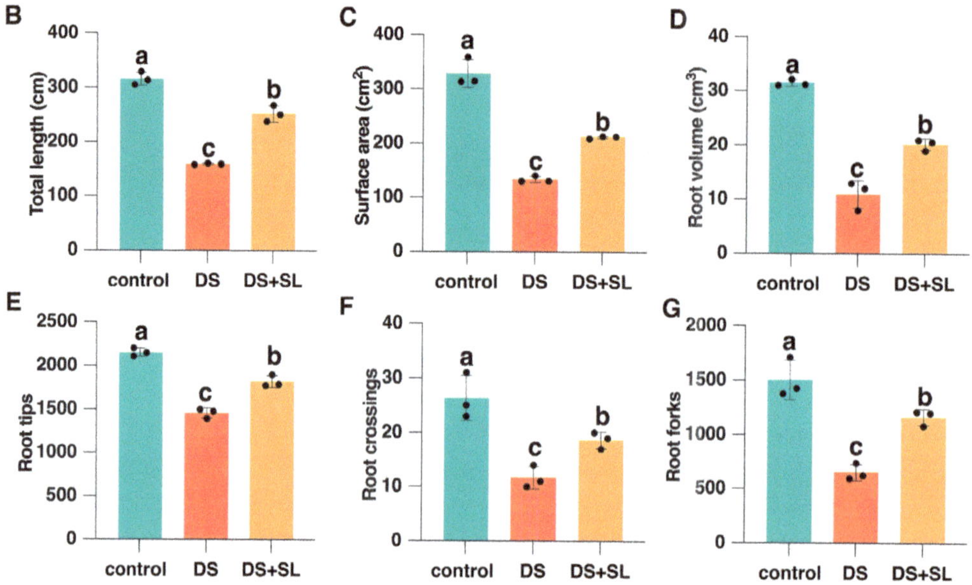

Figure 1. Phenotypic changes in pepper seedlings. (**A**) The phenotypic of pepper plants and roots under different treatments. (**B**) Total root length, (**C**) root surface area, (**D**) root volume, (**E**) root tips, (**F**) root crossing, and (**G**) root forks. Control: under normal control conditions, DS: drought conditions, and DS + SL: drought conditions with SL application. The data are mean values and standard errors; the black dot indicates three biological replicates. Different letters indicate significant differences between different treatments ($p < 0.05$).

3.2. The Microscopic Structure of Leaf Tissue

We evaluated the stomatal characteristics of plants with various treatments to detect the effect of SL on their stomatal opening and closing ability under DS (Figure 2). SEM showed that compared to the control group, the plant stomata were completely closed in drought-only treatment (Figure 2B). In SL application along with DS conditions, stomatal closure was alleviated, demonstrating that exogenous SL could alter the stomata opening in response to DS (Figure 2C). Additionally, compared to the control group, the stomatal area decreased by 98%; however, the stomatal area increased by 57.45% with SL foliar application under drought stress (Figure 2D). SL foliar application may promote stomatal opening by regulating cell permeability to maintain water levels in leaves.

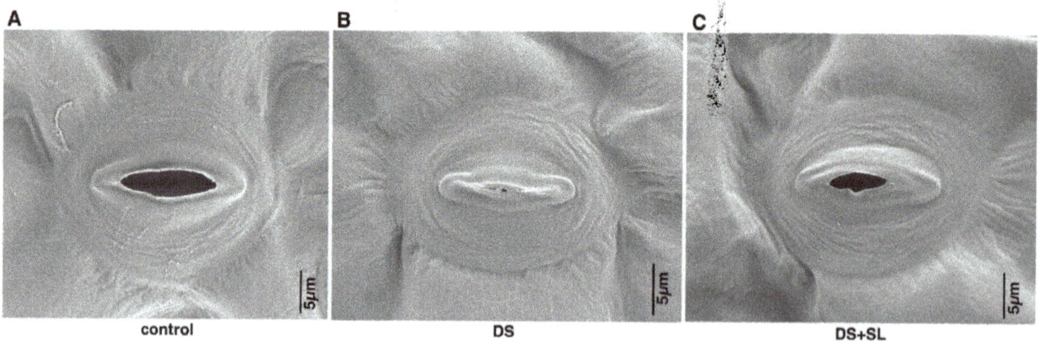

Figure 2. *Cont.*

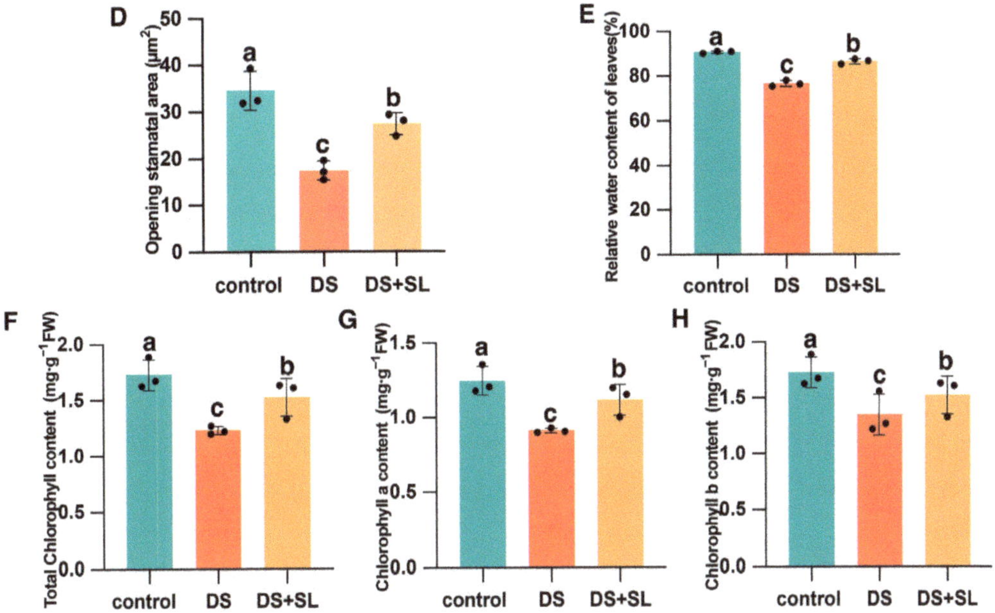

Figure 2. The combined effect of stomatal aperture performance and physiological analyses of pepper plants under control conditions, DS (drought conditions), and DS + SL (drought conditions with SL application). (**A–C**) Stomatal aperture performance; (**D**) analysis of stomatal area in different groups; (**E**) relative water content determined; (**F**) the total chlorophyll of the leaves; (**G**) chlorophyll a of the leaves; and (**H**) chlorophyll b of the leaves. The data are mean values and standard errors; the black dot indicates three biological replicates. Different letters indicate significant differences between different treatments ($p < 0.05$).

3.3. Relative Water Content and Chlorophyll Content of the Leaves

The drought strongly affected the RWC and chlorophyll content of the pepper seedlings. Only DS treatment has a lower RWC (76.57%) than the control group (90.61%). Remarkably, peppers treated with DS + SL have a significantly higher RWC in leaves (86.35%) than those grown under drought-only conditions (Figure 2E). Drought treatment significantly reduced the chlorophyll content in the leaves of pepper seedlings. Compared to the control, the concentrations of total chlorophyll, chlorophyll a, and chlorophyll b in pepper seedlings were decreased by 28.8%, 26.82%, and 21.91%, respectively. However, DS + SL plants exhibit considerably higher chlorophyll content, with increases of 23.69%, 22.33%, and 12.78%, respectively, compared to DS treatment plants (Figure 2F–H).

3.4. Measurement of Antioxidant Enzyme Activity and Related Metabolites

Drought + SL treatments significantly affected the antioxidant enzyme (POD, CAT, and SOD) activity of pepper seedlings. DS had significantly increased POD, CAT, and SOD activity compared to the control group. However, pretreated SL inhibited enzyme activity in the leaves of pepper seedlings. For example, the POD, CAT, and SOD activity of DS was only 180.86%, 93.59%, and 107.02% higher than in the control group. In contrast, the DS + SL group had lower enzyme activity than the DS-only treatments (Figure 3A–C). Compared to the control group, the DS-only group presented an extreme increase in proline, MDA, and H_2O_2. On the contrary, DS + SL plants presented significantly decreased proline, MDA, and H_2O_2 content, with 88.39%, 57.89%, and 61.83%, respectively, compared with the DS group in pepper seedlings (Figure 3D–F). Under DS, there is an increase in soluble sugar content. In addition, no significant differences were seen in DS and DS + SL treatments (Figure 3G).

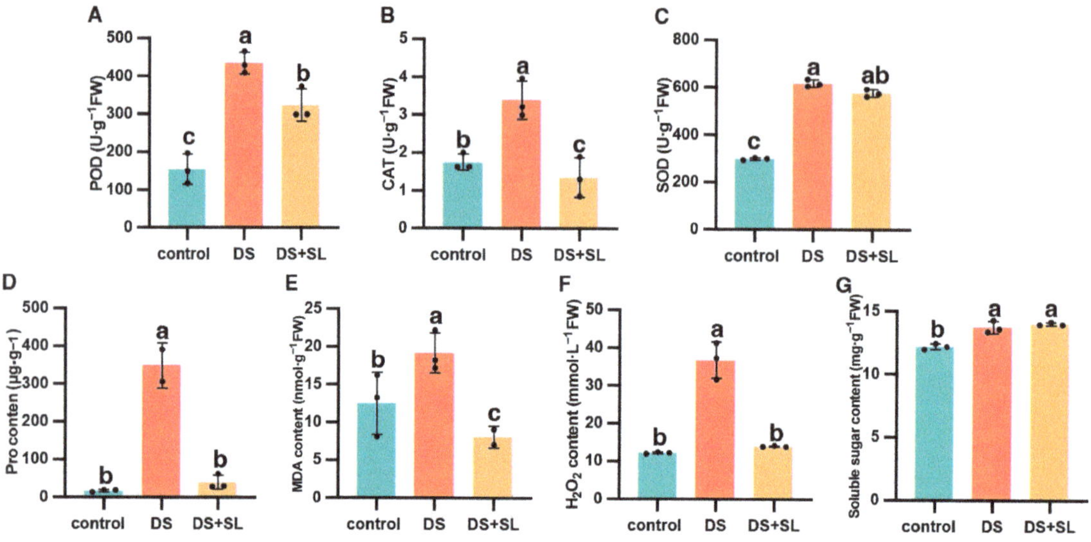

Figure 3. Effects of enzymatic and non-enzymatic antioxidant content in control, DS, and DS + SL pepper plant leaves. The antioxidant enzyme activity of (**A**) POD, (**B**) CAT, and (**C**) SOD; (**D**) Proline content, (**E**) MDA content, (**F**) H_2O_2 content, and (**G**) Soluble sugar content. The data are mean values and standard errors; the black dot indicates 3 biological replicates. Different letters indicate significant differences between different treatments ($p < 0.05$).

3.5. Assessment of RNA-Seq Data and Differentially Expressed Gene Analysis

A total of 40.32 million raw data were generated from transcriptome sequencing in three groups, which were filtered for low-quality reads, resulting in 38.69 million clean reads. It was found that 88.58–93.36% of reads could be successfully assigned to the *C. chinense* reference genome, and the unique mapping rate ranged from 85.35–89.02% (Table S2). Hierarchy analysis showed that different samples at the same treatment clustered together, and the DS + SL group demonstrated the most significant differences (Figure 4A). The transcriptome was divided into four clusters of DEGs, and the genes in each cluster had similar expression patterns, including 95, 1481, 159, and 25, respectively (Figure 4B). A total of 2097 DEGs were identified among the DS vs. control, DS + SL vs. control, and DS vs. DS + SL groups. The Venn diagram showed that only two genes were expressed in all groups, and the specific DEGs were higher than the common DEGs among all three groups (Figure 4C). Furthermore, there were 410 and 849 DEGs in group DS and DS + SL, compared with the control group, respectively. Moreover, there were 838 DEGs between the DS and DS + SL groups (Figure 4D).

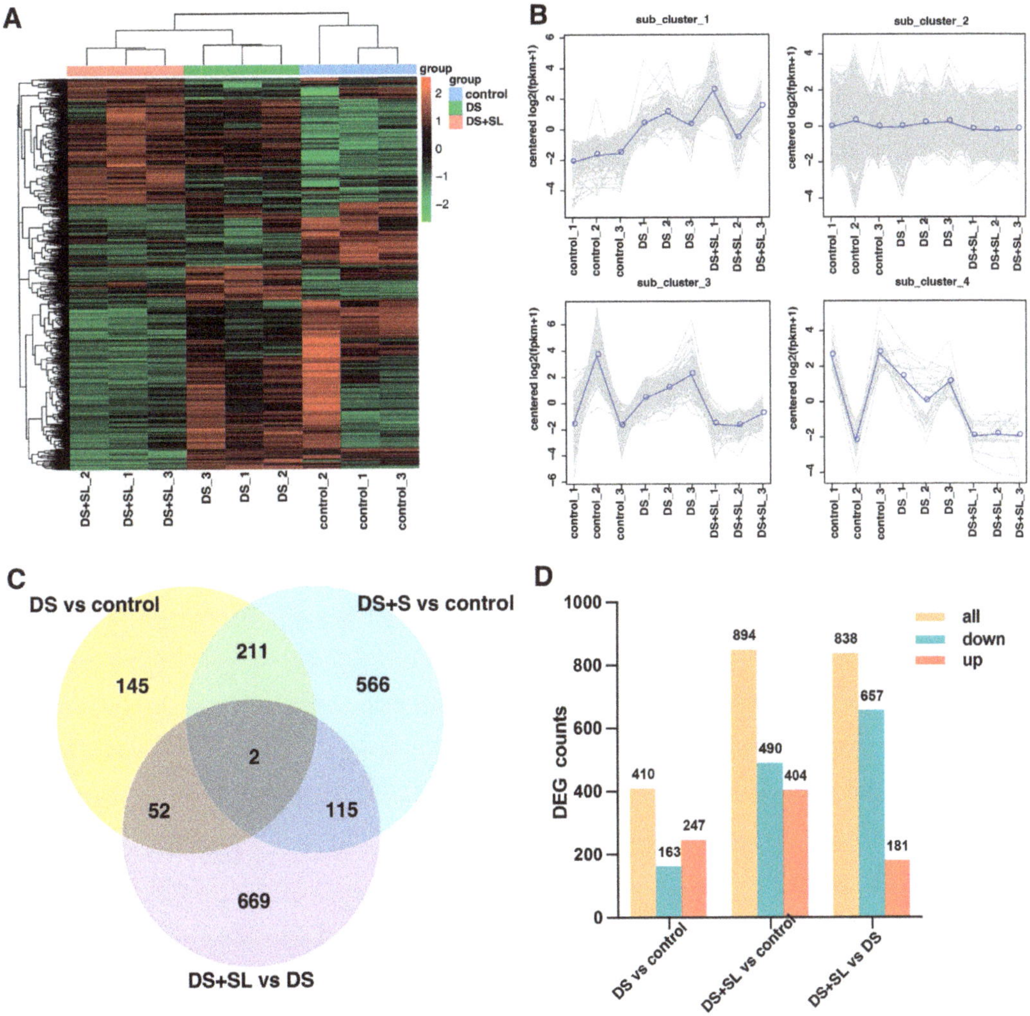

Figure 4. DEGs among three groups under different treatments. (**A**) Heatmap and cluster analysis expression of DEGs, (**B**) magnified regions of 4 subclusters of RNA–seq data. The gray lines represent the relative corrected gene expression levels of genes and the blue lines indicate consensus on relative corrected gene expression levels of all genes, (**C**) Venn diagram of three groups' DEGs, and (**D**) number of DEGs.

3.6. Enrichment Analysis of DEGs

Comparing DEGs between the control and DS group, galactose metabolism, plant hormone signal transduction, photosynthesis, and photosynthesis-antenna proteins were enriched in the latter. When control and DS + SL were compared, it was found that the biosynthesis of amino acids, brassinosteroid biosynthesis, carotenoid biosynthesis, and MAPK signaling were enriched in the latter. Intriguingly, the DS and DS + SL groups exhibited a significant enrichment in the ribosome pathway compared to the control group. Phenylpropanoid biosynthesis, terpenoid backbone biosynthesis, flavonoid biosynthesis, carotenoid biosynthesis, and MAPK signaling pathways were enriched in the DEGs identified between treatments with DS and DS + SL (Figure 5 and Figure S1 and Table S3). In addition, the genes of interest, as described by the molecular and biological functions of GO, were classified as molecular function (MF), cellular component (CC), and biological process (BP). GO functional annotation was performed to fully understand the roles of DEGs among the control, DS, and DS + SL groups (Figure S2).

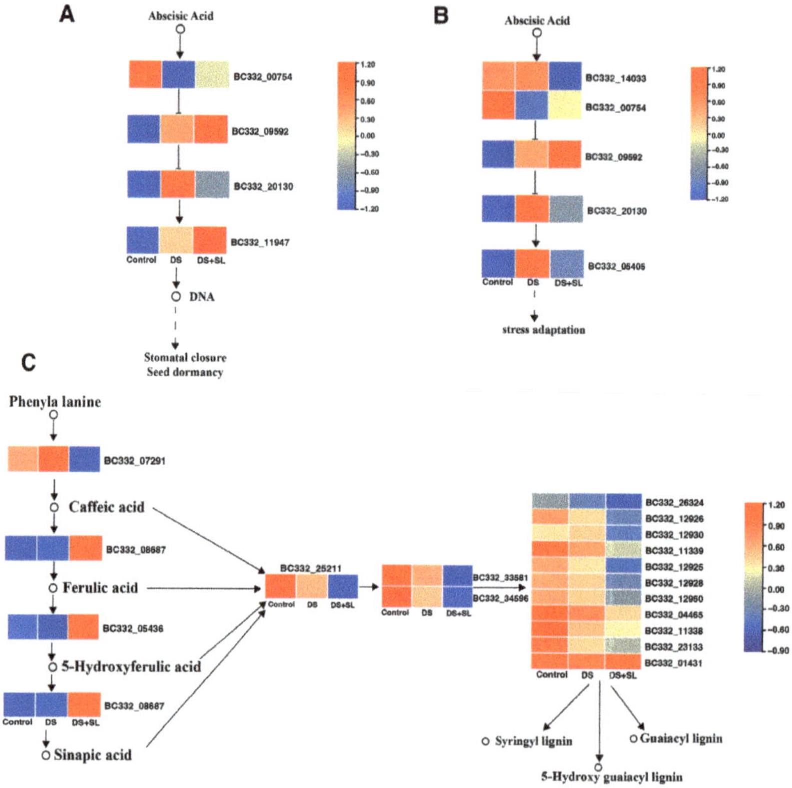

Figure 5. Enrichment pathways related to drought and SL treatments. (**A**) Plant hormone transduction, (**B**) MAPK signal, and (**C**) phenylpropanoid biosynthesis pathway. The heatmap colors show the log2 values of FPKM of control, DS, and DS + SL, from left to right.

3.7. Candidate Genes Involved in Drought and SL Treatment

Based on enrichment analysis, we identified 71 genes whose expression showed significant differences among the control, DS, and DS + SL groups, signifying that they may be responsive to DS or involved in the SL-induced alleviation of DS. Genes associated with the phenylpropanoid biosynthesis pathway were considerably enriched between DS alone and the DS + SL group, and 18 differentially expressed genes were identified. Interestingly, the expression of the genes related to phenylpropanoid biosynthesis *BC332_05436* (cytochrome 84A1) and *BC332_01431*(*peroxidase 15*) were notably higher in DS + SL group than in DS group plants. On the contrary, *BC332_25211*(*4-coumarate--CoA ligase 2*), *BC332_33581*(*caffeoyl-CoA O-methyltransferase*), and *BC332_34596* (*caffeoyl-CoA O-methyltransferase*) show substantially lower expression in DS + SL than in drought-only treatment. Furthermore, compared with the control group, in plants under DS alone and DS + SL treatment, genes related to the plant hormone signal transduction and MAPK pathways were significantly enriched. *BC332_22503* (*indole-3-acetic acid-induced protein ARG7*) and *BC332_09592* (*BRI1 kinase inhibitor 1*) were distinctly expressed in the DS + SL group, and had low expression in DS. Additionally, we identified 15 SL biosynthesis-related genes from DEGs. The expression profiles of genes involved in SL biosynthesis showed significant differences among the control, DS, and DS + SL groups. Compared with the DS group, *BC332_23620* (*SMAX1-LIKE 6*), *BC332_15224* (*SMAX1-LIKE 7*), and *BC332_18735* (*SMAX1-LIKE 4*) were expressed higher in DS + SL. Furthermore, 23 genes were identified in the ABC transporter pathway. Compared to the control treatment, the genes *BC332_30010*, *BC332_33666*, and *BC332_16040* were up-regulated in DS + SL treatment but down-regulated in DS-treated plants (Figure 6). Furthermore, we acquired mRNA levels of proline and antioxidant enzyme biosynthetic genes, which further elucidated the candidate genes at the molecular level (Figure S3).

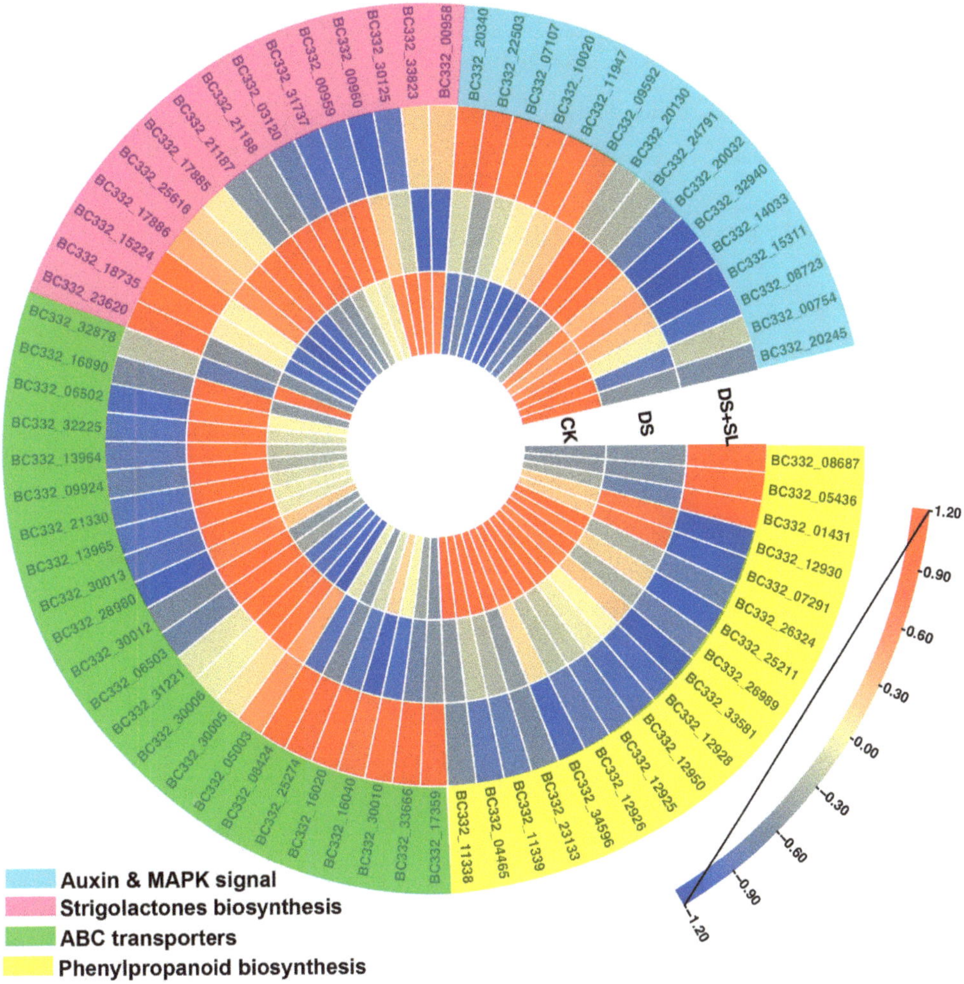

Figure 6. Expression of DEGs related to drought and SL treatments in control, DS, and DS + SL, respectively. Heatmap colors show the log2 values of FPKM.

3.8. qRT-PCR Validation of Gene Expression

Fifteen candidate genes closely associated with DS were selected for analysis of expression via RNA-Seq (Table S4). The control group was used as a control, while the

DS and DS + SL groups were investigated to validate the reliability of the RNA-seq data obtained in this study (Figure 7).

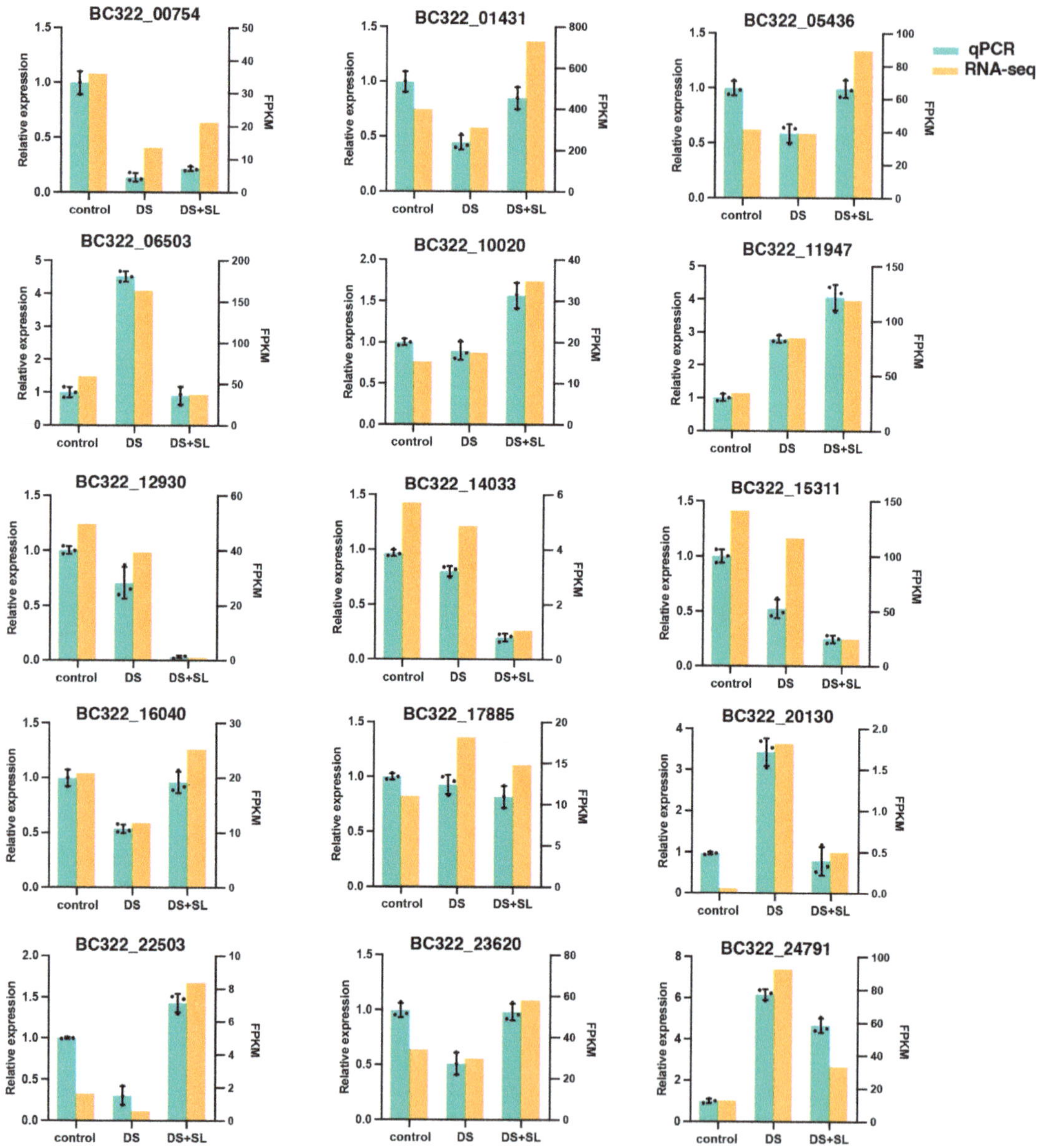

Figure 7. qRT-PCR validation of the relative expression levels of DEGs. Data are mean values and standard errors; the black dot indicates 3 biological replicates.

4. Discussion

Drought is well recognized as a primary abiotic stress factor that impedes plant growth and development [18]. The crucial role of several plant hormones in regulating the response to plant stress is well recognized [19]. SL has a crucial function in the regulation of plant development and the alleviation of environmental stressors [20]. Furthermore, the foliar application of SL enhances plant growth under DS conditions by improving photosynthetic characteristics and antioxidant enzyme activity [21]. However, little information about its effects on pepper abiotic stress responses is available.

DS considerably inhibited plant growth [22]. In this study, the roots and leaves of pepper seedlings were adversely affected under DS conditions. Pretreatment with SL can alleviate the drought symptoms of peppers. In the drought-stress treatment, the young wilted leaves showed signs of drooping and shrinking and most mature leaves had begun to curl, whereas the leaves of the DS + SL treatment only showed mild symptoms (Figures 1 and 2E). Hence, the results indicated that the application of SL reduced the apparent drought damage of pepper seedlings under DS. Chlorophyll content in plants is a critical indicator of photosynthetic activity, but this characteristic is adversely affected under stressful conditions [23]. The chlorophyll content significantly reduced by DS [24]. Additionally, SL plays a vital role in regulating plant photosynthetic efficiency [25]. Our research results show that the exogenous application of SL significantly alleviates the decrease in chlorophyll content under DS (Figure 2F–H). Consequently, this suggests that SL is an effective substance with which pepper can cope with DS; similar results have been reported in grapes [26]. Thus, applying SL is very possibly a method that can be used to alleviate the adverse effects of DS on pepper seedlings.

Proline is considered an important metabolite synthesized within plant cells under environmental stress conditions. It is assumed to have a significant impact on the protective function of plants grown under stress, which is attributed to its ability as an osmotic protector, membrane stabilizer, and ROS scavenger [27,28]. Previous literature suggested that proline accumulation may help improve water status and reduce oxidative damage caused by abiotic stress [18]. Similarly, we observed that DS significantly increased proline accumulation in pepper plants. In contrast, the application of SL significantly reduced the proline concentration in pepper seedlings under DS (Figure 3D). This demonstrates that SL has a potential efficiency effect on osmotic regulation during DS, which may help plants maintain growth and function. The proline biosynthesis gene (*AtP5CS1*) is pivotal in increasing proline biosynthesis under abiotic stress conditions [29]. We found that the mRNA level of *P5CS* (*BC332_15138*) was increased in peppers exposed to drought with or without SL (Figure S3). Captivatingly, the application of SL resulted in a significant decrease in proline content, which demonstrated that the up-regulation of *P5CS* transcription levels might contribute to an increase in proline concentration. However, the mechanism by which SL affects proline is not yet clear, and requires further exploration.

Maintaining redox homeostasis is of the utmost importance in mitigating the excessive generation of ROS and minimizing cellular membrane impairment in plants subjected to environmental stressors [30]. In the present study, there was a significant increase in the content of H_2O_2 and MDA under DS, while SL reversed the oxidative damage in DS-induced pepper seedlings by inhibiting the accumulation of H_2O_2 and MDA content (Figure 3E,F). The SL-mediated drought response may involve a complex interaction in H_2O_2 content and stomatal closure [8,31]. It may be necessary for plants to maintain the expression of some genes to maintain stomatal opening at specific levels, and subsequently balance CO_2 inflow and water loss under water scarcity conditions [32]. We perceived that the leaf stomata were firmly closed during DS. However, SL application had a remarkable effect in relieving this stomatal closure (Figure 2C), suggesting that SL plays a crucial role in regulating the plant's water balance, leading to the reopening of the stomata under DS. This is a complex molecular mechanism that still needs further research. To cope with oxidative damage, plants have developed a proficient antioxidant defense mechanism [33]. It was stated that SL can enhance the activity of antioxidant enzymes for ROS detoxification, thus endowing crop tolerance [34]. Compared with the DS group, lower levels of POD, CAT, and SOD activity were detected in the DS + SL group, signifying that SL can improve ROS clearance efficiency and potentially protect plants from the adverse effects of DS, consequently alleviating the oxidative stress caused by DS (Figure 3A–C). Furthermore, we assessed the expression levels of *peroxidase4* (*POD4*) and *peroxidase* (*POD15*), which are genes involved in the phenylpropane biosynthetic pathway. In drought-exposed seedlings, the expression of *POD15* decreased, whereas SL application significantly increased the expression levels of these genes, consistent with the transcriptomic results. Conversely, *POD4* showed almost no expression after SL treatment (Figure 7). Whether the transcription levels of *POD* genes contribute to improved drought tolerance or hinder drought resistance warrants further investigation.

SL effectively regulates plant growth in responses to various environmental conditions [35]. The physiological and molecular analysis of *Arabidopsis* suggests that plant hormone signaling pathways are critical to regulating drought or water-deficit responses [36]. *OSRK1* (*SAPK6*) is an upstream regulatory factor for stress signaling in rice roots, playing a significant role in ABA and hypertonic stress signaling [37]. Our research found that plants pretreated with SL exhibited significant gene changes associated with plant hormone signal transduction and the MAPK signal transduction pathway during DS (Figures 5 and 6). Specifically, we observed changes in the expression of auxin-responsive proteins from the Aux/IAA family, that are known to play a key role in plant stress responses [38]. An increase in ROS concentration exerts a negative regulatory influence on IAA turnover [39]. Interestingly, in this study, the expression level of *ARG7* (*BC332_22503*) was found to be decreased by 97.9% in the DS group, while it significantly increased in the DS + SL group. Furthermore, *BC332_20130* (*SAPK6*) encoding kinase was found to be up-regulated in response to DS without SL treatment, but significantly down-regulated in the group with SL pretreatment. We hypothesize that these genes may mitigate DS through the action of SL.

ABC transporters are now recognized to be involved in many physiological processes that enable plants to adapt environmental changes [40]. Multipotent drug-resistance (PDR) subfamily ABC proteins are found in many plants, and cope with biotic and abiotic stresses. A previous study demonstrated that the overexpression of the ABC transporter protein *AtABCG36/AtPDR8* made plants more resistant to drought and salt stress than wild-type plants. In contrast, knockout lines are more sensitive to DS than wild-type plants [41]. We identified that *BC332_16040* (PDR4) was significantly up-regulated after SL pretreatment under DS, implying that *PDR4* may play a crucial role in the response to DS and SL. Previous studies have demonstrated that the interaction between *MAX2* and *DWARF14* triggers the signal transduction of SL, while *SMXL6* (*SUPPRESSOR OF MAX1-Like 6*) functions as an inhibitor of SL signal transduction [42,43]. The *MAX2* mutant exhibits high sensitivity to DS [44]. A recent study has discovered that *SsMAX2* overexpression in *Arabidopsis* significantly enhances resistance to drought, osmotic, and salt stresses [11]. In addition, *SMXL6*, *SMXL7*, and *SMXL8* play a negative regulatory role during DS in *Arabidopsis* [45]. We found that the expression level of *BC332_17885* (*CcSMAX1*) was up-regulated under DS but down-regulated with SL treatment (Figure 7), indicating its negative regulatory role in response to DS. On the contrary, the expression of *CcSMXL6* showed the opposite trend. The investigation of the mechanism of action is elusive, necessitating further investigation.

5. Conclusions

In summary, our comprehensive investigation, which includes phenotypic observations, physiological measurements, and gene expression analysis, demonstrates the multifaceted positive impact of SL on pepper plants in a DS environment. SL effectively mitigates the adverse effects of DS and promotes the growth of pepper seedlings under such challenging conditions. The application of SL is a powerful tool to counteract drought-induced oxidative damage and enhance tolerance to DS, ultimately facilitating the overall healthy development of pepper plants. Furthermore, our findings highlight that DEGs critical to DS responses are predominantly related to the MAPK and plant hormone signal transduction pathways. These insights contribute to a deeper understanding of the molecular and physiological mechanisms underlying the ability of pepper to withstand DS and provide a valuable conceptual framework for elucidating how SL intricately interacts with diverse signaling pathways in pepper's adaptive response to drought stress. This research lays a solid foundation for future investigations into enhancing crop resilience and sustainability in the face of environmental challenges (Figure 8).

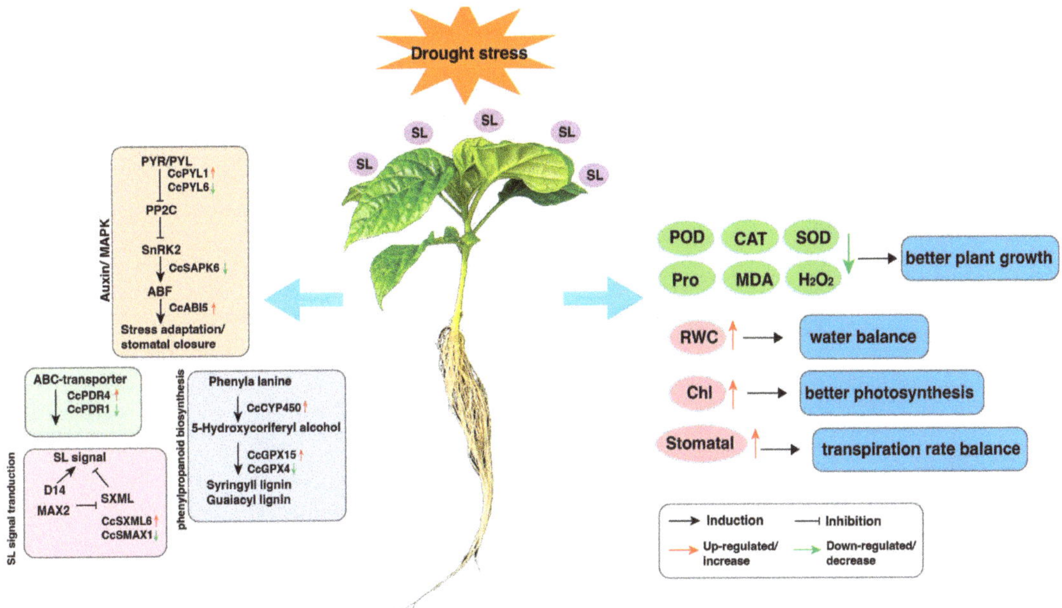

Figure 8. Summary of the mechanism to tolerance to drought stress induced by SL in pepper seedlings.

Supplementary Materials: The following supporting information can be downloaded at: https://www.mdpi.com/article/10.3390/antiox12122019/s1, Supplementary Figure S1: KEGG enrichment analysis. (A) DS vs. control, (B) DS + SL vs. control, and (C) DS + SL vs. DS; Figure S2: GO enrichment analysis. (A) DS vs. control, (B) DS + SL vs. control, and (C) DS + SL vs. DS; Figure S3: Heat-map representing the transcription level of differentially expressed proline and antioxidant enzymes related genes in pepper seedlings under drought stress. Heatmap colors show the log2FoldChange values; Table S1: Primers of qRT-PCR; Table S2: Quality assessment and mapping of sequencing data; Table S3: KEGG pathway assignments; Table S4: Information of candidate genes.

Author Contributions: Conceptualization, H.S.; Methodology, H.S., M.A.A., and H.F.; Software, H.S. and H.F.; Validation, X.L.; Formal analysis, N.M.; Investigation, M.A.A. and X.L.; Resources, G.Z. and S.C.; Data curation, N.M.; Writing—original draft, H.S.; Writing—review and editing, Z.W.; Visualization, H.S., G.Z., and S.C.; Supervision, Z.W.; Funding acquisition, Z.W. All authors have read and agreed to the published version of the manuscript.

Funding: This research was funded by the Collaborative Innovation Center of Nanfan and High-Efficiency Tropical Agriculture (XTCX2022NYB03), Hainan University, and Hainan Province Science and Technology Special Fund (ZDYF2023XDNY028).

Institutional Review Board Statement: Not applicable.

Informed Consent Statement: Not applicable.

Data Availability Statement: The original data can be found in NCBI (PRJNA957909).

Conflicts of Interest: The authors declare no conflict of interest.

References

1. Nadeem, M.; Li, J.; Yahya, M.; Sher, A.; Ma, C.; Wang, X.; Qiu, L. Research progress and perspective on drought stress in legumes: A review. *Int. J. Mol. Sci.* **2019**, *20*, 2541. [CrossRef] [PubMed]
2. Giordano, D.; Provenzano, S.; Ferrandino, A.; Vitali, M.; Pagliarani, C.; Roman, F.; Cardinale, F.; Castellarin, S.D.; Schubert, A. Characterization of a multifunctional caffeoyl-CoA O-methyltransferase activated in grape berries upon drought stress. *Plant Physiol. Biochem.* **2016**, *101*, 23–32. [CrossRef]
3. Qin, C.; Yu, C.; Shen, Y.; Fang, X.; Chen, L.; Min, J.; Cheng, J.; Zhao, S.; Xu, M.; Luo, Y.; et al. Whole-genome sequencing of cultivated and wild peppers provides insights into *Capsicum* domestication and specialization. *Proc. Natl. Acad. Sci. USA* **2014**, *111*, 5135–5140. [CrossRef] [PubMed]
4. Manivannan, A.; Kim, J.H.; Yang, E.Y.; Ahn, Y.K.; Lee, E.S.; Choi, S.; Kim, D.S. Next-Generation Sequencing Approaches in Genome-Wide Discovery of Single Nucleotide Polymorphism Markers Associated with Pungency and Disease Resistance in Pepper. *Biomed Res. Int.* **2018**, *2018*, 5646213. [CrossRef]
5. Gomez-Roldan, V.; Fermas, S.; Brewer, P.B.; Puech-Pagès, V.; Dun, E.A.; Pillot, J.P.; Letisse, F.; Matusova, R.; Danoun, S.; Portais, J.C.; et al. Strigolactone inhibition of shoot branching. *Nature* **2008**, *455*, 189–194. [CrossRef]
6. Waters, M.T.; Gutjahr, C.; Bennett, T.; Nelson, D.C. Strigolactone Signaling and Evolution. *Annu. Rev. Plant Biol.* **2017**, *68*, 291–322. [CrossRef] [PubMed]
7. Chesterfield, R.J.; Vickers, C.E.; Beveridge, C.A. Translation of strigolactones from plant hormone to agriculture: Achievements, future perspectives, and challenges. *Trends. Plant. Sci.* **2020**, *25*, 1087–1106. [CrossRef]
8. Min, Z.; Li, R.; Chen, L.; Zhang, Y.; Li, Z.; Liu, M.; Ju, Y.; Fang, Y. Alleviation of drought stress in grapevine by foliar-applied strigolactones. *Plant Physiol. Biochem.* **2019**, *135*, 99–110. [CrossRef]
9. Visentin, I.; Vitali, M.; Ferrero, M.; Zhang, Y.; Ruyter-Spira, C.; Novák, O.; Strnad, M.; Lovisolo, C.; Schubert, A.; Cardinale, F. Low levels of strigolactones in roots as a component of the systemic signal of drought stress in tomato. *New Phytol.* **2016**, *212*, 954–963. [CrossRef]
10. Ha, C.V.; Leyva-González, M.A.; Osakabe, Y.; Tran, U.T.; Nishiyama, R.; Watanabe, Y.; Tanaka, M.; Seki, M.; Yamaguchi, S.; Dong, N.V.; et al. Positive regulatory role of strigolactone in plant responses to drought and salt stress. *Proc. Natl. Acad. Sci. USA* **2014**, *111*, 851–856. [CrossRef] [PubMed]
11. Wang, Q.; Ni, J.; Shah, F.; Liu, W.; Wang, D.; Yao, Y.; Hu, H.; Huang, S.; Hou, J.; Fu, S.; et al. Overexpression of the stress-Inducible SsMAX2 promotes drought and dalt resistance via the regulation of redox homeostasis in *Arabidopsis*. *Int. J. Mol. Sci.* **2019**, *20*, 837. [CrossRef] [PubMed]
12. Altaf, M.A.; Hao, Y.; He, C.; Mumtaz, M.A.; Shu, H.; Fu, H.; Wang, Z. Physiological and biochemical responses of Ppepper (*Capsicum annuum* L.) seedlings to nickel toxicity. *Front. Plant Sci.* **2022**, *13*, 950392. [CrossRef] [PubMed]
13. Kim, D.; Langmead, B.; Salzberg, S.L. HISAT: A fast spliced aligner with low memory requirements. *Nat. Methods* **2015**, *12*, 357–360. [CrossRef] [PubMed]
14. Love, M.I.; Huber, W.; Anders, S. Moderated estimation of fold change and dispersion for RNA-seq data with DESeq2. *Genome Biol.* **2014**, *15*, 550. [CrossRef]
15. Young, M.D.; Wakefield, M.J.; Smyth, G.K.; Oshlack, A. Gene ontology analysis for RNA-seq: Accounting for selection bias. *Genome Biol.* **2010**, *11*, R14. [CrossRef]
16. Kanehisa, M.; Goto, S. KEGG: Kyoto encyclopedia of genes and genomes. *Nucleic Acids Res.* **2000**, *28*, 27–30. [CrossRef]
17. Shu, H.; Zhang, Y.; He, C.; Altaf, M.A.; Hao, Y.; Liao, D.; Li, L.; Li, C.; Fu, H.; Cheng, S.; et al. Establishment of in vitro regeneration system and molecular analysis of early development of somatic callus in *Capsicum chinense* and *Capsicum baccatum*. *Front. Plant Sci.* **2022**, *13*, 1025497. [CrossRef]
18. Kaya, C.; Ashraf, M.; Wijaya, L.; Ahmad, P. The putative role of endogenous nitric oxide in brassinosteroid-induced antioxidant defence system in pepper (*Capsicum annuum* L.) plants under water stress. *Plant Physiol. Biochem.* **2019**, *143*, 119–128. [CrossRef]
19. Salvi, P.; Manna, M.; Kaur, H.; Thakur, T.; Gandass, N.; Bhatt, D.; Muthamilarasan, M. Phytohormone signaling and crosstalk in regulating drought stress response in plants. *Plant Cell Rep.* **2021**, *40*, 1305–1329. [CrossRef]
20. Saeed, W.; Naseem, S.; Ali, Z. Strigolactones biosynthesis and their role in abiotic stress resilience in plants: A critical review. *Front. Plant Sci.* **2017**, *8*, 1487. [CrossRef] [PubMed]
21. Liu, B.; Zhang, Y.; Wang, S.; Wang, W.; Xu, X.; Wu, J.; Fang, Y.; Ju, Y. Effects of strigolactone and abscisic acid on the quality and antioxidant activity of grapes (*Vitis vinifera* L.) and wines. *Food Chem. X* **2022**, *16*, 100496. [CrossRef] [PubMed]
22. Mahmood, T.; Khalid, S.; Abdullah, M.; Ahmed, Z.; Shah, M.K.N.; Ghafoor, A.; Du, X. Insights into drought Stress dignaling in plants and the molecular genetic Basis of cotton drought tolerance. *Cells* **2019**, *9*, 105. [CrossRef] [PubMed]
23. Kalaji, H.M.; Jajoo, A.; Oukarroum, A.; Brestic, M.; Zivcak, M.; Samborska, I.A.; Cetner, M.D.; Łukasik, I.; Goltsev, V.; Ladle, R.J. Chlorophyll a fluorescence as a tool to monitor physiological status of plants under abiotic stress conditions. *Acta Physiol. Plant.* **2016**, *38*, 1–11. [CrossRef]
24. Ozturk, M.; Turkyilmaz Unal, B.; García-Caparrós, P.; Khursheed, A.; Gul, A.; Hasanuzzaman, M. Osmoregulation and its actions during the drought stress in plants. *Physiol. Plant* **2021**, *172*, 1321–1335. [CrossRef]
25. Müller, M.; Munné-Bosch, S. Hormonal impact on photosynthesis and photoprotection in plants. *Plant Physiol.* **2021**, *185*, 1500–1522. [CrossRef] [PubMed]

26. Wang, W.N.; Min, Z.; Wu, J.R.; Liu, B.C.; Xu, X.L.; Fang, Y.L.; Ju, Y.L. Physiological and transcriptomic analysis of Cabernet Sauvginon (*Vitis vinifera* L.) reveals the alleviating effect of exogenous strigolactones on the response of grapevine to drought stress. *Plant Physiol. Biochem.* **2021**, *167*, 400–409. [CrossRef]
27. Szabados, L.; Savouré, A. Proline: A multifunctional amino acid. *Trends. Plant Sci.* **2010**, *15*, 89–97. [CrossRef]
28. Bandurska, H.; Niedziela, J.; Pietrowska-Borek, M.; Nuc, K.; Chadzinikolau, T.; Radzikowska, D. Regulation of proline biosynthesis and resistance to drought stress in two barley (*Hordeum vulgare* L.) genotypes of different origin. *Plant Physiol. Biochem.* **2017**, *118*, 427–437. [CrossRef] [PubMed]
29. Fichman, Y.; Gerdes, S.Y.; Kovács, H.; Szabados, L.; Zilberstein, A.; Csonka, L.N. Evolution of proline biosynthesis: Enzymology, bioinformatics, genetics, and transcriptional regulation. *Biol. Rev. Camb. Philos. Soc.* **2015**, *90*, 1065–1099. [CrossRef]
30. Sharma, A.; Zheng, B. Melatonin mediated regulation of drought stress: Physiological and molecular aspects. *Plants* **2019**, *8*, 190. [CrossRef]
31. Yao, Y.; Liu, X.; Li, Z.; Ma, X.; Rennenberg, H.; Wang, X.; Li, H. Drought-induced H_2O_2 accumulation in subsidiary cells is involved in regulatory signaling of stomatal closure in maize leaves. *Planta* **2013**, *238*, 217–227. [CrossRef]
32. Ding, Z.J.; Yan, J.Y.; Xu, X.Y.; Yu, D.Q.; Li, G.X.; Zhang, S.Q.; Zheng, S.J. Transcription factor WRKY46 regulates osmotic stress responses and stomatal movement independently in Arabidopsis. *Plant J.* **2014**, *79*, 13–27. [CrossRef]
33. Ahmad, S.; Kamran, M.; Ding, R.; Meng, X.; Wang, H.; Ahmad, I.; Fahad, S.; Han, Q. Exogenous melatonin confers drought stress by promoting plant growth, photosynthetic capacity and antioxidant defense system of maize seedlings. *PeerJ* **2019**, *7*, e7793. [CrossRef]
34. Qiu, C.W.; Zhang, C.; Wang, N.H.; Mao, W.; Wu, F. Strigolactone GR24 improves cadmium tolerance by regulating cadmium uptake, nitric oxide signaling and antioxidant metabolism in barley (*Hordeum vulgare* L.). *Environ. Pollut.* **2021**, *273*, 116486. [CrossRef]
35. Gururani, M.A.; Mohanta, T.K.; Bae, H. Current understanding of the interplay between phytohormones and photosynthesis under environmental stress. *Int. J. Mol. Sci.* **2015**, *16*, 19055–19085. [CrossRef] [PubMed]
36. Gupta, A.; Rico-Medina, A.; Caño-Delgado, A.I. The physiology of plant responses to drought. *Science* **2020**, *368*, 266–269. [CrossRef]
37. Nam, M.H.; Huh, S.M.; Kim, K.M.; Park, W.J.; Seo, J.B.; Cho, K.; Kim, D.Y.; Kim, B.G.; Yoon, I.S. Comparative proteomic analysis of early salt stress-responsive proteins in roots of SnRK2 transgenic rice. *Proteome Sci.* **2012**, *10*, 25. [CrossRef] [PubMed]
38. Luo, J.; Zhou, J.J.; Zhang, J.Z. Aux/IAA gene family in plants: Molecular structure, regulation, and function. *Int. J. Mol. Sci.* **2018**, *19*, 259. [CrossRef]
39. Nguyen, H.T.; Umemura, K.; Kawano, T. Indole-3-acetic acid-induced oxidative burst and an increase in cytosolic calcium ion concentration in rice suspension culture. *Biosci. Biotechnol. Biochem.* **2016**, *80*, 1546–1554. [CrossRef] [PubMed]
40. Kretzschmar, T.; Burla, B.; Lee, Y.; Martinoia, E.; Nagy, R. Functions of ABC transporters in plants. *Essays Biochem.* **2011**, *50*, 145–160. [CrossRef]
41. Kim, D.Y.; Jin, J.Y.; Alejandro, S.; Martinoia, E.; Lee, Y. Overexpression of AtABCG36 improves drought and salt stress resistance in *Arabidopsis*. *Physiol. Plant* **2010**, *139*, 170–180. [CrossRef] [PubMed]
42. Wang, L.; Wang, B.; Jiang, L.; Liu, X.; Li, X.; Lu, Z.; Meng, X.; Wang, Y.; Smith, S.M.; Li, J. Strigolactone signaling in *Arabidopsis* regulates shoot development by targeting D53-Like SMXL repressor proteins for ubiquitination and degradation. *Plant Cell* **2015**, *27*, 3128–3142. [CrossRef]
43. Yao, R.; Ming, Z.; Yan, L.; Li, S.; Wang, F.; Ma, S.; Yu, C.; Yang, M.; Chen, L.; Chen, L.; et al. DWARF14 is a non-canonical hormone receptor for strigolactone. *Nature* **2016**, *536*, 469–473. [CrossRef] [PubMed]
44. Bu, Q.; Lv, T.; Shen, H.; Luong, P.; Wang, J.; Wang, Z.; Huang, Z.; Xiao, L.; Engineer, C.; Kim, T.H.; et al. Regulation of drought tolerance by the F-box protein MAX2 in *Arabidopsis*. *Plant Physiol.* **2014**, *164*, 424–439. [CrossRef] [PubMed]
45. Yang, T.; Lian, Y.; Kang, J.; Bian, Z.; Xuan, L.; Gao, Z.; Wang, X.; Deng, J.; Wang, C. The SUPPRESSOR of MAX2 1 (SMAX1)-Like SMXL6, SMXL7 and SMXL8 act as negative regulators in response to drought stress in *Arabidopsis*. *Plant Cell Physiol.* **2020**, *61*, 1477–1492. [CrossRef]

antioxidants

MDPI

Article

Caffeine Produced in Rice Plants Provides Tolerance to Water-Deficit Stress

Youngchul Yoo [1], Yo-Han Yoo [2], Dong Yoon Lee [3], Ki-Hong Jung [3], Sang-Won Lee [3,*,†] and Jong-Chan Park [4,*,†]

1 Advanced Radiation Technology Institute (ARTI), Korea Atomic Energy Research Institute (KAERI), Jeongeup 56212, Republic of Korea; yooyc84@kakao.com
2 Central Area Crop Breeding Division, Department of Central Area Crop Science, National Institute of Crop Science, RDA, Suwon 16429, Republic of Korea; yohan04@korea.kr
3 Graduate School of Green-Bio Science, Crop Biotech Institute, Kyung Hee University, Yongin 17104, Republic of Korea; ehddbs0502@khu.ac.kr (D.Y.L.); khjung2010@khu.ac.kr (K.-H.J.)
4 Plant Systems Engineering Research Center, Korea Research Institute of Bioscience & Biotechnology (KRIBB), Daejeon 34141, Republic of Korea
* Correspondence: swlee6803@khu.ac.kr (S.-W.L.); reredfox@gmail.com (J.-C.P.); Tel.: +82-31-201-3473 (S.-W.L.); +82-42-860-4296 (J.-C.P.)
† These authors contributed equally to this work.

Abstract: Exogenous or endogenous caffeine application confers resistance to diverse biotic stresses in plants. In this study, we demonstrate that endogenous caffeine in caffeine-producing rice (CPR) increases tolerance even to abiotic stresses such as water deficit. Caffeine produced by CPR plants influences the cytosolic Ca^{2+} ion concentration gradient. We focused on examining the expression of Ca^{2+}-dependent protein kinase genes, a subset of the numerous proteins engaged in abiotic stress signaling. Under normal conditions, CPR plants exhibited increased expressions of seven *OsCPKs* (*OsCPK10*, *OsCPK12*, *OsCPK21*, *OsCPK25*, *OsCPK26*, *OsCPK30*, and *OsCPK31*) and biochemical modifications, including antioxidant enzyme (superoxide dismutase, catalase, peroxidase, and ascorbate peroxidase) activity and non-enzymatic antioxidant (ascorbic acid) content. CPR plants exhibited more pronounced gene expression changes and biochemical alterations in response to water-deficit stress. CPR plants revealed increased expressions of 16 *OsCPKs* (*OsCPK1*, *OsCPK2*, *OsCPK3*, *OsCPK4*, *OsCPK5*, *OsCPK6*, *OsCPK9*, *OsCPK10*, *OsCPK11*, *OsCPK12*, *OsCPK14*, *OsCPK16*, *OsCPK18*, *OsCPK22*, *OsCPK24*, and *OsCPK25*) and 8 genes (*OsbZIP72*, *OsLEA25*, *OsNHX1*, *OsRab16d*, *OsDREB2B*, *OsNAC45*, *OsP5CS*, and *OsRSUS1*) encoding factors related to abiotic stress tolerance. The activity of antioxidant enzymes increased, and non-enzymatic antioxidants accumulated. In addition, a decrease in reactive oxygen species, an accumulation of malondialdehyde, and physiological alterations such as the inhibition of chlorophyll degradation and the protection of photosynthetic machinery were observed. Our results suggest that caffeine is a natural chemical that increases the potential ability of rice to cope with water-deficit stress and provides robust resistance by activating a rapid and comprehensive resistance mechanism in the case of water-deficit stress. The discovery, furthermore, presents a new approach for enhancing crop tolerance to abiotic stress, including water deficit, via the utilization of a specific natural agent.

Keywords: caffeine; water-deficit stress tolerance; calcium-dependent protein kinase; antioxidants; transcription factors

1. Introduction

Throughout their life cycle, plants encounter a broad spectrum of biotic and abiotic stresses. Among these challenges, drought stress emerges as a particularly formidable and pressing issue, primarily due to its extensive occurrence on a significant portion of the world's land [1,2]. This widespread phenomenon is largely attributed to the recent and rapid changes in our climate [3,4]. Drought stress exerts a profound influence on the physiological, metabolic, and molecular functions of crops, ultimately leading to substantial reductions in crop yields [5–7]. This, in turn, poses a significant threat to global food security. Of particular concern is rice, the second most widely cultivated food crop globally, following maize [8–10]. The centrality of rice in the global food supply underscores its significance. Nevertheless, the water-intensive cultivation methods employed for rice make it especially vulnerable to water-deficit stress, as it necessitates two to three times more water for cultivation compared to other crops [11–14]. Consequently, there is a pressing need to accelerate the development of water-deficit-tolerant rice varieties and innovative technologies via a multifaceted approach.

Plants have developed defense mechanisms against biotic and abiotic stresses. One of these defense strategies involves the production of secondary metabolites that can directly or indirectly counteract the stressor [15,16]. An illustrative example of this mechanism is the accumulation of secondary metabolites, such as lignin. Numerous studies have provided evidence that certain genes, including *OsCCR10* [17] and *OsTF1L* [18] in rice, *PeLAC10* [19] in Arabidopsis, and *PoCCoAOMT* [20] in tobacco, play pivotal roles in enhancing drought tolerance by regulating the accumulation of lignin in response to drought stress. To date, an extensive array of over 200,000 secondary metabolites, spanning diverse biosynthetic families such as alkaloids, terpenes, and phenolics, have been identified in various plant species [15,21]. Caffeine (1, 3, 7-trimethylxanthine) is a major secondary metabolite of the purine alkaloid family and is synthesized in a few plant species, including coffee, cacao, and tea [22]. Although it is a very familiar compound now widely consumed in beverages such as coffee and tea, it originally played an important role in plants. Caffeine has allelopathic effects on other plants, where it can inhibit their growth and development, helping the caffeine-producing plant to gain an advantage in competing for resources [23]. Furthermore, several studies have shown that caffeine plays a role in protecting host plants from pathogens and herbivores, either directly or indirectly [22]. Such properties of caffeine have been employed in transgenic plants. Transgenic tobacco plants that synthesize caffeine are highly resistant to fungal pathogens, viruses, and herbivores [24–28]. In our previous study, we demonstrated that caffeine synthesized in transgenic rice (CPR, caffeine-producing rice) directly inhibits the activity of phosphodiesterase, resulting in an influx of calcium ions (Ca^{2+}) into the cell. This causes alterations in the Ca^{2+} signals, which eventually drive the synthesis and accumulation of salicylic acid, providing resistance to diverse biotic stresses [27].

Ca^{2+} is a crucial element for plant growth and development [29,30]. It is essential for maintaining cell wall structure, cell division, and elongation [31]. Furthermore, changes in cytosolic free calcium concentration are among the swiftest intracellular responses to both abiotic and biotic stresses [32–34]. This transient increase in cytosolic calcium ion concentration generates a specific calcium signature, which can be detected by various calcium sensors, thereby initiating downstream effects [33,34]. Representatively, Ca^{2+}-dependent protein kinases (CPKs) and calcium/calmodulin-regulated kinases (CAMKs) are activated via specific calcium signatures, subsequently triggering changes in gene expression patterns, production of reactive oxygen species (ROS), and the reinforcement of cell walls [35]. Nevertheless, excessive fluctuations in calcium levels can trigger programmed cell death or apoptosis within the hypersensitive response (HR) [36–38]. In our previous study, despite the production of caffeine in rice resulting in intracellular calcium changes, we did not observe severe HR responses or programmed cell death, even when the plants were subjected to various biotic stresses [27]. This phenomenon has not been previously reported in either rice or transgenic plants producing caffeine [25,26]. The reactions triggered by changes in intracellular Ca^{2+} concentration are also deeply involved in the response to abiotic stresses such as salinity, drought, cold, and heat [39–41]. Among them, drought stress impacts plant photosynthesis and overall ROS production via the signaling of the plant hormone abscisic acid (ABA), resulting in ROS accumulation and oxidative stress [42–44]. ROS molecules, previously believed only to have detrimental effects, are essential for long-distance signaling, response to stress, and the growth and development of plants [45]. However, the excessive production of ROS, on the other hand, can cause oxidative damage to essential cellular components such as nucleic acids, proteins, and cell membranes. Hence, the balance between ROS generation and detoxification is strictly controlled [46]. Changes in cytosolic Ca^{2+} concentration caused by water deficits play a role in the drought stress response mechanism in an ABA-dependent manner by activating CPKs and releasing ABA from the cell [47]. These mechanisms encompass cellular functions, including stomatal closure [48]. The ABA signals induce the activity and accumulation of enzymatic antioxidants such as superoxide dismutase (SOD), catalase (CAT), peroxidase (POX), and ascorbate peroxidase (APX), as well as non-enzymatic antioxidants such as ascorbic acid, carotenoids, and glutathione [49–51]. In addition, the expression of numerous genes is affected under drought stress, many of which play important roles in other abiotic stress responses and tolerance. Several genes with expressions induced under drought stress are also induced under salt stress, indicating a significant interaction between these two stresses [52,53].

We hypothesized that caffeine synthesized in CPR not only confers resistance to biotic stresses, as shown in our previous report [27], but may also contribute to providing tolerance to abiotic stresses, including water deficit. To prove this, the present study aimed to assess the efficacy of caffeine produced in transgenic rice as a tolerance-inducing agent to water-deficit stress and elucidate the mechanism underlying its stress tolerance properties. Our results indicated involvement in the expression changes in OsCPKs and transcription factors related to stress tolerance, leading to the enforcement of ROS detoxification and ultimately maintaining the stability of physiological factors related to photosynthesis. Our finding indicates that caffeine possesses the ability to enhance and augment resistance to water deficit, providing novel perspectives on strategies for developing tolerance to abiotic stresses.

2. Materials and Methods

2.1. Plant Growth Conditions and Water-Deficit Stress Treatment

Wild-type rice (WT, *Oryza sativa. japonica* cv. Dongjin) and two homozygous T_5 transgenic rice lines (caffeine-producing transgenic rice, CPR #1 and CPR #2) were used in these experiments. Wild-type rice was obtained from the Institute of Crop Biotechnology at Kyung Hee University in Yongin, Republic of Korea. To generate the CPR lines, the full-length coding sequence (CDS) of *CaMXMT1* (AB048794), *CaXMT1* (AB048793), and *CaDXMT1* (AB084125) from *Coffea arabica* were amplified and cloned into vector pBIN-NMT 777 under the control of the 35s promoter. The constructed vector pBIN-NMT 777 was introduced into rice (Dongjin) using the Agrobacterium-mediated early infection method [54]. The detailed methodology has been previously reported [27]. Wild-type and transgenic rice plants were grown in a growth chamber (28 °C, 60% relative humidity, 14/10 h light/dark photoperiod, light intensity of 200 $\mu mol\ m^{-2}\ s^{-1}$) for 28 days.

To minimize the potential variations in stress effects, we cultivated both wild-type and transgenic rice plants in the same pots. Each pot contained five wild-type and five CPR plants. Stress treatments were conducted on a total of nine wild-type/CPR1 pots and nine wild-type/CPR2 pots to ensure biological replicates. Wild-type and transgenic rice plants were grown in a growth chamber with normal watering for 28 days. Subsequently, they were transferred to a greenhouse, where water supply was then stopped for 10 days. Following the conclusion of the water-deficit stress treatment, water was re-provided, and the phenotypic analysis was monitored for 14 days. We employed a temperature and humidity data logger, RC-4HC (Elitech Technology, San Jose, CA, USA), to measure soil moisture, providing confirmation of the water-deficit stress. Samples for gene transcript expression analyses and antioxidant enzyme activity assays were collected every 2 days over the course of the 10-day water-deficit treatment.

2.2. Determination of Caffeine Contents in Rice Leaves

To extract caffeine from transgenic rice leaves, 1 g of the sample was homogenized in a 50% methanol solution and incubated with stirring at 60 °C for 1 h. The extract was then centrifuged at 14,000× *g* for 10 min at 4 °C. An HPLC system (Agilent Technologies 1200 series, CA, USA) equipped with a reverse-phase column (Eclipse XDB-C18 column, 4.6 mm × 150 mm, 5 µm; Agilent Technologies, Santa Clara, CA, USA) was used to separate caffeine. The ChemStation software ver. B.04.03 (Agilent Technologies, CA, USA) was used to calculate the results on an HPLC system equipped with a UV-VIS spectrophotometer. The mobile phase A consisted of trifluoroacetic acid and water (0.05:99.95, *v*/*v*), and the mobile phase B was acetonitrile. The gradient elution profile was as follows: 0–21 min, a linear gradient from 2% B to 9% B; 21–32 min, a linear gradient from 9% B to 23% B; 32–45 min, 23% B. The elution was carried out at a flow rate of 0.8 mL min^{-1}. The caffeine peaks were identified by comparing their retention times with caffeine standards (Agilent Technologies, USA) at 280 nm. Standard curves for quantification were constructed using the quadratic fitting of the relationship between the area sum and the concentration of the peaks corresponding to each sample.

2.3. RNA Isolation and Quantitative Real-Time PCR (qRT-PCR) Analysis

Total RNA was extracted from rice leaf samples using RNAiso (Takara, Kusatsu, Japan) following the manufacturer's instructions. DNase I treatment (Qiagen, Hilden, Germany) was performed to remove any residual DNA. The isolated RNA was quantified using a NanoDrop ND-1000 spectrophotometer (Thermo Fisher Scientific, Waltham, MA, USA). First-strand complementary DNA (cDNA) was synthesized from 2 µg of total RNA using the SuPrime-Script cDNA Synthesis Kit (GeNet Bio, Daejeon, Republic of Korea) according to the manufacturer's instructions. The qRT-PCR was performed using 2X Prime Q-Mastermix SYBR Green I (GeNet Bio, Republic of Korea) reagent on a Rotor-Gene Q instrument system (Qiagen, Germany). The $2^{-\Delta\Delta Ct}$ method [55] was used to determine the relative quantification of gene expression. The fold change in gene expression was normalized to the internal control, *OsUBQ1* (AK121590, Os03g13170), and converted to the fold change using the data from the wild-type. The experiment was performed with three biological replicates. The primers used in the experiment are listed in Table S1.

2.4. Determination of Lipid Peroxidation and Reactive Oxygen Species Levels

Malondialdehyde (MDA) levels in rice leaf samples were determined using the Lipid Peroxidation (MDA) Assay Kit (Sigma-Aldrich, St. Louis, MO, USA) according to the manufacturer's instructions. Ten milligrams of rice leaves were homogenized with MDA lysis buffer and centrifuged at 13,000× *g* for 10 min, and the residue was discarded. To 200 µL of the sample, 600 µL of thiobarbituric acid was added and incubated at 95 °C for 1 h, followed by cooling on ice for 10 min. The MDA content in each sample was quantified as absorbance at 532 nm using the equation derived from the MDA standard curve. To determine the H_2O_2 content, 1 g of rice leaf samples was homogenized in 3 mL of 100 mM sodium phosphate buffer (pH 6.8). The homogenate was filtered through four layers of cheesecloth and centrifuged at 14,000× *g* for 20 min at 4 °C. The H_2O_2 content of the supernatant was measured via the modified Bernt and Bergmeyer [56] method using peroxidase. For this, 0.5 mL of supernatant was mixed with 2.5 mL of peroxide solution containing 83 mM sodium phosphate (pH 7.0), 0.005% *w*/*v* o-dianisidine, and 40 mg of peroxidase and incubated at 30 °C for 10 min. The reaction was terminated by the addition of 0.5 mL of 1 N perchloric acid, and the sample was then centrifuged at 5000× *g* for 5 min. The amount of H_2O_2 in each sample was quantified as the absorbance at 436 nm based on the equation derived from the H_2O_2 standard curve.

2.5. Activity Assays of Antioxidant Enzyme

For the determination of superoxide dismutase (SOD, EC: 1.15.1.1), catalase (CAT, EC: 1.11.1.6), and peroxidase (POX, EC: 1.11.1.7) activities, total proteins were extracted from 1 g of rice leaves via homogenization in 100 mM potassium phosphate buffer (pH 7.8) containing 0.1 mM EDTA, 1% *w/v* polyvinyl-pyrrolidone, and 0.5% *v/v* Triton X-100 at 4 °C. For the assay of ascorbate peroxidase (APX, EC: 1.11.1.11) activity, 1 g of rice leaves was homogenized in 100 mM sodium phosphate buffer (pH 7.0) containing 5 mM ascorbate and 1 mM EDTA. The homogenate was filtered through four layers of cheesecloth and centrifuged at 14,000× *g* at 4 °C for 25 min. The total protein content was determined according to the Bradford method [57]. The activity of SOD was determined according to the method of Beyer and Fridovich [58]. The reaction mixture consisted of 50 mM potassium phosphate (pH 7.8), 9.9 mM methionine, 57 μM nitro blue tetrazolium (NBT), and 40 μL protein extract. The reaction was initiated via light illumination. One unit of SOD is defined as the amount of enzyme that causes 50% inhibition of SOD-inhibitable NBT reduction. CAT activity was determined from the rate of H_2O_2 decomposition (extinction coefficient 39.4 mM/L cm^{-1}), measured as the decrease in absorbance at 240 nm, according to the method of Aebi [59]. The reaction mixture contained 50 mM potassium phosphate buffer (pH 7.0) and 40 μL of the protein extract. The reaction was initiated by the addition of 10 mM H_2O_2. Peroxidase activity was determined by monitoring the formation of guaiacol dehydrogenation product (extinction coefficient 6.39 mM/L cm^{-1}) at 436 nm according to the Pütter method [60]. The reaction mixture contained 100 mM potassium phosphate (pH 7.0), 0.3 mM guaiacol, and 50 μL protein extract. The reaction was initiated by the addition of 0.1 mM H_2O_2. APX activity was determined by measuring the decrease in absorbance at 290 nm (extinction coefficient 2.8 mM cm^{-1}) according to the method of Chen and Asada [61]. The reaction mixture contained 50 mM potassium phosphate (pH 7.0), 0.5 mM ascorbate, 0.2 mM H_2O_2, and 10 μL protein extract.

2.6. Measurement of Chlorophyll, Carotenoid, Ascorbic Acid, and Proline Contents

Chlorophyll and carotenoid contents were determined using the method described by Lichtenthaler and Wellburn [62]. Freshly collected leaves (0.1 g) were homogenized with 80% (*v/v*) chilled acetone and centrifuged at 10,000× *g* at 4 °C for 10 min. After filtration, the optical density was measured using a UV-1800 spectrophotometer (Shimadzu, Kyoto, Japan) at 663, 645, and 440 nm. The pigment levels were calculated using the following equations:

$$\text{Total chlorophyll (mg g}^{-1}\text{)} = 20.2 \times \text{Abs}_{645} + 8.05 \times \text{Abs}_{663}$$

$$\text{Chlorophyll a (mg g}^{-1}\text{)} = 12.7 \times \text{Abs}_{663} - 2.69 \times \text{Abs}_{645}$$

$$\text{Chlorophyll b (mg g}^{-1}\text{)} = 22.9 \times \text{Abs}_{645} - 4.68 \times \text{Abs}_{663}$$

$$\text{Carotenoids (mg g}^{-1}\text{)} = 4.7 \times \text{Abs}_{440} - (1.38 \times \text{Abs}_{663} + 2.69 \times \text{Abs}_{645})$$

The ascorbic acid content of the plants was measured using the protocol of Mukherjee and Choudhuri [63]. A sample of 0.5 g of fresh leaves was homogenized with 10 mL of trichloroacetic acid and then filtered. One drop of 10% thiourea, 2 mL of 2% diphenylhydrazine, and 5 mL of 80% H_2SO_4 were added to 4 mL of the supernatant of the plant leaf extract in test tubes. Absorbance was measured at 530 nm using a UV-1800 spectrophotometer. The ascorbic acid content in each sample was quantified as the absorbance at 530 nm using the equation obtained from the ascorbic acid standard curve.

The proline content of the plant sample was determined using the method of Bates et al. [64]. Half a gram of fresh leaf sample was homogenized in 5 mL of 3% sulphosalicylic acid, filtered, and centrifuged at 10,000× *g* for 15 min. The collected 2 mL of supernatant was mixed with 2 mL of acid ninhydrin and 2 mL of glacial acetic acid. These test tubes were placed in a water bath at 80 °C for 1 h. The tubes were then kept in an ice bath to terminate the reaction. The developed chromophore was extracted with 4 mL of toluene. Using the equation from the proline standard curve, the proline content in each red-colored solution was quantified as the absorbance at 520 nm.

2.7. Measurement of Transient Chlorophyll a Fluorescence

Chlorophyll fluorescence was measured using a HANDY PEA fluorimeter (Hansatech, King's Lynn, London, UK) on the longest leaves at their apex, midsection, and base. Prior to all measurements, the leaves were dark-adapted for a period of 30 min. The light intensity was 3500 μmol photons m^{-2} s^{-1}, provided by an array of three high-intensity light-emitting diodes, focused on a spot with a diameter of 5 mm. The measurements were recorded for 1 s at a resolution of 12-bits. The fluorescence parameters Fv/Fm were obtained using HANDY PEA software (version 1.31).

2.8. Statistical Analysis

All data are presented as mean ± standard deviation, as indicated in each figure. Student's *t*-test (** $p < 0.01$) was used to compare each data point with the control to test for significant differences. In addition, one-way analysis of variance (ANOVA) followed by Duncan's test was used to statistically analyze differences between data points. Statistical significance was set at $p < 0.01$ using SPSS software ver. 22.0 (IBM, New York, NY, USA).

3. Results

3.1. Endogenous Caffeine in CPR Plants Alters OsCPKs Expression

In our previous study, transgenic rice plants producing caffeine were prepared by introducing three N-methyltransferase genes for caffeine synthesis into rice plants, and we investigated the role of endogenous caffeine in response to various biotic stresses [27]. In this study, we used the transgenic rice plants of the T_5 generation (the next generation of the plants that we used for the previous study), specifically the independent transgenic lines CPR #1 and CPR #2, and confirmed the expression of the introduced genes via RT-PCR (Figure 1A). Using HPLC, we detected the accumulation of caffeine with a content of approximately 0.545 μg g^{-1} fresh weight in CPR #1 and 0.65 μg g^{-1} fresh weight in CPR #2 at the four-leaf stage (Figure 1B), which was similar to the T_4 generation of CPR plants [27], indicating stable caffeine synthesis across generations. In CPR plants, the production of caffeine via a series of processes leads to cytosolic calcium accumulation, resulting in changes in the expression of calcium sensors, CPKs, which have various effects on calcium signaling [27]. We hypothesized that even though the synthesis of caffeine is relatively small at this four-leaf stage, it would have a significant effect on CPK expression. To test this, we analyzed the expression of 31 *OsCPKs* via qRT-PCR and found that *OsCPK10* (LOC_Os03g57450), *OsCPK12* (LOC_Os04g47300), *OsCPK21* (LOC_Os08g42750), *OsCPK25* (LOC_Os11g04170), *OsCPK26* (LOC_Os12g03970), *OsCPK30* (LOC_Os07g44710), and *OsCPK31* (AK110341) were more than 2-fold upregulated in CPR plants at the four-leaf stage compared with wild-type plants (Figure 1C). Other *OsCPKs* were either similarly expressed or downregulated in CPR plants compared to wild-type plants (Figure S1). Those *OsCPKs* are well known as positive regulators that enhance tolerance to abiotic stresses as well as to biotic stresses [65–70].

3.2. Endogenous Caffeine in CPR Plants Confers Water-Deficit Tolerance

The investigation of water-deficit tolerance was conducted using CPR plants (Figure 2A). Irrigation was withheld for 10 days on the four-leaf stage wild and CPR plants. To ensure that water-deficit stress was evenly distributed to both rice lines, we planted wild-type and CPR plants together in the same pot. Water-deficit stress treatment lasted 10 days, and soil moisture monitored for 3 days in each pot showed that each pot received the same amount of water-deficit stress (Figure 2B). In this period, leaves were withered in both wild-type and CPR plants, but it was more severe in the wild type (left side of the pot in the middle picture of Figure 2A). While the leaves of wild-type rice had turned yellow, the leaves of CPR had retained their green color to a significant extent even after withering. After 14 days of recovery with rewatering, the survival rate of CPR plants (CPR #1, 95%; CPR #2, 96.66%) was dramatically higher than that of wild-type plants (6.66%) (Figure 2C). During the 10-day period of water-deficit treatment, we conducted a comparative analysis of photosynthesis-related parameters between wild-type and CPR plants. The change in chlorophyll content was significantly different in wild-type and CPR plants. Chlorophyll content in wild-type plants distinguishably decreased after 6 days of withholding irradiation and was only about 1/6 after 10 days compared to CPR (Figure 2D). The Fv/Fm values, which represent the photochemical efficiency of photosystem (PS) II, exhibited a sharp decline in wild-type plants from the sixth day of the water-deficit stress treatment. In contrast, a slight decrease was observed in CPR plants (Figure 2E). These results suggest that caffeine in the CPR plants maintains a stable photosynthetic metabolism under water-deficit stress.

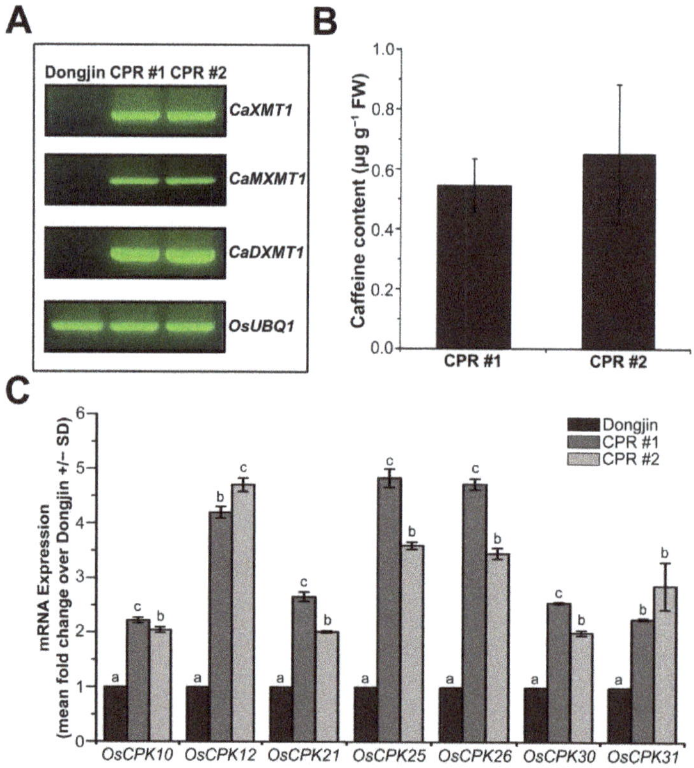

Figure 1. Caffeine synthesis and expression of *OsCPKs* in caffeine-producing transgenic rice (CPR, transgenic CPR line #1 and #2) and wild-type plants (Dongjin) at four-leaf stage under normal conditions. (**A**) Validation of the expression of the introduced *CaMXMT1*, *CaXMT1*, and *CaDXMT1* genes in CPR plants. Total RNA was extracted from leaves of wild-type and CPR plants at the four-leaf stage, and reverse transcription-polymerase chain reaction (RT-PCR) was performed using specific primers for each of the three introduced genes. (**B**) Validation of caffeine synthesis in leaves of CPR plants at the four-leaf stage. Caffeine was extracted from CPR plant leaves by methanol extraction and detected by high-performance liquid chromatography (HPLC). All experiments were repeated three times. Error bars represent the means ± SD (n = 3). (**C**) Expression of *OsCPKs* gene in Dongjin and CPR plants at the four-leaf stage under normal conditions. Total RNA was extracted from leaves of wild-type and CPR plants at the four-leaf stage, and quantitative RT-PCR (qRT-PCR) was performed using specific primers for each *OsCPK* gene. All experiments were repeated three times. Error bars represent the means ± SD (n = 3). Different letters above bars indicate statistically significant differences as determined by one-way analysis of variance (ANOVA), $p < 0.01$.

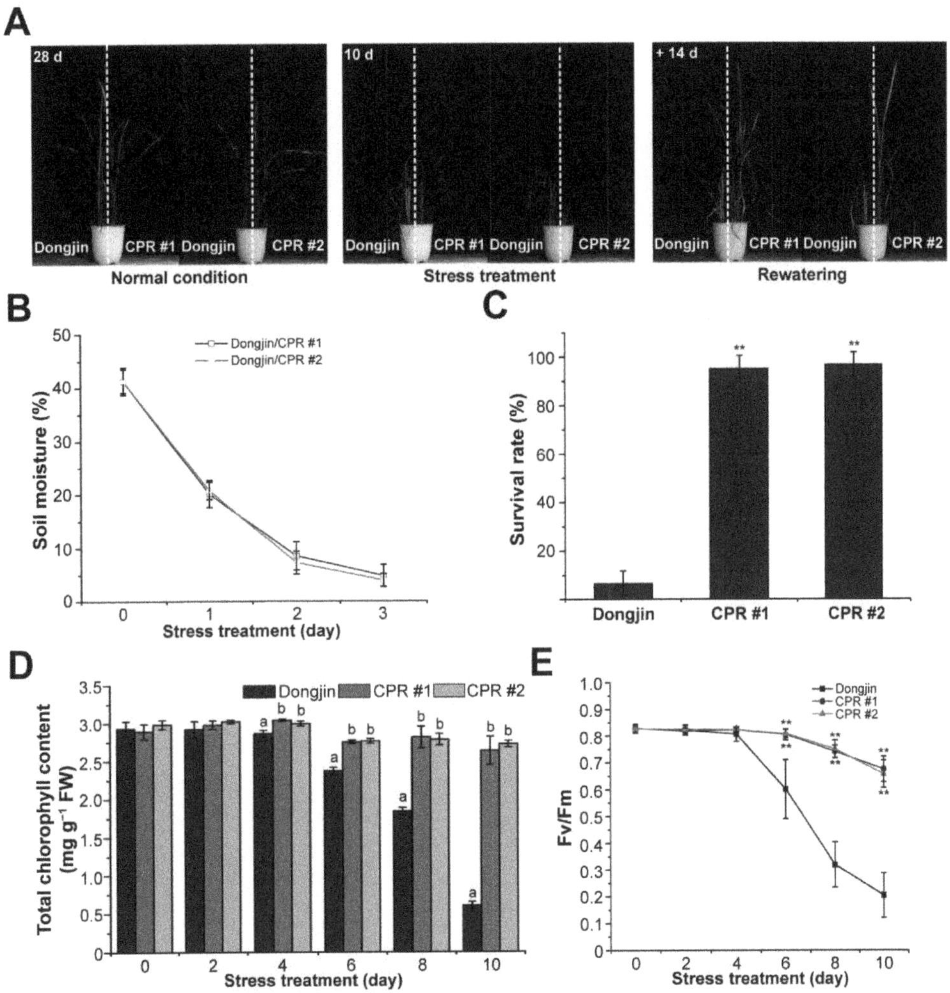

Figure 2. Caffeine production in CPR plants confers water-deficit stress tolerance. (**A**) Water-deficit stress tolerance phenotypes in CPR plants. Tolerance to water-deficit stress was assessed by growing CPR plants under normal conditions for 4 weeks, then withholding water for 10 days and rewatering for 14 days. (**B**) Measurement of soil moisture levels. All experiments were six biological replicates. Error bars represent the means ± SD (n = 6). (**C**) Survival of wild-type and CPR plants was measured 14 days after rewatering. All experiments were six biological replicates. Error bars represent the means ± SD (n = 6). Asterisks indicate statistically significant differences compared with wild-type (Dongjin) plants according to a Student's t-test (** $p < 0.01$). (**D**) Accumulation of chlorophyll content in the wild-type and CPR plants during water-deficit stress treatment. All experiments were repeated three times with three biological replicates. Error bars represent the means ± SD (n = 9). Different letters above bars indicate statistically significant differences as determined by one-way analysis of variance (ANOVA), $p < 0.01$. (**E**) Chlorophyll fluorescence (Fv/Fm) of leaves during wild-type and CPR plants during water-deficit stress treatment. All experiments were repeated three times with four biological replicates. Error bars represent the means ± SD (n = 12), ** $p < 0.01$ (Student's t-test).

3.3. Endogenous Caffeine in CPR Plants Alters OsCPKs Expression

To further investigate the role of endogenous caffeine in water-deficit tolerance, we examined the expression of calcium signaling-related *OsCPKs* (Figure 3A). In the absence of water deficit, it was observed that seven *OsCPKs* exhibited upregulation of more than 2-fold in CPR plants compared to wild-type plants (Figure 1C). Under water-deficit stress, 16 *OsCPKs (OsCPK1*, LOC_Os01g43410; OsCPK2, LOC_Os01g59360; OsCPK3, LOC_Os01g61590; OsCPK4, LOC_Os02g03410; OsCPK5, LOC_Os02g46090, OsCPK6, LOC_Os02g58520; OsCPK9, LOC_Os03g48270; OsCPK11, LOC_Os03g57510; OsCPK14, LOC_Os05g41270; OsCPK16, LOC_Os05g39090; OsCPK18, LOC_Os07g22710; OsCPK22, LOC_Os09g33910; and OsCPK24, LOC_Os11g07040), including three genes (*OsCPK10*, *OsCPK12*, and *OsCPk25*) that have already increased their expression in CPR without water-deficit treatment, have increased more than 2-fold in CPR plants compared to wild type (Figure 3A). The expression of the remaining 15 *OsCPK* genes was either not different from the wild type in expression under water-deficit stress, downregulated, or upregulated, but the level of gene expression was not significant (Figure S2). We further examined the expression of 20 genes known to be involved in responses to water-deficit and osmotic stress (Figreus 3B and S3). In total, 8 (*OsbZIP72*, LOC_Os09g28310; *OsLEA25*, LOC_Os11g26570; *OsNHX1*, LOC_Os07g47100; *OsRab16d*, LOC_Os11g26780; *OsDREB2B*, LOC_Os05g27930; *OsNAC45*, LOC_Os11g03370; *OsP5CS*, LOC_Os05g38150; *OsRSUS1*, LOC_Os06g09450) of the 20 genes were expressed more than 2-fold in CPR plants compared to wild-type plants, while the remaining 12 genes were not significantly different from wild-type in expression under water-deficit stress (Figure S3). These eight transcription factor genes have been well studied and confirmed to be involved in tolerance to water-deficit and osmotic stress [71–78]. These results suggest that endogenous caffeine in CPR plants induces the expression of genes encoding CPKs and transcription factors that cause (or may trigger) resistance to water-deficit stress.

3.4. Endogenous Caffeine Suppresses Water-Deficit Stress-induced Production of H_2O_2 and MDA in Rice Plants

In plants, including rice, H_2O_2 and malondialdehyde (MDA), which is a substance produced by membrane lipids in response to reactive oxygen species (ROS) such as H_2O_2, are suggested to be increased by drought stress [79–81]. The alterations in H_2O_2 and MDA amounts were investigated in CPR subjected to water-deficit stress (Figure 4). Under normal growth conditions, the H_2O_2 levels of CPR plants and wild-type plants were alike, but the changes in H_2O_2 levels caused by water-deficit stress in the two plants were different (Figure 4A). During the water-deficit treatment period, both Dongjin and CPR plants exhibited a continuous increase in H_2O_2 levels. However, the extent of accumulation varied significantly. After four days of water-deficit treatment, H_2O_2 increased by 84.48% in Dongjin but only by 8.9% (CPR#1) and 12.87% (CPR#2) in CPR plants. Notably, ten days after the initiation of the water-deficit treatment, when the H_2O_2 content reached its peak, it had risen by 184.77% in Dongjin and by 134.37% (CPR#1) and 115.58% (CPR#2) in CPR plants (Figure 4A). The MDA levels in CPR plants were somewhat lower than in wild-type rice under normal conditions. It was 85.06% in CPR #1 and 85.52% in CPR #2 compared to Dongjin. The MDA levels in both Dongjin and CPR plants increased throughout the water-deficit stress treatment. Notably, in Dongjin, the MDA level reached its peak six days after the initiation of the water-deficit stress treatment (77.77% in CPR#1 and 73.6% in CPR#2 compared to Dongjin), after which it gradually decreased. In contrast, the CPR plants exhibited a relatively mild increase in MDA levels during the water-deficit treatment period and then maintained homeostasis (Figure 4B). These results suggest that endogenous caffeine in CPR plants contributes to reducing ROS production and lipid peroxidation caused by water-deficit stress.

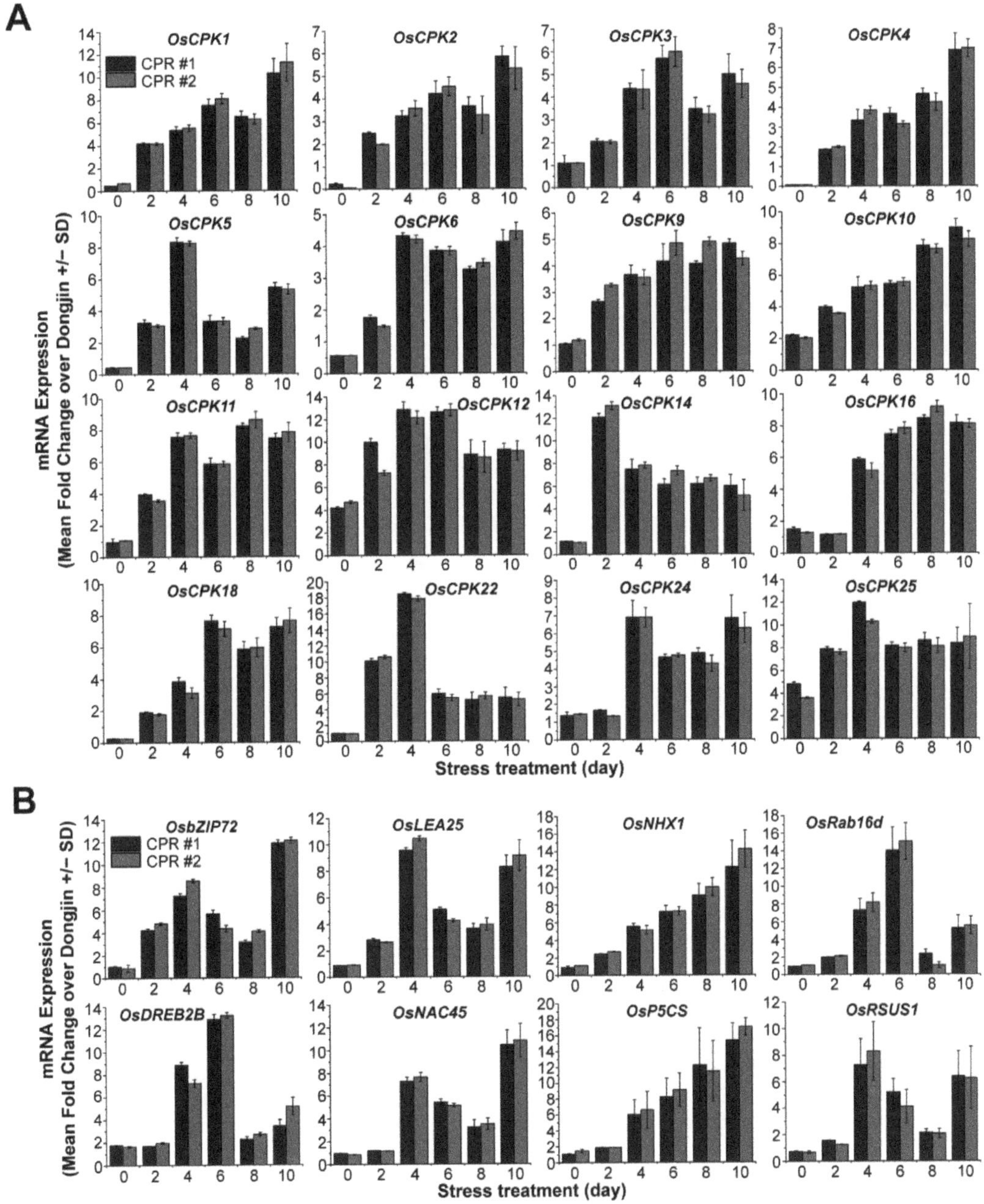

Figure 3. Gene expression analysis of *OsCPKs* and transcription factors under water-deficit stress treatment in wild-type and CPR plants. (**A**) Expression of *OsCPKs* gene in Dongjin and CPR plants under water-deficit stress treatment. (**B**) Expression of transcription factor genes involved in water-deficit tolerance under water deficit stress treatment. Total RNA was extracted from leaves of wild-type and CPR plants collected at 0, 1, 2, and 3 d after water-deficit stress treatment. The qRT-PCR was performed using specific primers for each gene. All experiments were repeated three times. Error bars represent the means ± SD ($n = 3$).

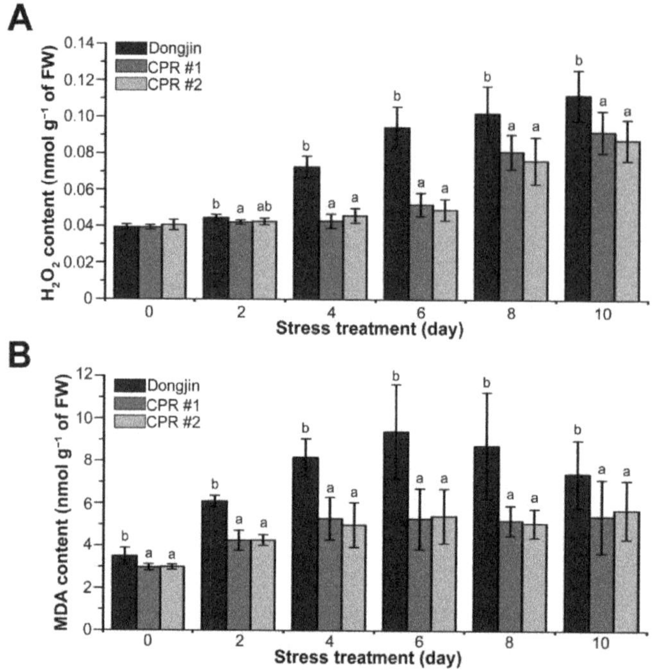

Figure 4. Analysis of changes in hydrogen peroxide and malondialdehyde levels during water-deficit stress treatment in wild-type and CPR plants. Quantification of H_2O_2 (**A**) and MDA levels (**B**) in wild-type and CPR plants under water-deficit stress. H_2O_2 and MDA were extracted from leaves of wild-type and CPR plants collected at 0, 2, 4, 6, 8, and 10 d after water-deficit stress treatment and quantified via spectrophotometry. All experiments were repeated three times with three biological replicates. Error bars represent the means ± SD (n = 9). Different letters above bars indicate statistically significant differences as determined by one-way analysis of variance (ANOVA), $p < 0.05$.

3.5. Endogenous Caffeine Enhances Activity of Antioxidant Enzyme in CPR Plants under Water-Deficit Stress

Based on the results shown in Figures 2E and 4, we hypothesized that endogenous caffeine enhances the function of antioxidant enzymes. To validate this hypothesis, we measured the activity of antioxidant enzymes w/wo water-deficit stress. CPR plants had higher activities of SOD, CAT, POX, and APX compared to wild-type plants without water-deficit stress (Figure 5). In the case of POX, enzyme activity in CPR was nearly three times higher than that of Dongjin. During water deficit stress, SOD activity in Dongjin decreased to 11.56% after 4 days of water-deficit treatment, followed by a subsequent upward trend. In contrast, CPR plants showed a continuous increase in SOD activity throughout the entire period of water-deficit stress (Figure 5A). In Dongjin, CAT activity increased during 4 days of water-deficit stress treatment, decreased after 6 days, and then increased more. In contrast, CPR plants exhibited a continuous and moderate increase in CAT activity. The most substantial increase in CAT activity for Dongjin occurred at 4 days post-stress, rising by 16.46% compared to 0 days. In comparison, CPR plants showed increases of 16.97% (CPR#1) and 17.52% (CPR#2), respectively, during the same period (Figure 5B). The POX activity exhibited a gradual increase throughout the water-deficit stress period and showed a notable steepening ten days after the stress treatment. Notably, CPR plants displayed a 2.35- (CPR#1) and 2.30-fold (CPR#2) difference in POX activity compared to the wild-type plants, respectively (Figure 5C). Dongjin plants exhibited the highest APX activity four days after the water-deficit stress treatment, followed by a gradual decline. In contrast, CPR plants maintained high levels of APX activity until eight days after the water-deficit stress treatment. Notably, four days after the water-deficit stress treatment, when APX activity was at its peak, CPR plants displayed 32.91% (CPR#1) and 36.76% (CPR#2) higher activity compared to the wild type, respectively (Figure 5D). Overall, these results suggest that endogenous caffeine in CPR plants helps to mitigate the damage caused by high ROS due to water-deficit stress by increasing antioxidant enzyme activity.

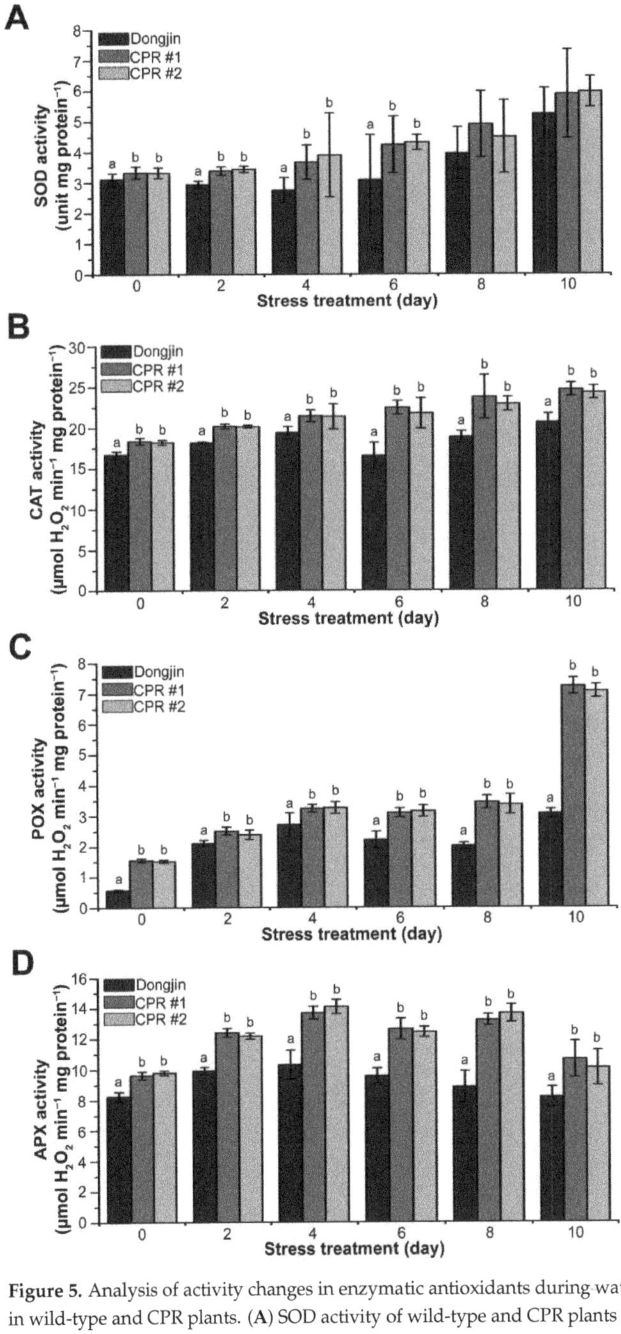

Figure 5. Analysis of activity changes in enzymatic antioxidants during water-deficit stress treatment in wild-type and CPR plants. (**A**) SOD activity of wild-type and CPR plants under water-deficit stress.

(**B**) CAT activity of wild-type and CPR plants under water-deficit stress. (**C**) POX activity of wild-type and CPR plants under water-deficit stress. (**D**) APX activity of wild-type and CPR plants under water-deficit stress. Antioxidant enzymes were extracted from leaves of wild-type and CPR plants collected at 0, 2, 4, 6, 8, and 10 d after water-deficit stress treatment. The activities of the extracted antioxidant enzymes were measured by colorimetric spectrophotometry. All experiments were repeated three times with three biological replicates. Error bars represent the means ± SD ($n = 9$). Different letters above bars indicate statistically significant differences as determined via one-way analysis of variance (ANOVA), $p < 0.05$.

3.6. Endogenous Caffeine Increases Non-Enzymatic Antioxidant Content of CPR Plants under Water-Deficit Stress

To further explore the role of endogenous caffeine in mitigating the damage caused by water-deficit stress, we analyzed changes in the levels of non-enzymatic antioxidants and osmoprotectants. Under normal growth conditions, CPR plants accumulated 1.8–2.7% less carotenoids than wild-type plants. Under water-deficit stress, carotenoid content declined dramatically in wild-type plants, whereas the content remained stable in CPR plants, with the difference in content between wild-type and CPR plants being 8.61 to 8.77-fold (Figure 6A). For ascorbic acid content, CPR plants accumulated 22–26% more than the wild-type plants under normal growth conditions. During the water-deficit stress treatment, there was a tendency for ascorbic acid accumulation to increase, particularly up to 4 days after the initiation of the stress treatment. CPR plants, in particular, accumulated 28.31% (CPR#1) and 28.07% (CPR#2) more ascorbic acid than the wild-type plants, respectively. Beginning four days after the water-deficit stress treatment, both the wild-type and CPR plants exhibited a trend of decreased ascorbic acid accumulation (Figure 6B). Under normal conditions, there was minimal difference in proline content between wild-type and CPR plants. The proline content in both wild-type and CPR plants exhibited an increase under the water-deficit stress treatment. However, during the water-deficit stress treatment, measurements taken at 2-day intervals revealed higher proline content in CPR plants, increasing by 4.28% (CPR#1), and 7.74% (CPR#2) two days after the stress treatment, 9.98% (CPR#1) and 12.16% (CPR#2) four days after the stress treatment, 10.21% (CPR#1) and 13.88% (CPR#2) six days after the stress treatment, 14.23% (CPR#1) and 20.81% (CPR#2) eight days after the stress treatment, and 21.11% (CPR#1) and 29.51% (CPR#2) in CPR plants ten days after the stress treatment, respectively (Figure 6C). These results indicate that endogenous caffeine plays a role in modulating the levels of non-enzymatic antioxidants and osmoprotectants, which are responsible for enhancing the tolerance of CPR plants to water-deficit stress.

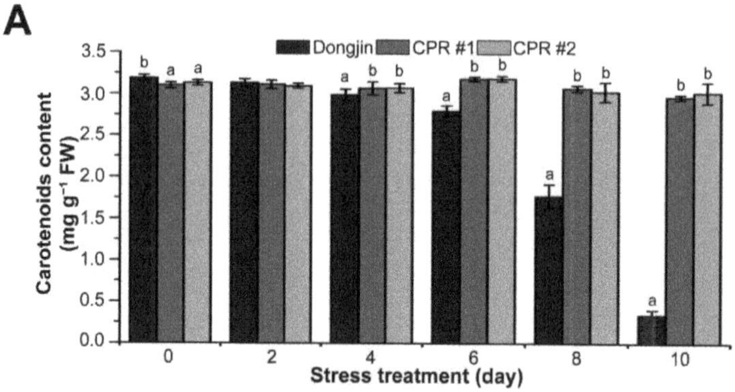

Figure 6. *Cont.*

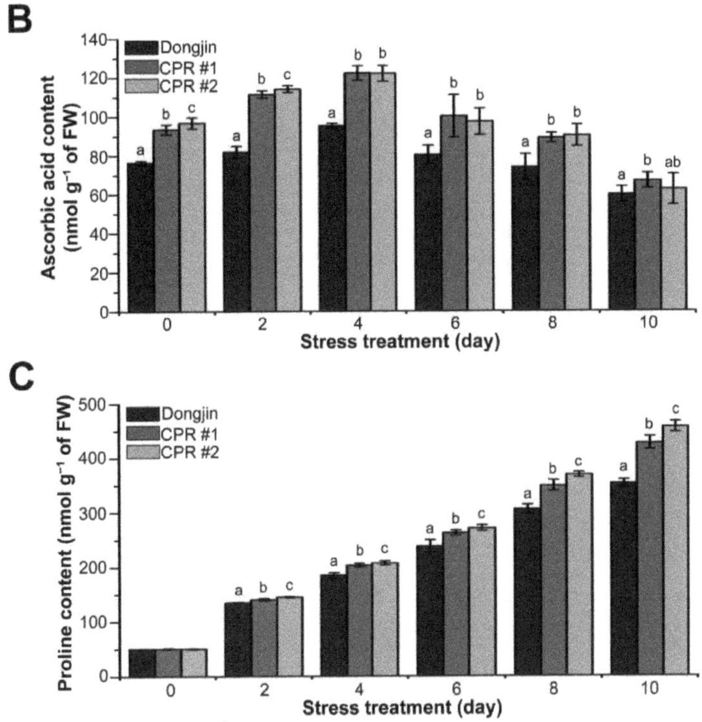

Figure 6. Analysis of changes in non-enzymatic antioxidants (carotenoids and ascorbic acid) and osmoprotectant (proline) contents during water-deficit stress treatment in wild-type and CPR plants. Quantification of carotenoids (**A**), ascorbic acid (**B**), and proline levels (**C**) in wild-type and CPR plants under water-deficit stress. The carotenoids, ascorbic acid, and proline were extracted from leaves of wild-type and CPR plants collected at 0, 2, 4, 6, 8, and 10 d after water-deficit stress treatment and quantified via spectrophotometry. All experiments were repeated three times with three biological replicates. Error bars represent the means ± SD ($n = 9$). Different letters above bars indicate statistically significant differences as determined via one-way analysis of variance (ANOVA), $p < 0.05$.

4. Discussion

Plants must constantly adapt to a variety of challenges, including biotic and abiotic stresses. The outcome of these interactions with stresses is crucial for determining a plant's survival. Especially in cereal plants, rapid and adequate responses to mitigate biotic and/or abiotic stresses are very important for crop productivity. It is now estimated that nearly 800 million people face hunger [82]; minimizing the loss in crop products due to stresses is, therefore, one of the effective ways to solve this food problem of the world. In order to achieve this objective, a lot of effort is dedicated to the identification of diverse secondary metabolites originating from plants as well as the investigation of their function in relation to stress resistance. Several studies, including our previous report, have shown that caffeine, a well-known secondary metabolite, confers biotic stress resistance directly or indirectly [24–27,83]. Caffeine acts directly as an antimicrobial and repellent and indirectly as a triggering agent for resistance signaling. Caffeine increases salicylic acid levels via calcium signaling, making plants more resistant to variable biotic stresses [27]. However, the knowledge of the role of caffeine in abiotic stress tolerance is very limited. Recently, exogenous caffeine application to spinach has been shown to improve resistance to cadmium toxicity [84]. We now report that this study provides evidence that caffeine also has a positive effect on water-deficit tolerance in rice. Wild-type rice, Dongjin, exhibited a very poor survival rate due to the disorders of physiological and biochemical metabolism caused by the water-deficit stress (Figures 2–6). On the other hand, CPR under water-deficit stress can express more water-deficit tolerance-related genes such as particular *CPKs* and transcription factors, has higher activities of antioxidant enzymes, and has increased non-enzymatic antioxidant accumulation (Figures 3, 5 and 6). These alterations are able to alleviate water-deficit stress-induced damage in rice plants, as shown in our working model (Figure 7). All the changes shown in this present study are most likely due to Ca^{2+} signaling triggered by caffeine. Although we did not repeat the investigation of the change in intracellular Ca^{2+} concentration in this study, the CPR has much a higher Ca^{2+} concentration in cytosol than Dongjin [27].

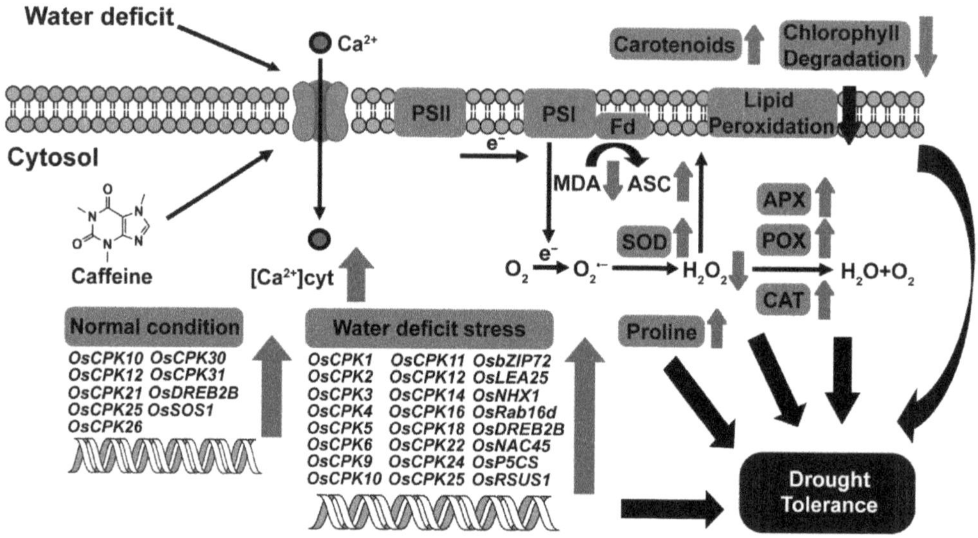

Figure 7. A model of endogenous caffeine-mediated tolerance regulation that induces water-deficit stress tolerance in CPR plants. Endogenous caffeine produced by CPR plants triggers an intracellular influx of calcium ions, which under normal growth conditions, upregulates the expression of OsCPK genes (*OsCPK10, OsCPK12, OsCPK21, OsCPK25, OsCPK26, OsCPK30, OsCPK31*) and water-deficit tolerance-related transcription factors (*OsDREB2B, OsSOS1*). Under water-deficit stress, these molecular level changes result in upregulation of OsCPK genes (*OsCPK1, OsCPK2, OsCPK3, OsCPK4, OsCPK5, OsCPK6, OsCPK9, OsCPK10, OsCPK11, OsCPK12, OsCPK14, OsCPK16, OsCPK18, OsCPK22, OsCPK24, OsCPK25*) and water-deficit tolerance-related transcription factors (*OsbZIP72, OsLEA25, OsNHX1, OsRab16d, OsDREB2B, OsNAC45, OsP5CS, OsRSUS1*), leading to biochemical level changes such as reduced accumulation of ROS and MDA, increased activity of antioxidant enzymes (SOD, CAT, POX, APX), and increased accumulation of non-enzymatic antioxidants (carotenoids, ASA, proline). These biochemical changes then lead to physiological changes such as inhibition of lipid peroxidation, reduction of chlorophyll degradation, and protection of photosynthetic machinery, ultimately resulting in acquired tolerance to water-deficit stress. Abbreviations: APX, ascorbate peroxidase; CAT, catalase; ASC, ascorbic acid; Fd, ferredoxin; MDA, malondialdehyde; POX, peroxidase; PSI/II, photosystem I/II; SOD, superoxide dismutase.

Previous studies have shown that the amount of caffeine synthesized in CPR plants depends on the amount of its precursor, xanthosine, and that HPLC-detectable caffeine is synthesized in CPR plants from the two-leaf stage onward [27]. Caffeine, as a type of methylxanthine, acts as a non-selective inhibitor of phosphodiesterase [85]. The enzyme catalyzes a hydrolysis reaction from 3′, 5′-cAMP/3′, 5′-cGMP to AMP/GMP [86], thereby causing an abnormal accumulation of cAMP/cGMP in CPR plants [27]. High levels of cAMP/cGMP, which are involved in signal transduction, bind to cyclic nucleotide-gated channels in the cell membrane and activate the channels [87], leading to Ca^{2+} accumulation in the cytosol [27]. The high intracellular concentration of Ca^{2+} is recognized by calcium-binding proteins that act as calcium sensors, initiating an immediate calcium-mediated signal transduction in plants [40]. Ca^{2+} signaling has been known to play a broad and diverse role in plants, including the growth, development, stress perception, and regulation of cell responses [40,88]. In particular, a specific CPK plays a critical role in the Ca^{2+} signaling pathway that mediates responses to biotic and abiotic stresses by regulating important processes such as ROS generation and hypersensitive reactions [40,89]. We analyzed the transcript levels of 31 *OsCPK* genes to identify alterations in Ca^{2+} signaling induced by endogenous caffeine at the four-leaf stage. We found that the expression levels of seven *OsCPK* genes were significantly increased in CPR plants under normal growth conditions (Figure 1). Although two *OsCPKs* (*OsCPK30* and *OsCPK31*) have not been mentioned in relation to abiotic stress tolerance, the five CPKs are known to be involved in abiotic stress resistance. The overexpression of *OsCPK10* resulted in decreased accumulation of H_2O_2 and MDA levels, reducing lipid peroxidation and ultimately demonstrating resistance to drought stress by phosphorylating CAT-A [66]. The overexpression of *OsCPK12* increased the expression levels of *OsAPX2* and *OsAPX8*, which encode APX, resulting in increased ROS detoxification, thus demonstrating resistance to salt stress [67]. The overexpression of *OsCPK21* increased the expression levels of stress-related transcription factors such as *OsP5CS*, *OsLEA3*, *OsNAC6*, *OsbZIP23*, *OsNHX1*, and *OsSOS1*, demonstrating resistance to salt stress [68]. The detailed function of water-deficit tolerance remains unclear, *OsCPK26* is a gene that is upregulated in drought tolerance isogenic lines under water-deficient conditions [90]. Under water-deficit conditions, the expression of many CPKs in CPR plants was surprisingly enhanced. The expression of three (*OsCPK10*, *OsCPK12*, and *OsCPK25*) of the seven *OsCPKs* that were already increased in CPR without water-deficit treatment was further enhanced with water-deficit stress. However, the expression of the rest of the four *OsCPKs* (*OsCPK21*, *OsCPK26*, *OsCPK30*, and *OsCPK31*) declined with water-deficit treatment and did not show a big difference with those of the wild type at three days after water-deficit treatment. In addition to the 3 *OsCPKs*, the expression of the 13 *OsCPKs* (*OsCPK1*, *OsCPK2*, *OsCPK3*, *OsCPK4*, *OsCPK5*, *OsCPK6*, *OsCPK9*, *OsCPK11*, *OsCPK14*, *OsCPK16*, *OsCPK18*, *OsCPK22*, and *OsCPK24*) markedly enhanced compared to wild-type rice plants with water-deficit treatment (Figure 2A). Some of these genes have been reported to play a role in response to abiotic stress, especially drought, but five of them (*OsCPK3*, *OsCPK5*, *OsCPK16*, *OsCPK18*, and *OsCPK24*) have not yet. The *OsCPK1* gene is known to be activated by sucrose and involved in drought tolerance mechanism [91]. CPK1 negatively regulates GA biosynthesis and activates 14-3-3 protein expression to prevent damage by drought stress in the germination stage. To date, there have been no reports investigating the role of CPK2 in rice and its connection to drought stress. However, it is worth noting that barley CPK2 has been found to function as a negative regulator of the drought stress response [92]. *OsCPK4* expression is induced by salt, drought, and ABA. The overexpression of *OsCPK4* confers significant tolerance to salt and drought stress via the protection of cellular membranes from stress-induced oxidative damage [93]. OsCPK6 is a gene that is upregulated under drought stress [70]. OsCPK9 plays a positive role in drought stress tolerance and spikelet fertility. OsCPK9 enhances stomatal closure and improves osmotic adjustment ability in plants [94]. The investigation of wild-type and heterozygous mutants of *OsCPK11* suggested that *OsCPK11* is likely involved in salt, cold, and drought stress signaling [95]. *OsCPK14* interacts with *OsDi19-4*, one of the drought-induced 19 transcription factor family, with an overexpression that enhances drought tolerance in *Arabidopsis* [96]. OsCPK14 is, therefore, suggested as an indirect regulator of drought stress. Using knockout and overexpression mutants, the role of OsCPK24 as a positive regulator of cold stress has been elucidated [97]. In addition, the expression of the eight transcription factors related to abiotic stress resistance was also altered in CPR plants (Figure 3B). One member of the basic leucine zipper (bZIP) family, *OsbZIP72*, is hypersensitive to ABA and plays a positive role in drought stress resistance by increasing the expression levels of late embryogenesis abundant (LEA) genes, which are ABA-responsive genes that enhance drought tolerance [72]. The exogenous application of melatonin confers salt stress tolerance in rice plants, and under these conditions, *OsLEA25* is upregulated as a responsive gene [71]. The overexpression of rice Na^+/H^+ exchanger 1 (*OsNHX1*) enhances resistance to both drought and salt stress [73,74]. *OsRab16d* is associated with ABA signaling pathways and increases stress tolerance in response to drought and salt stress [98]. The overexpression of rice dehydration-responsive element 2B (*OsDREB2B*) confers strong drought tolerance [76], and the overexpression of *OsDREB2B* in Arabidopsis increases resistance to drought and heat-shock stress [99]. The overexpression of rice *OsNAC45* mediates POX activity and plays an important role in drought and salt stress tolerance by responding to ABA [75,100]. The *OsPC5S* gene contributes to drought stress tolerance by enhancing proline accumulation and the activity of antioxidant enzymes [78]. The expression of rice sucrose synthase1 (*OsRSUS1*) is upregulated as a responsive gene to enhance drought stress tolerance [101,102]. Although we are not able to explain how water-deficit signals mediated and distributed by particular CPKs are delivered to these transcription factors in the signaling pathways of rice, it is clear that caffeine synthesized in CPR plants triggers significant changes that can cope with abiotic stresses including water deficit.

Drought stress accelerates the aging process of plant leaves, causing a rapid breakdown of chlorophyll and a reduction in photosynthesis that significantly impacts the lifespan of plants. This process leads to an excessive accumulation of ROS within the plant, which can cause oxidative damage to cell membranes, essential enzymes, proteins, nucleic acids, and more [103–107]. Like the leaves of the Dongjin, the leaves of the CPR plants wilted from the water-deficit treatment but remained greener than the wild-type plants and recovered nearly 100% with watering (Figure 2). This is because CPR plants excellently retain chlorophyll content, photosystem II photochemical efficiency (Figure 2), H_2O_2, and MDA content under water-deficit stress (Figure 4), whereas wild types do not. These caffeine-induced changes are attributed to protecting the CPR plants from oxidative damage caused by water-deficit stress.

Plants exhibit homeostasis by effectively regulating the amount of reactive oxygen species (ROS) to maintain optimal levels. However, abiotic stress disrupts regulatory mechanisms, leading to excessive ROS accumulation and oxidative stress [108]. Meanwhile, plants have well-developed antioxidant defense systems, including enzymatic and non-enzymatic antioxidants [108,109]. According to a recent report, the application of exogenous caffeine enhances the capacity of the antioxidant defense system. Exogenous caffeine reduces cadmium stress-induced ROS accumulation by increasing the enzymatic activities of SOD, CAT, and POX and increasing ascorbic acid content. These biochemical changes induced by caffeine conferred resistance to cadmium in plants [84]. In CPR plants, endogenous caffeine activates the antioxidant defense system. Under no water-deficit stress, all four enzymes (SOD, CAT, POX, and APX) scavenging ROS have higher activity than Dongjin, and the activity significantly increased by water-deficit treatment, except SOD (Figure 5). While SOD activity persisted in CPR plants, it diminished in wild-type rice under water-deficit stress (Figure 5A). These results suggest that SOD functions normally under water-deficit stress in CPR but not in wild-type plants. Non-enzymatic antioxidants in CPR plants, such as carotenoids, ascorbic acid, and proline, also showed significant differences in content compared to wild-type rice (Figure 6). Carotenoids are pigments that give fruits and vegetables their color. They have antioxidant properties that can protect cells from free radical damage by donating electrons to free radicals or scavenging singlet oxygen, a highly reactive form of oxygen [110]. The antioxidant properties of ascorbic acid are attributed to its ability to donate hydrogen atoms to lipid radicals, neutralize singlet oxygen, and eliminate molecular oxygen [111]. Proline accumulates in plant cells under stress conditions, allowing the maintenance of plant macromolecules and structure from stress damage via ROS detoxification. Additionally, it acts as an effective osmo-protectant that maintains cell turgor pressure balance, increasing plant resistance to osmotic stress [81,104,112]. Proline accumulation and enhanced drought stress tolerance are closely associated with the *OsP5CS* gene product [113]. *OsP5CS* encodes an enzyme called Δ1-pyrroline-5-carboxylate synthetase, which is involved in the biosynthesis of proline, an amino acid that helps plants cope with drought and salt stress [114]. In CPR plants, the expression of *OsP3CS* also increased with water-deficit treatment (Figure 3). Taken together, caffeine uses both enzymatic and non-enzymatic antioxidants present in plants to strengthen antioxidant defense mechanisms and lessen the oxidative damage water-deficit stress causes to rice.

We previously reported that caffeine as a defense inducer confers multi-resistance against variable biotic stresses, including bacterial, fungal, and insect pathogens, in rice without fitness cost [27]. Caffeine directly functions as an antibacterial and insect repellent agent and as a signal molecule to trigger Ca^{2+}-mediated defense responses. In the present study, we report the novel discovery of the additional properties of caffeine as an enhancer conferring tolerance to abiotic stress in rice plants. The rice plants producing caffeine exhibited significant resistance to water-deficit stress and demonstrated a remarkable survival rate upon re-irrigation following a severe water-deficit treatment. At the molecular level, as shown in the working model (Figure 7), there were notable disparities in the expression patterns of *OsCPKs* and some genes implicated in the plant abiotic stress response between CPR and wild-type rice. CPR also robustly maintains photosynthetic capacity under water-deficit stress. Furthermore, the activity of ROS-detoxifying enzymes (SOD, CAT, POX, and APX) and the production of non-enzymatic antioxidants (carotenoid, ascorbic acid, and proline) are enhanced in CPR, resulting in tolerance to water deficit-induced oxidative damage. Although this study could not provide evidence of how the Ca^{2+} signals decoded via CPKs are delivered to the response molecules, these biochemical changes induced by endogenous caffeine are sufficient to obtain strong resistance in rice plants. In this study, following on from our previous report [27], we suggest again how versatile and effective caffeine is as a priming agent and confidence its application in agriculture will contribute to exceptional rice productivity.

5. Conclusions

In conclusion, the endogenous production of caffeine in CPR plants resulted in a stress-ready state, enhancing their tolerance to water-deficit stress, a significant threat to growth and survival. The stable production of caffeine in transgenic CPR plants over generations led to the upregulation of seven OsCPK genes (*OsCPK10, OsCPK12, OsCPK21, OsCPK25, OsCPK26, OsCPK30, OsCPK31*) and two stress tolerance-related transcription factor genes (*OsDREB2B, OsSOS1*) even under normal growth conditions. These molecular changes led to biochemical changes, such as decreased MDA accumulation, increased antioxidant enzyme activity (SOD, CAT, POX, APX), and accumulation of the non-enzymatic antioxidant ascorbic acid, equipping CPR plants with defense against water-deficit stress. Under water-deficit stress, rice suffers from oxidative stress that accelerates chlorophyll degradation and reduces photosynthesis due to increased ROS and MDA accumulation. However, endogenous caffeine produced in CPR plants upregulated the expression of 16 OsCPK genes (*OsCPK1, OsCPK2, OsCPK3, OsCPK4, OsCPK5, OsCPK6, OsCPK9, OsCPK10, OsCPK11, OsCPK12, OsCPK14, OsCPK16, OsCPK18, OsCPK22, OsCPK24, OsCPK25*) and 8 stress tolerance-related transcription factor genes (*OsbZIP72, OsLEA25, OsNHX1, OsRab16d, OsDREB2B, OsNAC45, OsP5CS, OsRSUS1*), leading to increased activities of antioxidant enzymes (SOD, CAT, POX, APX) and reduced ROS and MDA accumulation via non-enzymatic antioxidant (carotenoids, ascorbic acid, proline) accumulation. These molecular and biochemical changes resulted in significant tolerance to water-deficit stress, ultimately inhibiting chlorophyll degradation and protecting photosynthetic machinery under water-deficit stress. Our results provide insights into metabolic engineering approaches using plant secondary metabolites for stress tolerance and offer a potential avenue for the discovery of novel molecular breeding sources.

Supplementary Materials: The following supporting information can be downloaded at: https://www.mdpi.com/article/10.3390/antiox12111984/s1, Table S1: Primers used for RT-PCR and qRT-PCR analysis. Figure S1: Gene expression analysis of *OsCPKs* in wild-type and CPR plants at the four-leaf stage under normal conditions. Figure S2: Gene expression analysis of *OsCPKs* under water-deficit stress treatment in wild-type and CPR plants. Figure S3: Gene expression analysis of transcription factors under water-deficit stress treatment in wild-type and CPR plants.

Author Contributions: Conceptualization, J.-C.P. and S.-W.L.; methodology, J.-C.P., Y.Y., Y.-H.Y. and D.Y.L.; validation, J.-C.P. and Y.Y.; formal analysis, J.-C.P., Y.Y. and Y.-H.Y.; writing—original draft preparation, J.-C.P. and Y.Y.; writing—review and editing, Y.Y., K.-H.J. and S.-W.L.; supervision, S.-W.L. and J.-C.P. All authors have read and agreed to the published version of the manuscript.

Funding: This work was supported by the Ministry of Education of the Republic of Korea and the National Research Foundation of Korea (NRF-2023R1A2C1005706).

Institutional Review Board Statement: Not applicable.

Informed Consent Statement: Not applicable.

Data Availability Statement: The data are contained within this article.

Acknowledgments: We thank Kyung Hee University for providing the rice paddy fields and greenhouse.

Conflicts of Interest: The authors declare no conflict of interest.

References

1. Etesami, H. Can interaction between silicon and plant growth promoting rhizobacteria benefit in alleviating abiotic and biotic stresses in crop plants? *Agric. Ecosyst. Environ.* **2018**, *253*, 98–112. [CrossRef]
2. Grillakis, M.G. Increase in severe and extreme soil moisture droughts for Europe under climate change. *Sci. Total Environ.* **2019**, *660*, 1245–1255. [CrossRef] [PubMed]
3. Lesk, C.; Rowhani, P.; Ramankutty, N. Influence of extreme weather disasters on global crop production. *Nature* **2016**, *529*, 84–87. [CrossRef] [PubMed]
4. Helman, D.; Bonfil, D.J. Six decades of warming and drought in the world's top wheat-producing countries offset the benefits of rising CO_2 to yield. *Sci. Rep.* **2022**, *12*, 7921. [CrossRef] [PubMed]
5. Mickelbart, M.V.; Hasegawa, P.M.; Bailey-Serres, J. Genetic mechanisms of abiotic stress tolerance that translate to crop yield stability. *Nat. Rev. Genet.* **2015**, *16*, 237–251. [CrossRef] [PubMed]
6. Hussain, M.; Farooq, S.; Hasan, W.; Ul-Allah, S.; Tanveer, M.; Farooq, M.; Nawaz, A. Drought stress in sunflower: Physiological effects and its management through breeding and agronomic alternatives. *Agric. Water Manag.* **2018**, *201*, 152–166. [CrossRef]
7. Leng, G.; Hall, J. Crop yield sensitivity of global major agricultural countries to droughts and the projected changes in the future. *Sci. Total Environ.* **2019**, *654*, 811–821. [CrossRef] [PubMed]
8. Prasad, R.; Shivay, Y.S.; Kumar, D. Current status, challenges, and opportunities in rice production. *Rice Prod. Worldw.* **2017**, 1–32. [CrossRef]
9. FAOSTAT. FAOSTAT Crops and Livestock Products Database. Rice Production Statistics. Available online: https://www.fao.org/faostat/en/#data/QCL/visualize (accessed on 3 March 2023).
10. International Rice Research Institute (IRRI). World Rice Statistics Database. Rice Consumption. Available online: https://ricestat.irri.org/wrs/ (accessed on 3 March 2023).
11. Zu, X.; Lu, Y.; Wang, Q.; Chu, P.; Miao, W.; Wang, H.; La, H. A new method for evaluating the drought tolerance of upland rice cultivars. *Crop J.* **2017**, *5*, 488–498. [CrossRef]
12. Tuong, P.; Bouman, B.A.M.; Mortimer, M. More Rice, Less Water—Integrated Approaches for Increasing Water Productivity in Irrigated Rice-Based Systems in Asia. *Plant Prod. Sci.* **2005**, *8*, 231–241. [CrossRef]
13. Zhang, Q. Strategies for developing Green Super Rice. *Proc. Natl. Acad. Sci. USA* **2007**, *104*, 16402–16409. [CrossRef] [PubMed]
14. Barnabás, B.; Jäger, K.; Fehér, A. The effect of drought and heat stress on reproductive processes in cereals. *Plant Cell Environ.* **2008**, *31*, 11–38. [CrossRef] [PubMed]
15. Douglas, A.E. Strategies for Enhanced Crop Resistance to Insect Pests. *Annu. Rev. Plant Biol.* **2018**, *69*, 637–660. [CrossRef] [PubMed]
16. Kessler, A.; Kalske, A. Plant Secondary Metabolite Diversity and Species Interactions. *Annu. Rev. Ecol. Evol. Syst.* **2018**, *49*, 115–138. [CrossRef]
17. Bang, S.W.; Choi, S.; Jin, X.; Jung, S.E.; Choi, J.W.; Seo, J.S.; Kim, J.K. Transcriptional activation of rice Cinnamoyl-CoA Reductase 10 by OsNAC5, contributes to drought tolerance by modulating lignin accumulation in roots. *Plant Biotechnol. J.* **2022**, *20*, 736–747. [CrossRef] [PubMed]

18. Bang, S.W.; Lee, D.K.; Jung, H.; Chung, P.J.; Kim, Y.S.; Choi, Y.D.; Suh, J.W.; Kim, J.K. Overexpression of OsTF1L, a rice HD-Zip transcription factor, promotes lignin biosynthesis and stomatal closure that improves drought tolerance. *Plant Biotechnol. J.* **2019**, *17*, 118–131. [CrossRef] [PubMed]
19. Li, L.; Yang, K.; Wang, S.; Lou, Y.; Zhu, C.; Gao, Z. Genome-wide analysis of laccase genes in moso bamboo highlights PeLAC10 involved in lignin biosynthesis and in response to abiotic stresses. *Plant Cell Rep.* **2020**, *39*, 751–763. [CrossRef] [PubMed]
20. Zhao, D.; Luan, Y.; Shi, W.; Zhang, X.; Meng, J.; Tao, J. A Paeonia ostii caffeoyl-CoA O-methyltransferase confers drought stress tolerance by promoting lignin synthesis and ROS scavenging. *Plant Sci.* **2021**, *303*, 110765. [CrossRef]
21. Gleadow, R.M.; Møller, B.L. Cyanogenic glycosides: Synthesis, physiology, and phenotypic plasticity. *Annu. Rev. Plant Biol.* **2014**, *65*, 155–185. [CrossRef]
22. Ashihara, H.; Sano, H.; Crozier, A. Caffeine and related purine alkaloids: Biosynthesis, catabolism, function and genetic engineering. *Phytochemistry* **2008**, *69*, 841–856. [CrossRef]
23. Pham, V.T.T.; Ismail, T.; Mishyna, M.; Appiah, K.S.; Oikawa, Y.; Fujii, Y. Caffeine: The Allelochemical Responsible for the Plant Growth Inhibitory Activity of Vietnamese Tea (*Camellia sinensis* L. Kuntze). *Agronomy* **2019**, *9*, 396. [CrossRef]
24. Kim, Y.S.; Sano, H. Pathogen resistance of transgenic tobacco plants producing caffeine. *Phytochemistry* **2008**, *69*, 882–888. [CrossRef] [PubMed]
25. Kim, Y.S.; Lim, S.; Yoda, H.; Choi, Y.E.; Sano, H. Simultaneous activation of salicylate production and fungal resistance in transgenic Chrysanthemum producing caffeine. *Plant Signal Behav.* **2011**, *6*, 409–412. [CrossRef]
26. Kim, Y.S.; Uefuji, H.; Ogita, S.; Sano, H. Transgenic tobacco plants producing caffeine: A potential new strategy for insect pest control. *Transgenic Res.* **2006**, *15*, 667–672. [CrossRef] [PubMed]
27. Park, J.-C.; Yoo, Y.; Lim, H.; Yun, S.; Win, K.T.Y.S.; Kim, K.-M.; Lee, G.-S.; Cho, M.-H.; Lee, T.H.; Sano, H.; et al. Intracellular Ca^{2+} accumulation triggered by caffeine provokes resistance against a broad range of biotic stress in rice. *Plant Cell Environ.* **2022**, *45*, 1049–1064. [CrossRef] [PubMed]
28. Kim, Y.-S.; Choi, Y.-e.; Sano, H. Plant vaccination: Stimulation of defense system by caffeine production in planta. *Plant Signal. Behav.* **2010**, *5*, 489–493. [CrossRef] [PubMed]
29. Demarty, M.; Morvan, C.; Thellier, M. Calcium and the cell wall. *Plant Cell Environ.* **1984**, *7*, 441–448. [CrossRef]
30. Ketehouli, T.; Nguyen Quoc, V.H.; Dong, J.; Do, H.; Li, X.; Wang, F. Overview of the roles of calcium sensors in plants' response to osmotic stress signalling. *Funct. Plant Biol.* **2022**, *49*, 589–599. [CrossRef]
31. Hepler, P.K. Calcium: A Central Regulator of Plant Growth and Development. *Plant Cell* **2005**, *17*, 2142–2155. [CrossRef]
32. Kudla, J.; Batistič, O.; Hashimoto, K. Calcium signals: The lead currency of plant information processing. *Plant Cell* **2010**, *22*, 541–563. [CrossRef]
33. McAinsh, M.R.; Pittman, J.K. Shaping the calcium signature. *N. Phytol.* **2009**, *181*, 275–294. [CrossRef] [PubMed]
34. Plieth, C. Signal percolation through plants and the shape of the calcium signature. *Plant Signal. Behav.* **2010**, *5*, 379–385. [CrossRef] [PubMed]
35. Aldon, D.; Mbengue, M.; Mazars, C.; Galaud, J.P. Calcium Signalling in Plant Biotic Interactions. *Int. J. Mol. Sci.* **2018**, *19*, 665. [CrossRef] [PubMed]
36. Ren, H.; Zhao, X.; Li, W.; Hussain, J.; Qi, G.; Liu, S. Calcium Signaling in Plant Programmed Cell Death. *Cells* **2021**, *10*, 1089. [CrossRef] [PubMed]
37. O'Brien, I.; Ferguson, I. Calcium signalling in programmed cell death in plants. In *Proceedings of the Plant Nutrition for Sustainable Food Production and Environment: Proceedings of the XIII International Plant Nutrition Colloquium, Tokyo, Japan, 13–19 September 1997*; pp. 99–103.
38. Levine, A.; Pennell, R.I.; Alvarez, M.E.; Palmer, R.; Lamb, C. Calcium-mediated apoptosis in a plant hypersensitive disease resistance response. *Curr. Biol.* **1996**, *6*, 427–437. [CrossRef] [PubMed]
39. Sanders, D.; Pelloux, J.; Brownlee, C.; Harper, J.F. Calcium at the crossroads of signaling. *Plant Cell* **2002**, *14*, S401–S417. [CrossRef]
40. Choi, W.-G.; Hilleary, R.; Swanson, S.J.; Kim, S.-H.; Gilroy, S. Rapid, Long-Distance Electrical and Calcium Signaling in Plants. *Annu. Rev. Plant Biol.* **2016**, *67*, 287–307. [CrossRef]
41. Harmon, A.C.; Gribskov, M.; Harper, J.F. CDPKs—A kinase for every Ca^{2+} signal? *Trends Plant Sci.* **2000**, *5*, 154–159. [CrossRef]
42. Davies, W.J.; Zhang, J. Root Signals and the Regulation of Growth and Development of Plants in Drying Soil. *Annu. Rev. Plant Physiol. Plant Mol. Biol.* **1991**, *42*, 55–76. [CrossRef]
43. Mittler, R. Oxidative stress, antioxidants and stress tolerance. *Trends Plant Sci.* **2002**, *7*, 405–410. [CrossRef]
44. Jiang, F.; Hartung, W. Long-distance signalling of abscisic acid (ABA): The factors regulating the intensity of the ABA signal. *J. Exp. Bot.* **2008**, *59*, 37–43. [CrossRef] [PubMed]
45. Mhamdi, A.; Van Breusegem, F. Reactive oxygen species in plant development. *Development* **2018**, *145*, dev164376. [CrossRef] [PubMed]
46. Tripathy, B.C.; Oelmüller, R. Reactive oxygen species generation and signaling in plants. *Plant Signal Behav.* **2012**, *7*, 1621–1633. [CrossRef]
47. Zou, J.J.; Li, X.D.; Ratnasekera, D.; Wang, C.; Liu, W.X.; Song, L.F.; Zhang, W.Z.; Wu, W.H. Arabidopsis Calcium-Dependent Protein Kinase8 and Catalase3 Function in Abscisic Acid-Mediated Signaling and H_2O_2 Homeostasis in Stomatal Guard Cells under Drought Stress. *Plant Cell* **2015**, *27*, 1445–1460. [CrossRef] [PubMed]

48. Bharath, P.; Gahir, S.; Raghavendra, A.S. Abscisic acid-induced stomatal closure: An important component of plant defense against abiotic and biotic stress. *Front. Plant Sci.* **2021**, *12*, 615114. [CrossRef] [PubMed]
49. Miao, Y.; Lv, D.; Wang, P.; Wang, X.C.; Chen, J.; Miao, C.; Song, C.P. An Arabidopsis glutathione peroxidase functions as both a redox transducer and a scavenger in abscisic acid and drought stress responses. *Plant Cell* **2006**, *18*, 2749–2766. [CrossRef] [PubMed]
50. Ahmad, P.; Jaleel, C.A.; Salem, M.A.; Nabi, G.; Sharma, S. Roles of enzymatic and nonenzymatic antioxidants in plants during abiotic stress. *Crit. Rev. Biotechnol.* **2010**, *30*, 161–175. [CrossRef]
51. Wang, W.; Zhang, H.; Wei, X.; Yang, L.; Yang, B.; Zhang, L.; Li, J.; Jiang, Y.Q. Functional characterization of calcium-dependent protein kinase (CPK) 2 gene from oilseed rape (*Brassica napus* L.) in regulating reactive oxygen species signaling and cell death control. *Gene* **2018**, *651*, 49–56. [CrossRef]
52. Li, B.; Liu, Y.; Cui, X.Y.; Fu, J.D.; Zhou, Y.B.; Zheng, W.J.; Lan, J.H.; Jin, L.G.; Chen, M.; Ma, Y.Z.; et al. Genome-Wide Characterization and Expression Analysis of Soybean TGA Transcription Factors Identified a Novel TGA Gene Involved in Drought and Salt Tolerance. *Front. Plant Sci.* **2019**, *10*, 549. [CrossRef]
53. Sadau, S.B.; Mehari, T.G.; Ahmad, A.; Tajo, S.M.; Ibrahim, S.; Iqbal, M.S.; Elasad, M.; Zhang, J.; Wei, H.; Yu, S. Genome wide identification and characterization of MAPK genes reveals their potential in enhancing drought and salt stress tolerance in *Gossypium hirsutum*. *J. Cotton Res.* **2022**, *5*, 23. [CrossRef]
54. Toki, S.; Hara, N.; Ono, K.; Onodera, H.; Tagiri, A.; Oka, S.; Tanaka, H. Early infection of scutellum tissue with Agrobacterium allows high-speed transformation of rice. *Plant J.* **2006**, *47*, 969–976. [CrossRef] [PubMed]
55. Livak, K.J.; Schmittgen, T.D. Analysis of relative gene expression data using real-time quantitative PCR and the 2(-Delta Delta C(T)) Method. *Methods* **2001**, *25*, 402–408. [CrossRef]
56. Bernt, E.; Bergmeyer, H.U. Inorganic Peroxides. In *Methods of Enzymatic Analysis*; Bergmeyer, H.U., Ed.; Verlag Chemie-Academic Press: London, UK, 1974; Volume 14, pp. 2246–2248.
57. Bradford, M.M. A rapid and sensitive method for the quantitation of microgram quantities of protein utilizing the principle of protein-dye binding. *Anal. Biochem.* **1976**, *72*, 248–254. [CrossRef] [PubMed]
58. Beyer, W.F.; Fridovich, I. Assaying for superoxide dismutase activity: Some large consequences of minor changes in conditions. *Anal. Biochem.* **1987**, *161*, 559–566. [CrossRef] [PubMed]
59. Aebi, H. *Catalase, In Methods of Enzymatic Analysis*, 2nd ed.; Bergmeyer, H.U., Ed.; Academic Press: New York, NY, USA, 1974; pp. 673–684.
60. Pütter, J. Peroxidases. In *Methods of Enzymatic Analysis*; Elsevier: New York, NY, USA, 1974; pp. 685–690.
61. Chen, G.-X.; Asada, K. Ascorbate Peroxidase in Tea Leaves: Occurrence of Two Isozymes and the Differences in Their Enzymatic and Molecular Properties. *Plant Cell Physiol.* **1989**, *30*, 987–998. [CrossRef]
62. Lichtenthaler, H.K.; Wellburn, A.R. Determinations of total carotenoids and chlorophylls a and b of leaf extracts in different solvents. *Biochem. Soc. Trans.* **1983**, *11*, 591–592. [CrossRef]
63. Mukherjee, S.P.; Choudhuri, M.A. Implications of water stress-induced changes in the levels of endogenous ascorbic acid and hydrogen peroxide in Vigna seedlings. *Physiol. Plant.* **1983**, *58*, 166–170. [CrossRef]
64. Bates, L.S.; Waldren, R.P.; Teare, I.D. Rapid determination of free proline for water-stress studies. *Plant Soil.* **1973**, *39*, 205–207. [CrossRef]
65. Fu, L.; Yu, X.; An, C. Overexpression of constitutively active OsCPK10 increases Arabidopsis resistance against *Pseudomonas syringae* pv. tomato and rice resistance against *Magnaporthe grisea. Plant Physiol. Biochem.* **2013**, *73*, 202–210. [CrossRef]
66. Bundó, M.; Coca, M. Calcium-dependent protein kinase OsCPK10 mediates both drought tolerance and blast disease resistance in rice plants. *J. Exp. Bot.* **2017**, *68*, 2963–2975. [CrossRef]
67. Asano, T.; Hayashi, N.; Kobayashi, M.; Aoki, N.; Miyao, A.; Mitsuhara, I.; Ichikawa, H.; Komatsu, S.; Hirochika, H.; Kikuchi, S.; et al. A rice calcium-dependent protein kinase OsCPK12 oppositely modulates salt-stress tolerance and blast disease resistance. *Plant J.* **2012**, *69*, 26–36. [CrossRef] [PubMed]
68. Asano, T.; Hakata, M.; Nakamura, H.; Aoki, N.; Komatsu, S.; Ichikawa, H.; Hirochika, H.; Ohsugi, R. Functional characterisation of OsCPK21, a calcium-dependent protein kinase that confers salt tolerance in rice. *Plant Mol. Biol.* **2011**, *75*, 179–191. [CrossRef] [PubMed]
69. Tang, W.; Luo, C. Overexpression of Zinc Finger Transcription Factor ZAT6 Enhances Salt Tolerance. *Open Life Sci.* **2018**, *13*, 431–445. [CrossRef] [PubMed]
70. Wan, B.; Lin, Y.; Mou, T. Expression of rice Ca(2+)-dependent protein kinases (CDPKs) genes under different environmental stresses. *FEBS Lett.* **2007**, *581*, 1179–1189. [CrossRef] [PubMed]
71. Xie, Z.; Wang, J.; Wang, W.; Wang, Y.; Xu, J.; Li, Z.; Zhao, X.; Fu, B. Integrated Analysis of the Transcriptome and Metabolome Revealed the Molecular Mechanisms Underlying the Enhanced Salt Tolerance of Rice Due to the Application of Exogenous Melatonin. *Front. Plant Sci.* **2021**, *11*, 618680. [CrossRef] [PubMed]
72. Lu, G.; Gao, C.; Zheng, X.; Han, B. Identification of OsbZIP72 as a positive regulator of ABA response and drought tolerance in rice. *Planta* **2009**, *229*, 605–615. [CrossRef] [PubMed]
73. Wang, F.-b.; Wan, C.-z.; Niu, H.-f.; Qi, M.-y.; Li, G.; Zhang, F.; Hu, L.-b.; Ye, Y.-x.; Wang, Z.-x.; Pei, B.-l.; et al. OsMas1, a novel maspardin protein gene, confers tolerance to salt and drought stresses by regulating ABA signaling in rice. *J. Integr. Agric.* **2023**, *22*, 341–359. [CrossRef]

74. Liu, S.; Zheng, L.; Xue, Y.; Zhang, Q.; Wang, L.; Shou, H. Overexpression of OsVP1 and OsNHX1 increases tolerance to drought and salinity in rice. *J. Plant Biol.* **2010**, *53*, 444–452. [CrossRef]
75. Yu, S.; Huang, A.; Li, J.; Gao, L.; Feng, Y.; Pemberton, E.; Chen, C. OsNAC45 plays complex roles by mediating POD activity and the expression of development-related genes under various abiotic stresses in rice root. *Plant Growth Regul.* **2018**, *84*, 519–531. [CrossRef]
76. Chen, J.-Q.; Meng, X.-P.; Zhang, Y.; Xia, M.; Wang, X.-P. Over-expression of OsDREB genes lead to enhanced drought tolerance in rice. *Biotechnol. Lett.* **2008**, *30*, 2191–2198. [CrossRef]
77. Yu, N.; Liang, Y.; Wang, Q.; Peng, X.; He, Z.; Hou, X. Transcriptomic analysis of OsRUS1 overexpression rice lines with rapid and dynamic leaf rolling morphology. *Sci. Rep.* **2022**, *12*, 6736. [CrossRef] [PubMed]
78. Saddique, M.A.B.; Ali, Z.; Sher, M.A.; Farid, B.; Ikram, R.M.; Ahmad, M.S. Proline, total antioxidant capacity, and OsP5CS gene activity in radical and plumule of rice are efficient drought tolerance indicator traits. *Int. J. Agron.* **2020**, *2020*, 8862792. [CrossRef]
79. Duan, Z.-Q.; Bai, L.; Zhao, Z.-G.; Zhang, G.P.; Cheng, F.-M.; Jiang, L.-X.; Chen, K.-M. Drought-Stimulated Activity of Plasma Membrane Nicotinamide Adenine Dinucleotide Phosphate Oxidase and Its Catalytic Properties in Rice. *J. Integr. Plant Biol.* **2009**, *51*, 1104–1115. [CrossRef] [PubMed]
80. Zhang, Y.; Luan, Q.; Jiang, J.; Li, Y. Prediction and Utilization of Malondialdehyde in Exotic Pine Under Drought Stress Using Near-Infrared Spectroscopy. *Front. Plant Sci.* **2021**, *12*, 735275. [CrossRef] [PubMed]
81. Seleiman, M.F.; Al-Suhaibani, N.; Ali, N.; Akmal, M.; Alotaibi, M.; Refay, Y.; Dindaroglu, T.; Abdul-Wajid, H.H.; Battaglia, M.L. Drought Stress Impacts on Plants and Different Approaches to Alleviate Its Adverse Effects. *Plants* **2021**, *10*, 259. [CrossRef] [PubMed]
82. FAO; IFAD; UNICEF; WFP; WHO. *The State of Food Security and Nutrion in the World 2023*; Urbanization, Agrifood Systems Transformation and Healthy Diets Across the Rural–Urban Continuum; FAO: Rome, Italy, 2023; pp. 6–25.
83. Hollingsworth, R.G.; Armstrong, J.W.; Campbell, E. Caffeine as a repellent for slugs and snails. *Nature* **2002**, *417*, 915–916. [CrossRef] [PubMed]
84. Emanuil, N.; Akram, M.S.; Ali, S.; Majrashi, A.; Iqbal, M.; El-Esawi, M.A.; Ditta, A.; Alharby, H.F. Exogenous Caffeine (1,3,7-Trimethylxanthine) Application Diminishes Cadmium Toxicity by Modulating Physio-Biochemical Attributes and Improving the Growth of Spinach (*Spinacia oleracea* L.). *Sustainability* **2022**, *14*, 2806. [CrossRef]
85. Essayan, D.M. Cyclic nucleotide phosphodiesterases. *J. Allergy Clin. Immunol.* **2001**, *108*, 671–680. [CrossRef]
86. Assmann, S.M. Cyclic AMP as a second messenger in higher plants (status and future prospects). *Plant Physiol.* **1995**, *108*, 885. [CrossRef]
87. Kaplan, B.; Sherman, T.; Fromm, H. Cyclic nucleotide-gated channels in plants. *FEBS Lett.* **2007**, *581*, 2237–2246. [CrossRef]
88. Wilkins, K.A.; Matthus, E.; Swarbreck, S.M.; Davies, J.M. Calcium-mediated abiotic stress signaling in roots. *Front. Plant Sci.* **2016**, *7*, 1296. [CrossRef] [PubMed]
89. Dvořák, P.; Krasylenko, Y.; Zeiner, A.; Šamaj, J.; Takáč, T. Signaling toward reactive oxygen species-scavenging enzymes in plants. *Front. Plant Sci.* **2021**, *11*, 2178. [CrossRef] [PubMed]
90. Moumeni, A.; Satoh, K.; Venuprasad, R.; Serraj, R.; Kumar, A.; Leung, H.; Kikuchi, S. Transcriptional profiling of the leaves of near-isogenic rice lines with contrasting drought tolerance at the reproductive stage in response to water deficit. *BMC Genom.* **2015**, *16*, 1110. [CrossRef] [PubMed]
91. Ho, S.L.; Huang, L.F.; Lu, C.A.; He, S.L.; Wang, C.C.; Yu, S.P.; Chen, J.; Yu, S.M. Sugar starvation- and GA-inducible calcium-dependent protein kinase 1 feedback regulates GA biosynthesis and activates a 14-3-3 protein to confer drought tolerance in rice seedlings. *Plant Mol. Biol.* **2013**, *81*, 347–361. [CrossRef] [PubMed]
92. Cieśla, A.; Mituła, F.; Misztal, L.; Fedorowicz-Strońska, O.; Janicka, S.; Tajdel-Zielińska, M.; Marczak, M.; Janicki, M.; Ludwików, A.; Sadowski, J. A Role for Barley Calcium-Dependent Protein Kinase CPK2a in the Response to Drought. *Front. Plant Sci.* **2016**, *7*, 1550. [CrossRef] [PubMed]
93. Campo, S.; Baldrich, P.; Messeguer, J.; Lalanne, E.; Coca, M.; San Segundo, B. Overexpression of a Calcium-Dependent Protein Kinase Confers Salt and Drought Tolerance in Rice by Preventing Membrane Lipid Peroxidation. *Plant Physiol.* **2014**, *165*, 688–704. [CrossRef] [PubMed]
94. Wei, S.; Hu, W.; Deng, X.; Zhang, Y.; Liu, X.; Zhao, X.; Luo, Q.; Jin, Z.; Li, Y.; Zhou, S.; et al. A rice calcium-dependent protein kinase OsCPK9 positively regulates drought stress tolerance and spikelet fertility. *BMC Plant Biol.* **2014**, *14*, 133. [CrossRef]
95. Kim, H.-M.; Kim, S.-H. Expression Analysis of OsCPK11 by ND0001 oscpk11 Mutants of *Oryza sativa* L. under Salt, Cold and Drought Stress Conditions. *J. Life Sci.* **2021**, *31*, 115–125.
96. Li, G.-Z.; Li, H.-X.; Xu, M.-J.; Wang, P.-F.; Xiao, X.-H.; Kang, G.-Z. Functional characterization and regulatory mechanism of wheat CPK34 kinase in response to drought stress. *BMC Genom.* **2020**, *21*, 577. [CrossRef]
97. Liu, Y.; Xu, C.; Zhu, Y.; Zhang, L.; Chen, T.; Zhou, F.; Chen, H.; Lin, Y. The calcium-dependent kinase OsCPK24 functions in cold stress responses in rice. *J. Integr. Plant Biol.* **2018**, *60*, 173–188. [CrossRef]
98. Wang, F.; Niu, H.; Xin, D.; Long, Y.; Wang, G.; Liu, Z.; Li, G.; Zhang, F.; Qi, M.; Ye, Y.; et al. OsIAA18, an Aux/IAA Transcription Factor Gene, Is Involved in Salt and Drought Tolerance in Rice. *Front. Plant Sci.* **2021**, *12*, 738660. [CrossRef] [PubMed]
99. Matsukura, S.; Mizoi, J.; Yoshida, T.; Todaka, D.; Ito, Y.; Maruyama, K.; Shinozaki, K.; Yamaguchi-Shinozaki, K. Comprehensive analysis of rice DREB2-type genes that encode transcription factors involved in the expression of abiotic stress-responsive genes. *Mol. Genet. Genom.* **2010**, *283*, 185–196. [CrossRef] [PubMed]

100. Zhang, X.; Long, Y.; Huang, J.; Xia, J. OsNAC45 is involved in ABA response and salt tolerance in rice. *Rice* **2020**, *13*, 79. [CrossRef] [PubMed]
101. Hirose, T.; Scofield, G.N.; Terao, T. An expression analysis profile for the entire sucrose synthase gene family in rice. *Plant Sci.* **2008**, *174*, 534–543. [CrossRef]
102. Wang, X.; Li, B.-B.; Ma, T.-T.; Sun, L.-Y.; Tai, L.; Hu, C.-H.; Liu, W.-T.; Li, W.-Q.; Chen, K.-M. The NAD kinase OsNADK1 affects the intracellular redox balance and enhances the tolerance of rice to drought. *BMC Plant Biol.* **2020**, *20*, 11. [CrossRef] [PubMed]
103. Cui, G.; Zhao, X.; Liu, S.; Sun, F.; Zhang, C.; Xi, Y. Beneficial effects of melatonin in overcoming drought stress in wheat seedlings. *Plant Physiol. Biochem.* **2017**, *118*, 138–149. [CrossRef] [PubMed]
104. Demidchik, V. Mechanisms of oxidative stress in plants: From classical chemistry to cell biology. *Environ. Exp. Bot.* **2015**, *109*, 212–228. [CrossRef]
105. Mahmood, T.; Khalid, S.; Abdullah, M.; Ahmed, Z.; Shah, M.K.N.; Ghafoor, A.; Du, X. Insights into Drought Stress Signaling in Plants and the Molecular Genetic Basis of Cotton Drought Tolerance. *Cells* **2020**, *9*, 105. [CrossRef]
106. Farooq, M.; Wahid, A.; Lee, D.-J.; Cheema, S.A.; Aziz, T. DROUGHT STRESS: Comparative Time Course Action of the Foliar Applied Glycinebetaine, Salicylic Acid, Nitrous Oxide, Brassinosteroids and Spermine in Improving Drought Resistance of Rice. *J. Agron. Crop Sci.* **2010**, *196*, 336–345. [CrossRef]
107. Laxa, M.; Liebthal, M.; Telman, W.; Chibani, K.; Dietz, K.-J. The Role of the Plant Antioxidant System in Drought Tolerance. *Antioxidants* **2019**, *8*, 94. [CrossRef]
108. Nadarajah, K.K. ROS Homeostasis in Abiotic Stress Tolerance in Plants. *Int. J. Mol. Sci.* **2020**, *21*, 5208. [CrossRef]
109. Hasanuzzaman, M.; Bhuyan, M.H.M.B.; Zulfiqar, F.; Raza, A.; Mohsin, S.M.; Mahmud, J.A.; Fujita, M.; Fotopoulos, V. Reactive Oxygen Species and Antioxidant Defense in Plants under Abiotic Stress: Revisiting the Crucial Role of a Universal Defense Regulator. *Antioxidants* **2020**, *9*, 681. [CrossRef] [PubMed]
110. Stahl, W.; Sies, H. Antioxidant activity of carotenoids. *Mol. Asp. Med.* **2003**, *24*, 345–351. [CrossRef] [PubMed]
111. Akbari, A.; Jelodar, G.; Nazifi, S.; Sajedianfard, J. An Overview of the Characteristics and Function of Vitamin C in Various Tissues: Relying on its Antioxidant Function. *Zahedan J. Res. Med. Sci.* **2016**, *18*, e4037. [CrossRef]
112. Khaleghi, A.; Naderi, R.; Brunetti, C.; Maserti, B.E.; Salami, S.A.; Babalar, M. Morphological, physiochemical and antioxidant responses of *Maclura pomifera* to drought stress. *Sci. Rep.* **2019**, *9*, 19250. [CrossRef] [PubMed]
113. Silva-Ortega, C.O.; Ochoa-Alfaro, A.E.; Reyes-Agüero, J.A.; Aguado-Santacruz, G.A.; Jiménez-Bremont, J.F. Salt stress increases the expression of p5cs gene and induces proline accumulation in cactus pear. *Plant Physiol. Biochem.* **2008**, *46*, 82–92. [CrossRef] [PubMed]
114. Igarashi, Y.; Yoshiba, Y.; Sanada, Y.; Yamaguchi-Shinozaki, K.; Wada, K.; Shinozaki, K. Characterization of the gene for delta1-pyrroline-5-carboxylate synthetase and correlation between the expression of the gene and salt tolerance in *Oryza sativa* L. *Plant Mol. Biol.* **1997**, *33*, 857–865. [CrossRef]

antioxidants

MDPI

Review

The Regulation of ROS and Phytohormones in Balancing Crop Yield and Salt Tolerance

Lei Jiang [1,†], Minggang Xiao [2,†], Rongfeng Huang [1] and Juan Wang [1,*]

1 Biotechnology Research Institute, Chinese Academy of Agricultural Sciences, Beijing 100081, China; jianglei031197@163.com (L.J.); rfhuang@caas.cn (R.H.)
2 Biotechnology Research Institute, Heilongjiang Academy of Agricultural Sciences, Harbin 150028, China; xiaominggang@haas.cn
* Correspondence: wangjuan@caas.cn
† These authors contributed equally to this work.

Abstract: Salinity affects crop growth and productivity, and this stress can be increased along with drought or high temperature stresses and poor irrigation management. Cultivation of salt-tolerant crops plays a critical role in enhancing crop yield under salt stress. In the past few decades, the mechanisms of plant adaptation to salt stress have been described, especially relying on ionic homeostasis, reactive oxygen species (ROS) scavenging, and phytohormone signaling. The studies of these molecular mechanisms have provided a basis for breeding new salt-tolerant crop germplasm and have facilitated the entry into the era of molecular breeding of salt-tolerant crops. In this review, we outline the recent progress in the molecular regulations underlying crop salt tolerance, focusing on the double-edged sword effect of ROS, the regulatory role of phytohormones, and the trade-off effects of ROS and phytohormones between crop yield and salt tolerance. A future challenge is to identify superior alleles of key salt-tolerant genes that will accelerate the breeding of high-yield and salt-tolerant varieties.

Keywords: reactive oxygen species signaling; oxidative stress; phytohormone; salt stress; crop yield; superior alleles; molecular breeding

1. Introduction

The area of salt-affected land worldwide is approximately 950 million hectares and is still increasing, which seriously threatens the production of crops [1,2]. As an important reserve of cultivated land resources, the productivity of salt-affected land urgently needs to be enhanced. Therefore, it is of practical significance to improve the salt tolerance of crops and cultivate high-yield salt-tolerant crops. The salt-tolerant varieties generally have no advantage in terms of yield. Generally speaking, the rice variety with the strongest salt-stress tolerance displays the lowest grain yield, and this contradiction restricts the breeding process of high-yield salt-tolerant crops [3].

When plants are exposed to salt stress, the excessive accumulation of Na^+ disrupts the Na^+/K^+ homeostasis within plant cells, resulting in the inhibition of numerous crucial physiological processes [4]. During the adaptation to salt stress, plants have evolved a series of signaling pathways to deal with salt stress. The salt overly sensitive (SOS) signaling pathway is composed of the core components Na^+/H^+ antiporter SOS1, serine/threonine protein kinase SOS2, and calcium-binding protein SOS3/SCABP8 [5,6]. Recent studies have indicated that reducing high-affinity K^+ transporters (HKTs) and increasing vacuolar Na^+/H^+ exchanger 1 (NHX1) and the CHX-type ion transporter 1 (CHX1) can jointly

 https://doi.org/10.3390/antiox14010063

enhance Na^+ efflux, restrict Na^+ uptake, and sequester Na^+ in vacuoles to alleviate the damage caused by salt stress [7–9].

Salt stress not only induces ionic stress, but also causes the excessive accumulation of ROS, such as hydrogen peroxide (H_2O_2), superoxide radical ($O_2^{\bullet-}$), singlet oxygen ($^1O_2$), and hydroxyl radical (OH•), resulting in oxidative stress and nutrient deficiency. ROS are products under stresses and signaling molecules for growth and development, and ROS homeostasis is of great significance to plants [10]. Accumulating research has elucidated the signaling, generation, and metabolism of ROS [10–12]. Phytohormones are small chemical substances that play a vital role in the growth and development of plants. Moreover, phytohormones also play significant roles when plants cope with stresses. Increased or decreased levels of certain hormones could help plants adapt to salt stress, including abscisic acid (ABA), ethylene (ET), jasmonic acid (JA), salicylic acid (SA), auxin, cytokinin (CK), gibberellic acid (GA), and brassinosteroid (BR) [13]. Whether to regulate the normal growth of plants or mediate the responses to abiotic stress, the integration and coordination of multiple plant hormones are requisites for plants. Moreover, ROS homeostasis and phytohormone signaling pathways are interrelated and form a more complex regulatory network in plants [10,13], revealing the potential functions of ROS and phytohormones in plant growth and salt-stress responses, probably responsible for the key modulation of the collaborative balance between crop yield and salt tolerance. By searching published research in the National Center for Biotechnology Information (NCBI) PubMed https://www.ncbi.nlm.nih.gov/ (accessed on 1 October 2024) database using "ROS or hormones", "salt stress", and "crop yield" as key words, a tremendous number of reports were found, followed by a selection focusing on crops (rice, wheat, maize, soybean, potato, etc.). Here, we mainly review the latest research progress in the regulation of ROS homeostasis and phytohormones in crops under salt stress, especially the critical genes in the last ten years.

2. ROS Homeostasis Under Salt Stress

2.1. Signaling Function of ROS Under Salt Stress

ROS have important roles throughout the entire life cycle of plants. Under normal circumstances, ROS are maintained at a relatively low level and can serve as signaling molecules to regulate plant growth and development [14,15]. In the early stage of salt stress, ROS are produced "actively" or "passively" in plant cells [10]. The extracellular ROS burst is mediated by plasma-membrane-localized respiratory burst oxidase homologs (RBOHs), which utilize NADPH in the cytosol to generate $O_2^{\bullet-}$ in the apoplast and convert it into H_2O_2 through the superoxide dismutase [16]. The RBOH family in rice (*Oryza sativa*) contains nine members (*OsRbohA* to *I*) [17]. Research has disclosed that not every RBOH serves as the principal protein for ROS bursts in response to salt stress. OsRBOHA and OsRBOHI are the key factors in rice, whereas AtRBOHD and AtRBOHF are the crucial ones in *Arabidopsis* (*Arabidopsis thaliana*) [16,17]. The H_2O_2 generated by RBOHs is capable of mediating calcium ion signaling in peas, further activating RBOHD by binding to the EF-hand domain of RBOHD, thereby augmenting H_2O_2 production [18,19]. It has been discovered that H_2O_2 induced by salt stress in soybeans (*Glycine max*) promotes oxidative modification of the transcription factor GmNTL1 protein, facilitating its translocation to the nucleus to activate the expression of *GmRbohBs*, further enhancing the ROS level and thereby rapidly amplifying the ROS signal elicited by salt stress [20]. It has been reported that the trafficking of RBOHs is regulated by the epsin-like clathrin adaptor 4 (ECA4) [21]. Furthermore, RBOHs are also regulated by ubiquitylation, persulfidation, nitrosylation, and endocytosis [10].

Apart from the extracellular space, ROS are also produced in various cellular compartments, such as mitochondria, the cytosol, peroxisomes, and the endoplasmic reticulum [12,22]. Salt stress could also damage the ultrastructure of chloroplasts, resulting in the generation of H_2O_2 and $O_2^{\bullet -}$ in chloroplasts [23]. Due to the fact that aquaporins (AQPs) identified on the plasma membrane and organelle membranes promote the bidirectional transportation of H_2O_2, the H_2O_2 level within the cytoplasm can influence the H_2O_2 level in other cellular compartments, and vice versa [24].

During the process of salt stress, the interplay between the generation and scavenging of ROS in each cellular compartment gives rise to compartment-specific ROS signals [10]. These signals, in combination with other non-ROS signals, traverse the cytoplasm to reach the nucleus, modify nuclear ROS signals, and trigger defense and adaptation responses [25]. It has been discovered that ROS could regulate gene expression through mRNA stability and microRNA abundance during salt-stress responses. For instance, H_2O_2 could enhance the stability of SOS1 mRNA to facilitate the Na^+ efflux [26]. Stress-induced ROS inhibit the accumulation of *miR408* in maize (*Zea mays*), thereby elevating the transcription levels of its target genes *ZmLAC9* and *ZmLAC18* and promoting cell wall development and salt tolerance by regulating the polymerization of lignin monomers [27]. In addition to regulating transcription via epigenetic mechanisms during stress, ROS also influence the translocation of different redox-regulated transcriptional regulators. Non-expressor of pathogenesis-related genes 1 (NPR1), heat shock factors (HSFs), and membrane-bound NAC transcription factors have been determined to sense H_2O_2 signals in the cytoplasm and convey them to the nucleus to modulate the expression of salt-stress response genes [12,28,29]. These studies indicate that ROS signals play a crucial role in the salt-stress response.

2.2. Oxidative Stress Induced by ROS Under Salt Stress

Appropriate levels of ROS constitute an essential signaling molecule for activating salt stress tolerance; however, excessive accumulation of ROS could elicit oxidative stress in plants, disrupting the oxidation–reduction balance of cells and resulting in DNA damage, protein degradation, as well as peroxidation of carbohydrates and lipids [30]. At present, there have been detailed reviews discussing the molecular mechanisms of ROS scavenging under oxidative stress [12]. To counteract oxidative stress, there is a large number of antioxidant enzymes that clear ROS [31]. Enzymatic scavengers include superoxide dismutase (SOD), catalase (CAT), ascorbic acid peroxidase (APX), glutathione peroxidase (GPX), glutathione reductase (GR), dehydroascorbate reductase (DHAR), monodehydroascorbate reductase (MDHAR), and glutathione S-transferase (GST) [32–34]. In an S-type cytoplasmic male sterile maize line, the moderately expressed sterility gene open reading frame 355 (ZmORF355) in the mitochondria promotes salt tolerance by activating various antioxidant enzymes [35]. Biochemical analyses have revealed that hydroquinone could be transformed into arbutin under the catalysis of UDP-glucose transferases (UGTs). Moreover, the growth regulating factor 7 (OsGRF7) could directly bind to the promoters of OsUGT1 and OsUGT5 to stimulate their expression, thereby increasing arbutin biosynthesis and enlarging grains [36]. Inhibition of *ZmmiR169q* under salt stress contributes to salt tolerance by mediating ROS elimination, by increasing the expression of its target gene ZmNF-YA8 and enhancing the transcription of the antioxidant enzyme gene peroxidase 1 (ZmPER1) [37]. The salt-tolerant receptor-like cytoplasmic kinase 1 (STRK1) phosphorylates and activates catalase C (CatC), promoting the homeostasis of H_2O_2 in rice. Plants overexpressing OsSTRK1 exhibit a higher catalase activity, a lower H_2O_2 accumulation, and a greater salt tolerance compared to wild-type plants [38]. Further research has found that catalase 1 phosphatase (PC1)-mediated dephosphorylation of CAT can inhibit its activity [39]. These studies indicate that the activity regulation of many antioxidant enzymes under salt stress is a key regulatory node.

In addition to antioxidant enzymes, an increasing number of studies indicate that many non-enzymatic antioxidants are involved in mitigating ROS-induced damage under salt stress, such as ascorbic acid (AsA), anthocyanin, glutathione, and tocopherols [2]. It has been discovered that AsA confers salt tolerance to plants [40]. GDP-mannose pyrophosphorylase VTC1 is a key enzyme in the AsA biosynthesis pathway and positively regulates salt tolerance [41–43]. OsVTC1-1 plays a crucial role in rice salt tolerance by promoting the generation of AsA to scavenge excessive ROS [44]. Maize bronze 1 (ZmBZ1) enhances stress tolerance by mediating anthocyanin accumulation [45]. Branched-chain amino transferase 2 (OsBCAT2) governs the catabolism of branched-chain amino acids (BCAAs) and the synthesis of pantothenic acid (vitamin B5, VB5) in rice. Salt stress induces the expression of *OsBCAT2*, accelerating the intracellular degradation of BCAAs and augmenting the synthesis of the downstream product VB5, thereby enhancing the salt stress tolerance of rice [46]. *OsGSA1* encodes a UDP-glucosyltransferase, enabling plants to accumulate flavonoid glycosides, thereby protecting rice from abiotic stress [47], revealing that many antioxidants play a role in crop salt tolerance.

Thus, the studies in recent years have emphasized the significant role of ROS homeostasis in the response to salt stress. ROS show the double-edged sword effects in different crops in terms of their contribution to salt tolerance (Table 1). To prevent damage, it is essential for plants to strictly control ROS concentrations. Enhancing crop salt tolerance by recruiting enzymatic and non-enzymatic antioxidant systems to eliminate excessive ROS is an effective approach (Figure 1).

Table 1. Genes associated with ROS in crop salt–alkali tolerance.

Gene	ROS	Species	Salt–Alkali Tolerance	Ref.
Rbohs	ROS generation	*Oryza sativa, Glycine max*	Positive	[17] [20]
GmNTL1	ROS generation	*Glycine max*	Positive	[20]
miR408	Regulated by ROS signal	*Zea mays*	Positive	[27]
HSFB2b	Regulated by ROS signal	*Glycine max*	Positive	[29]
ZmORF355	ROS scavenging	*Zea mays*	Positive	[35]
OsGRF7 *OsUGT1/5*	ROS scavenging	*Oryza sativa*	Positive	[36]
ZmmiR169q	ROS scavenging	*Zea mays*	Positive	[37]
ZmNF-YA8 *ZmPER1*	ROS scavenging	*Zea mays*	Positive	[37]
STRK1	ROS scavenging	*Oryza sativa*	Positive	[38]
PC1	ROS scavenging	*Oryza sativa*	Negative	[39]
OsVTC1	ROS scavenging	*Oryza sativa*	Positive	[44]
ZmBZ1	ROS scavenging	*Zea mays*	Positive	[45]
OsBCAT2	ROS scavenging	*Oryza sativa*	Positive	[46]
OsGSA1	ROS scavenging	*Oryza sativa*	Positive	[47]
TaCHP	ROS scavenging	*Triticum aestivum*	Positive	[48]
AT1	ROS scavenging	*Sorghum bicolor* *Oryza sativa* *Zea mays,* *Triticum aestivum*	Negative	[49,50]

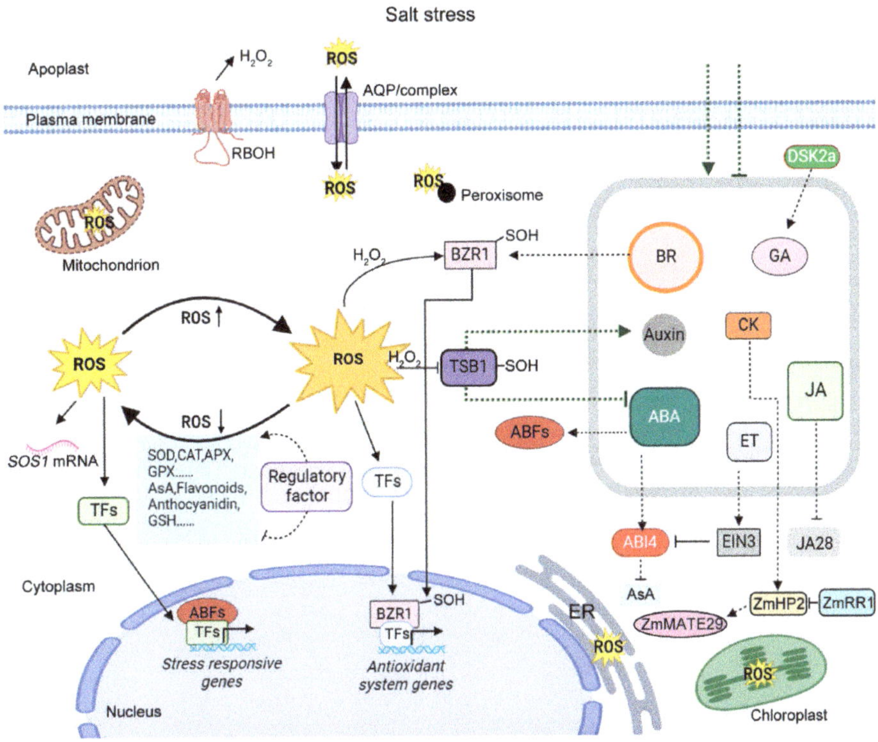

Figure 1. ROS and phytohormones in the context of salt stress. In the early phases of salt stress, RBOHs govern the outburst of ROS in organelles including chloroplasts, mitochondria, peroxisomes, etc. Due to the promotion of bidirectional transportation of H_2O_2 by AQP/complexes on the cell membrane and organelle membranes, the level of H_2O_2 in the cytoplasm could influence that in other organelles and vice versa. The ROS signal produced by salt stress regulates some transcription factors (TFs), facilitating the expression of some salt stress responsive genes. The stability of *SOS1* mRNA is also enhanced to regulate ionic homeostasis. Persistent salt stress could result in the excessive accumulation of ROS in plant cells when certain TFs are promoted to activate the expression of antioxidant system genes. Antioxidative enzymes, combined with non-enzymatic antioxidants, participate in the control of ROS homeostasis under salt stress. Meanwhile, phytohormone signals are activated or inhibited to regulate salt-stress responses and plant growth. Several regulators involved in phytohormone signals are also controlled by ROS homeostasis, which mediates the crosstalk between ROS and phytohormones under salt stress.

3. Modulation of Phytohormones in Salt-Stress Responses

The phytohormones facilitate plants' rapid responses to salt stress. Some hormones exert positive regulatory effects on plant salt tolerance, such as ABA and JA, whereas some have negative regulatory roles, such as GA [13]. Moreover, the regulatory function of some hormones in salt tolerance is opposite in different species. For instance, ethylene has a positive role in salt tolerance in *Arabidopsis*, while it negatively regulates salt tolerance in rice [51–53], demonstrating that phytohormones have complicated regulatory mechanisms.

When plants are exposed to salt stress, the accumulation level of ABA shows a positive correlation with the salt tolerance of plants [54,55]. ABA could activate a variety of downstream signaling factors, such as kinases, phosphatases, and G proteins. It has been discovered that the drought-induced protein GmDi19-5 could negatively regulate the responses of soybeans to drought and salt stress through the ABA-dependent pathway [56,57]. In rice, the transcription factor OsSGL is involved in inhibiting the gene expression of *OsNCED3*, encoding a key enzyme of ABA biosynthesis. Under salt stress, OsSGL undergoes acetylation modification and nuclear–cytoplasmic distributions, enabling a dynamic regulation of ABA content in plant cells [58,59]. Consequently, ABA, as a stress hormone, plays an important role in regulating salt tolerance in crops.

It has been discovered that ethylene treatments augment the salt sensitivity of rice seedlings, while, for *Arabidopsis* and soybeans, they enhance salt tolerance [51,60,61]. Salt intolerance 1 (OsSIT1), a lectin receptor-like kinase and an ethylene sensor, actively mediates salt sensitivity by activating the mitogen-activated protein kinase 3/6 (MPK3/6), thereby promoting ethylene biosynthesis, ROS accumulation, and sensitivity to salt stress in rice [62].

GA plays a crucial role in modulating growth under abiotic stress [63]. The genes involved in the GA catabolic pathway, i.e., gibberellin 2-oxidase 5/7 (*GA2ox5/7*), *OsMYB91*, and *OsCYP71D8L*, contribute to decreased GA levels and a retarded growth, thereby enhancing the salt tolerance of plants [64–66]. Moreover, OsDSK2a, functioning as a ubiquitin receptor and transporter, promotes the degradation of the GA-inactivating enzyme EUI in rice. The level of OsDSK2a is inhibited by salt stress, giving rise to the accumulation of EUI and the reduced level of biologically active GA, elucidating the mechanism underlying the inhibition of salt stress in the context of plant growth [67]. These studies indicate that growth inhibition under salt stress might be an active adaptive mechanism in crops.

JA plays a key role in plant growth and development, in the defense against insects and pathogens, and in secondary metabolism. It has also been unveiled that JA possesses significant functions in plant responses to salt stress [68]. When plants encounter high-concentration salt stress, degradation of the JA response factor JAZ8 relieves its inhibitory effect on the NF-YA1-YB2-YC9 transcription factor complex, promoting the expression of salt-stress response genes and enhancing salt tolerance in rice [69–71]. IbNAC087, a nuclear-localized transcriptional activator in sweet potatoes (*Ipomoea batatas*), could directly activate the expression of JA synthesis-related genes to increase JA content and, thus, enhance salt and drought resistance [72].

It has been found that the mutation of rice salt tolerance 1 (*RST1*) in rice, encoding the auxin response factor OsARF18, promotes nitrogen utilization and reduces the accumulation of $NH4^+$ under salt stress, thereby enhancing salt tolerance and grain yield [73]. Overexpression of rice big grain 1 (RBG1) leads to an increased grain size due to the elevated cell numbers by increasing auxin accumulation; meanwhile, RBG1 could upregulate stress proteins to enhance the tolerance to the heat and osmotic and salt stresses in rice [74,75]. Therefore, promoting growth to a certain extent is also necessary for improving salt tolerance in crops.

Some reports have indicated that cytokinin regulates salt tolerance in *Arabidopsis* [76–78]. Furthermore, a study on maize has determined that a type A response regulator, ZmRR1, modulates Cl^- exclusion and negatively regulates cytokinin signaling. ZmRR1 could interact with a key mediator of cytokinin signaling, ZmHP2, to inhibit Cl^- exclusion and salt tolerance [79]. This study demonstrates that cytokinin signaling has a positive role in salt tolerance through ionic homeostasis in crops.

Taken together, the roles of phytohormones in crop growth and development under salt stress are complicated (Table 2), and more in-depth research on some hormones in crops is needed. In particular, the hormones mainly regulating plant growth have a promising route for balancing salt tolerance and crop yield.

Table 2. Genes associated with phytohormones in crop salt tolerance.

Gene	Phytohormone	Species	Salt Tolerance	Ref.
GmDi19-5	ABA signal	*Glycine max*	Negative	[56]
OsSGL	ABA synthesis	*Oryza sativa*	Negative	[58]
OsNCED3	ABA synthesis	*Oryza sativa*	Positive	[58]
OsSIT1	ET synthesis	*Oryza sativa*	Negative	[62]
MPK3/6	ET synthesis	*Oryza sativa*	Negative	[62]
OsGA2ox5	GA metabolism	*Oryza sativa*	Positive	[65]
OsCYP71D8L	GA metabolism CK metabolism	*Oryza sativa*	Positive	[66]
OsDSK2a	GA metabolism	*Oryza sativa*	Positive	[67]
JAZ8	JA signal	*Oryza sativa*	Positive	[70]
IbNAC087	JA synthesis	*Ipomoea batatas*	Negative	[72]
OsARF18	Auxin signal	*Oryza sativa*	Negative	[73]
ZmRR1	CK signal	*Zea mays*	Negative	[79]
ZmHP2 *ZmMATE29*	CK signal	*Zea mays*	Positive	[79]

4. ROS and Phytohormones Jointly Balance Crop Yield and Salt Tolerance

ROS play a dual role in plant growth and stress responses, and phytohormones are also significant endogenous chemical signals that orchestrate plant growth and development in appropriate environments and under adverse conditions [10,13]. Owing to the high reactivity and oxidizing nature of ROS, ROS signaling is predominantly mediated by protein oxidative post-translational modifications. This characteristic enables ROS to target a broad spectrum of proteins and to dynamically and spatiotemporally regulate signaling transduction pathways under various stress conditions, a phenomenon which is similar to phytohormone signaling pathways [10]. It has been reported that phytohormones could be associated with ROS to jointly regulate plant growth, development, and adaptation to adverse circumstances.

First and foremost, phytohormones are capable of influencing the synthesis, metabolism, and signaling of ROS. Research has discovered that wheat (*Triticum aestivum*) brassinazole-resistant 1 (TaBZR1) could not only directly bind to the G-box motif in the promoter of the key ABA synthesis gene *TaNCED3* to activate its expression and ABA synthesis, but also bind to the promoters of the ROS scavenging-related genes *TaGPX2* and *TaGPX3* to promote their expression, thereby facilitating ROS scavenging [80]. It is interesting that ethylene inhibits but ABA promotes the accumulation of ROS in *Arabidopsis* [81]. Furthermore, AsA acts as a key factor in integrating the interaction of ethylene and ABA in the regulation of ROS levels. Ethylene and ABA antagonistically regulate AsA biosynthesis via ethylene-insensitive 3 (EIN3) and ABA insensitive 4 (ABI4), respectively, both of which are key factors in the ethylene and ABA signaling pathways, respectively [81]. In addition, EIN3 and ABI4 collaboratively modulate salt tolerance via AsA biosynthesis [82]. It is noteworthy that the GA signaling factor RGA1 (alpha subunit of heterotrimeric G protein) in rice negatively regulates salt tolerance by participating in ROS scavenging by modulating antioxidant enzymes protein abundance [83]. The most recent study has disclosed that, under salt stress, the downstream transcription factor OsMYB2 of ABA mediates the expression of the amino acid transporter gene ANT1, further enhancing the transportation of amino acids (Pro, Leu, Phe, Tyr) and regulating the balance between rice growth and salt tolerance through the accumulation of amino acids like Pro and Tyr and the reduction of ROS. This research provides novel insights into the function of plant amino acid transporter genes [84].

Secondly, ROS exert an influence on the synthesis, metabolism, and signaling of phytohormones. It has been discovered that GPX3 is not merely a ROS-scavenging enzyme highly sensitive to H_2O_2, and that the redox state of GPX3 could also influence the phosphatase activity of its interacting partner ABA insensitive 2 (ABI2), which is a negative component of the ABA signaling pathway and participates in stomatal closure [85]. H_2O_2 could sulfonate BZR1 to enhance its transcriptional activity, thereby positively regulating BR signaling and plant growth in *Arabidopsis* [86]. It is notable that H_2O_2 could affect both the contents of IAA and ABA through the tryptophan synthase B subunit 1 (TSB1). Salt-induced sulfoxidation of TSB1 increases ABA accumulation and facilitates salt stress tolerance in *Arabidopsis* [87]. Thus, ROS-mediated oxidative modification of phytohormone signaling factors confers to plants the ability to adapt to salt stress.

Furthermore, there exist certain transcription factors that could concurrently regulate both ROS and phytohormones. The mutation of the key regulatory factor ITPK4 in *Arabidopsis* leads to an augmented accumulation of ROS and decreased auxin levels in plants under salt stress [88]. In soybeans, the induced expression of *GmSIN1* under salt stress upregulates the cellular levels of ABA and ROS, giving rise to an amplification of the initial salt stress signal [89]. In brief, plants utilize multiple signaling molecules to coordinate developmental processes and responses to environmental stimuli [90] (Table 3).

Table 3. Genes associated with both ROS and phytohormones under salt stress.

Gene	ROS	Phytohormone	Species	Salt Tolerance	Ref.
BZR1	ROS scavenging	BR signal ABA synthesis	*Triticum Aestivum* L. *Arabidopsis thaliana*	Positive	[80,86]
ABI4	ROS scavenging	ABA signal ET signal	*Arabidopsis thaliana*	Negative	[81]
EIN3	ROS scavenging	ET signal ABA signal	*Arabidopsis thaliana*	Positive	[81]
RGA1	ROS scavenging	GA signal	*Oryza sativa*	Positive	[83]
OsMYB2	ROS scavenging	ABA signal	*Oryza sativa*	Positive	[84]
GPX3	ROS scavenging	ABA signal	*Arabidopsis thaliana*	Positive	[85]
TSB1	Regulated by ROS signal	ABA synthesis, auxin metabolism	*Arabidopsis thaliana*	Positive	[87]
ITPK4	ROS scavenging	Auxin metabolism	*Arabidopsis thaliana*	Positive	[88]
GmSIN1	ROS generation	ABA synthesis	*Glycine max*	Positive	[89]

In recent years, numerous reviews related to balancing crop growth and stress tolerance have been reported [3,91]. Regarding the balance between crop yield and salt tolerance, it seems to be distinct from other stresses, such as drought, heat, and cold, since the salt content in saline–alkali soil changes with the weather [91]. Recent studies have discovered that, in natural populations of wheat, there are many natural variations of the promoter of *TaCHP*, forming two main haplotypes, Hap1 and Hap2. The salt tolerance of the varieties carrying Hap1 is significantly higher than that of thosse carrying Hap2. TaCHP-Hap1 could stably increase the yield of wheat in saline–alkali soil without affecting the yield in normal soil. Therefore, the salt-tolerant haplotype of *TaCHP* has a strong application potential in molecular design breeding of wheat [48]. Under saline–alkaline conditions, high pH stress amplifies the harmful effects of salt stress on plants. Recent studies have shown that the *Alkaline tolerance 1* (*AT1*) gene in *Sorghum* negatively regulates crop tolerance to alkali by modulating ROS homeostasis [49,50]. It has been determined that *OsAT1* (known as grain size 3, OsGS3) is a high-yield gene that encodes an atypical G protein γ subunit [92]. Introducing the favorable allele of *OsAT1* into the high-quality rice variety Kongyu131 could increase the yield by 27.8% under alkaline stress conditions, indicating that this allele has a great potential value in cultivating alkali-tolerant rice [50]. Recent research has revealed that OsDSK2a, a co-regulatory factor for growth and salt tolerance in rice, could be phosphorylated by the conserved kinase SnRK1A, and natural variations exist at the phosphorylation sites. The varieties contained in the elite haplotype *OsDSK2a-G* have a higher grain yield than that of the varieties contained in *OsDSK2a-S* under salt stress [93]. Furthermore, a study has reported, for the first time, a grain length gene GL12 from the common wild rice (*Oryza rufipogon*) which is capable of concurrently enhancing the grain length and salt tolerance of rice. A G/T variation site in its promoter region influences its spatio-temporal expression pattern in the spikelets and roots, thereby balancing the yield and salt tolerance of rice [94]. Therefore, it is still necessary to explore the key salt-tolerant genes in crop plants, regardless of whether these genes are related to ROS or hormone pathways, and try to identify the associated excellent natural variations as much as possible to provide opportunities for the trade-off between salt tolerance and growth (Figure 2).

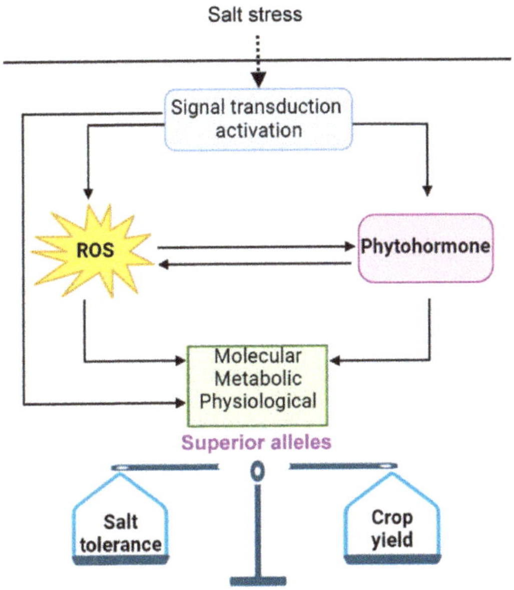

Figure 2. ROS and phytohormones mediate the balance between crop yield and salt tolerance. ROS and phytohormones display an intricate signal crosstalk and jointly govern the growth responses of plants to salt stress. Exploring the key salt-tolerant genes therein and identifying the associated excellent natural variations offer opportunities for striking a balance between salt tolerance and growth.

5. Conclusions and Future Prospects

Crop yield has been greatly improved after years of breeding, but the improvement in crop salt tolerance is still at a bottleneck. The identification of key salt-tolerant genes in plants and an in-depth exploration of their molecular regulatory mechanisms could provide significant theoretical foundations for the breeding of salt-tolerant crops. However, crop salt tolerance is a quantitative trait controlled by multiple genes. Due to the subtle effect of one or several genes, more genes need to work together to achieve synergistic effects under salt stress. Therefore, highly efficient methods for identifying salt-tolerant genes are a challenge for molecular breeding. Meanwhile, with the development of functional genomics and multiomics technology, a stable phenome identification system under salt stress is very important.

On the other hand, since the modulations of ROS and phytohormones in the context of plant growth and salt stress have been deeply researched in model plants, it is feasible to use these regulations to balance yield and salt tolerance in crops. Phytohormones could play a significant regulatory role in extremely small amounts, providing a potential way for balancing ROS homeostasis, salt tolerance, plant growth, and yield enhancement. GA, which was reduced to increase the harvest index in the Green Revolution, could also improve salt tolerance to a certain level. Identification of key genes referring to the crosstalk between ROS and phytohormones is helpful to control their dose effects.

During the prolonged breeding process, with yield traits as the main target orientation, the genetic diversity of crops has decreased, and superior alleles for salt tolerance have been lost or lacked directed selection. The discovery of superior alleles that can collaboratively improve both yield and salt tolerance remains an exciting challenge. To increase the productivity of salt–alkali land and improve grain yield, it is important to apply modern molecular breeding approaches for precise, rapid, and efficient improvement of crop salt tolerance.

Author Contributions: Conceptualization, J.W., R.H. and L.J.; figure software, L.J.; writing—original draft preparation, L.J.; writing—review and editing, L.J., M.X. and J.W.; supervision, J.W. and R.H. All authors have read and agreed to the published version of the manuscript.

Funding: This work was supported by the Biological Breeding-National Science and Technology Major Project (2023ZD04071), the National Natural Science Foundation of China (32030079, 32171949, 32472068), the Agricultural Science and Technology Innovation Program of the Chinese Academy of Agricultural Sciences (CAAS-ZDRW202407), and the Fundamental Research Funds for Central Non-profit Scientific Institution (1610392020004).

Acknowledgments: All the figures displayed in this review were created with BioRender.com.

Conflicts of Interest: The authors declare no conflicts of interest.

References

1. Ismail, A.M.; Horie, T. Genomics, Physiology, and molecular breeding approaches for improving salt tolerance. *Annu. Rev. Plant Biol.* **2017**, *68*, 405–434. [CrossRef]
2. Yang, Y.; Guo, Y. Elucidating the molecular mechanisms mediating plant salt-stress responses. *New Phytol.* **2018**, *217*, 523–539. [CrossRef] [PubMed]
3. Zhang, H.; Zhao, Y.; Zhu, J.K. Thriving under Stress: How plants balance growth and the stress response. *Dev. Cell* **2020**, *55*, 529–543. [CrossRef] [PubMed]
4. Munns, R.; Tester, M. Mechanisms of salinity tolerance. *Annu. Rev. Plant Biol.* **2008**, *59*, 651–681. [CrossRef] [PubMed]
5. Zhang, X.Y.; Tang, L.H.; Nie, J.W.; Zhang, C.R.; Han, X.; Li, Q.Y.; Qin, L.; Wang, M.H.; Huang, X.; Yu, F.; et al. Structure and activation mechanism of the rice Salt Overly Sensitive 1 (SOS1) Na(+)/H(+) antiporter. *Nat. Plants* **2023**, *9*, 1924–1936. [CrossRef]
6. Zhang, Y.; Zhou, J.; Ni, X.; Wang, Q.; Jia, Y.; Xu, X.; Wu, H.; Fu, P.; Wen, H.; Guo, Y.; et al. Structural basis for the activity regulation of Salt Overly Sensitive 1 in Arabidopsis salt tolerance. *Nat. Plants* **2023**, *9*, 1915–1923. [CrossRef] [PubMed]
7. Mottaleb, S.A.; Rodriguez-Navarro, A.; Haro, R. Knockouts of Physcomitrella patens CHX1 and CHX2 transporters reveal high complexity of potassium homeostasis. *Plant Cell Physiol.* **2013**, *54*, 1455–1468. [CrossRef]
8. Yang, L.; Wei, Y.; Li, N.; Zeng, J.; Han, Y.; Zuo, Z.; Wang, S.; Zhu, Y.; Zhang, Y.; Sun, J.; et al. Declined cadmium accumulation in Na(+)/H(+) antiporter (NHX1) transgenic duckweed under cadmium stress. *Ecotoxicol. Environ. Saf.* **2019**, *182*, 109397. [CrossRef] [PubMed]
9. Wang, X.; Shen, X.; Qu, Y.; Zhang, H.; Wang, C.; Yang, F.; Shen, H. Structural insights into ion selectivity and transport mechanisms of Oryza sativa HKT2; 1 and HKT2; 2/1 transporters. *Nat. Plants* **2024**, *10*, 633–644. [CrossRef] [PubMed]
10. Mittler, R.; Zandalinas, S.I.; Fichman, Y.; Van Breusegem, F. Reactive oxygen species signalling in plant stress responses. *Nat. Rev. Mol. Cell Biol.* **2022**, *23*, 663–679. [CrossRef]
11. Waszczak, C.; Carmody, M.; Kangasjarvi, J. Reactive oxygen species in plant signaling. *Annu. Rev. Plant Biol.* **2018**, *69*, 209–236. [CrossRef] [PubMed]

12. Wang, P.; Liu, W.C.; Han, C.; Wang, S.; Bai, M.Y.; Song, C.P. Reactive oxygen species: Multidimensional regulators of plant adaptation to abiotic stress and development. *J. Integr. Plant Biol.* **2024**, *66*, 330–367. [CrossRef]
13. Yu, Z.; Duan, X.; Luo, L.; Dai, S.; Ding, Z.; Xia, G. How plant hormones mediate salt stress responses. *Trends Plant Sci.* **2020**, *25*, 1117–1130. [CrossRef] [PubMed]
14. Evans, M.J.; Choi, W.G.; Gilroy, S.; Morris, R.J. A ROS-assisted calcium wave dependent on the AtRBOHD NADPH oxidase and TPC1 cation channel propagates the systemic response to salt stress. *Plant Physiol.* **2016**, *171*, 1771–1784. [CrossRef]
15. Zhang, M.; Smith, J.A.; Harberd, N.P.; Jiang, C. The regulatory roles of ethylene and reactive oxygen species (ROS) in plant salt stress responses. *Plant Mol. Biol.* **2016**, *91*, 651–659. [CrossRef] [PubMed]
16. Miller, G.; Suzuki, N.; Ciftci-Yilmaz, S.; Mittler, R. Reactive oxygen species homeostasis and signalling during drought and salinity stresses. *Plant Cell Environ.* **2010**, *33*, 453–467. [CrossRef] [PubMed]
17. Wang, G.F.; Li, W.Q.; Li, W.Y.; Wu, G.L.; Zhou, C.Y.; Chen, K.M. Characterization of Rice NADPH oxidase genes and their expression under various environmental conditions. *Int. J. Mol. Sci.* **2013**, *14*, 9440–9458. [CrossRef]
18. Ogasawara, Y.; Kaya, H.; Hiraoka, G.; Yumoto, F.; Kimura, S.; Kadota, Y.; Hishinuma, H.; Senzaki, E.; Yamagoe, S.; Nagata, K.; et al. Synergistic activation of the Arabidopsis NADPH oxidase AtrbohD by Ca2+ and phosphorylation. *J. Biol. Chem.* **2008**, *283*, 8885–8892. [CrossRef] [PubMed]
19. Zepeda-Jazo, I.; Velarde-Buendia, A.M.; Enriquez-Figueroa, R.; Bose, J.; Shabala, S.; Muniz-Murguia, J.; Pottosin, I.I. Polyamines interact with hydroxyl radicals in activating Ca(2+) and K(+)transport across the root epidermal plasma membranes. *Plant Physiol.* **2011**, *157*, 2167–2180. [CrossRef]
20. Zhang, W.; Zhi, W.; Qiao, H.; Huang, J.; Li, S.; Lu, Q.; Wang, N.; Li, Q.; Zhou, Q.; Sun, J.; et al. H2O2-dependent oxidation of the transcription factor GmNTL1 promotes salt tolerance in soybean. *Plant Cell* **2023**, *36*, 112–135. [CrossRef] [PubMed]
21. Lee, J.; Hanh, N.H.; Park, Y.; Lin, J.; Hwang, I. Spatial regulation of RBOHD via AtECA4-mediated recycling and clathrin-mediated endocytosis contributes to ROS accumulation during salt stress response but not flg22-induced immune response. *Plant J.* **2022**, *109*, 816–830. [CrossRef] [PubMed]
22. Janku, M.; Luhova, L.; Petrivalsky, M. On the origin and fate of reactive oxygen species in plant cell compartments. *Antioxidants* **2019**, *8*, 105. [CrossRef]
23. Yamane, K.; Taniguchi, M.; Miyake, H. Salinity-induced subcellular accumulation of H(2)O(2) in leaves of rice. *Protoplasma* **2012**, *249*, 301–308. [CrossRef] [PubMed]
24. Rodrigues, O.; Reshetnyak, G.; Grondin, A.; Saijo, Y.; Leonhardt, N.; Maurel, C.; Verdoucq, L. Aquaporins facilitate hydrogen peroxide entry into guard cells to mediate ABA- and pathogen-triggered stomatal closure. *Proc. Natl. Acad. Sci. USA* **2017**, *114*, 9200–9205. [CrossRef] [PubMed]
25. Van Breusegem, F.; Bailey-Serres, J.; Mittler, R. Unraveling the tapestry of networks involving reactive oxygen species in plants. *Plant Physiol.* **2008**, *147*, 978–984. [CrossRef] [PubMed]
26. Chung, J.S.; Zhu, J.K.; Bressan, R.A.; Hasegawa, P.M.; Shi, H. Reactive oxygen species mediate Na+-induced SOS1 mRNA stability in Arabidopsis. *Plant J.* **2008**, *53*, 554–565. [CrossRef] [PubMed]
27. Qin, R.; Hu, Y.; Chen, H.; Du, Q.; Yang, J.; Li, W.X. MicroRNA408 negatively regulates salt tolerance by affecting secondary cell wall development in maize. *Plant Physiol.* **2023**, *192*, 1569–1583. [CrossRef]
28. Jayakannan, M.; Bose, J.; Babourina, O.; Shabala, S.; Massart, A.; Poschenrieder, C.; Rengel, Z. The NPR1-dependent salicylic acid signalling pathway is pivotal for enhanced salt and oxidative stress tolerance in Arabidopsis. *J. Exp. Bot.* **2015**, *66*, 1865–1875. [CrossRef]
29. Bian, X.H.; Li, W.; Niu, C.F.; Wei, W.; Hu, Y.; Han, J.Q.; Lu, X.; Tao, J.J.; Jin, M.; Qin, H.; et al. A class B heat shock factor selected for during soybean domestication contributes to salt tolerance by promoting flavonoid biosynthesis. *New Phytol.* **2020**, *225*, 268–283. [CrossRef]
30. Noctor, G.; Foyer, C.H. Ascorbate and glutathione: Keeping active oxygen under control. *Annu. Rev. Plant Physiol. Plant Mol. Biol.* **1998**, *49*, 249–279. [CrossRef] [PubMed]
31. Gill, S.S.; Tuteja, N. Reactive oxygen species and antioxidant machinery in abiotic stress tolerance in crop plants. *Plant Physiol. Biochem.* **2010**, *48*, 909–930. [CrossRef] [PubMed]
32. Begara-Morales, J.C.; Sanchez-Calvo, B.; Chaki, M.; Mata-Perez, C.; Valderrama, R.; Padilla, M.N.; Lopez-Jaramillo, J.; Luque, F.; Corpas, F.J.; Barroso, J.B. Differential molecular response of monodehydroascorbate reductase and glutathione reductase by nitration and S-nitrosylation. *J. Exp. Bot.* **2015**, *66*, 5983–5996. [CrossRef]
33. Madeira, C.; Mendonca, V.; Leal, M.C.; Flores, A.; Cabral, H.N.; Diniz, M.S.; Vinagre, C. Environmental health assessment of warming coastal ecosystems in the tropics—Application of integrative physiological indices. *Sci. Total Environ.* **2018**, *643*, 28–39. [CrossRef]
34. Palma, J.M.; Mateos, R.M.; Lopez-Jaramillo, J.; Rodriguez-Ruiz, M.; Gonzalez-Gordo, S.; Lechuga-Sancho, A.M.; Corpas, F.J. Plant catalases as NO and H(2)S targets. *Redox Biol.* **2020**, *34*, 101525. [CrossRef]

35. Xiao, S.; Song, W.; Xing, J.; Su, A.; Zhao, Y.; Li, C.; Shi, Z.; Li, Z.; Wang, S.; Zhang, R.; et al. ORF355 confers enhanced salinity stress adaptability to S-type cytoplasmic male sterility maize by modulating the mitochondrial metabolic homeostasis. *J. Integr. Plant Biol.* **2023**, *65*, 656–673. [CrossRef]
36. Chen, Y.; Dan, Z.; Li, S. GROWTH REGULATING FACTOR 7-mediated arbutin metabolism enhances rice salt tolerance. *Plant Cell* **2024**, *36*, 2834–2850. [CrossRef]
37. Xing, L.; Zhu, M.; Luan, M.; Zhang, M.; Jin, L.; Liu, Y.; Zou, J.; Wang, L.; Xu, M. miR169q and NUCLEAR FACTOR YA8 enhance salt tolerance by activating PEROXIDASE1 expression in response to ROS. *Plant Physiol.* **2022**, *188*, 608–623. [CrossRef] [PubMed]
38. Zhou, Y.B.; Liu, C.; Tang, D.Y.; Yan, L.; Wang, D.; Yang, Y.Z.; Gui, J.S.; Zhao, X.Y.; Li, L.G.; Tang, X.D.; et al. The receptor-like cytoplasmic kinase STRK1 phosphorylates and activates CatC, thereby regulating H(2)O(2) homeostasis and improving salt tolerance in rice. *Plant Cell* **2018**, *30*, 1100–1118. [CrossRef]
39. Liu, C.; Lin, J.Z.; Wang, Y.; Tian, Y.; Zheng, H.P.; Zhou, Z.K.; Zhou, Y.B.; Tang, X.D.; Zhao, X.H.; Wu, T.; et al. The protein phosphatase PC1 dephosphorylates and deactivates CatC to negatively regulate H_2O_2 homeostasis and salt tolerance in rice. *Plant Cell* **2023**, *35*, 3604–3625. [CrossRef]
40. Zhang, Z.; Wang, J.; Zhang, R.; Huang, R. The ethylene response factor AtERF98 enhances tolerance to salt through the transcriptional activation of ascorbic acid synthesis in Arabidopsis. *Plant J.* **2012**, *71*, 273–287. [CrossRef]
41. Wang, J.; Yu, Y.; Zhang, Z.; Quan, R.; Zhang, H.; Ma, L.; Deng, X.W.; Huang, R. Arabidopsis CSN5B interacts with VTC1 and modulates ascorbic acid synthesis. *Plant Cell* **2013**, *25*, 625–636. [CrossRef] [PubMed]
42. Li, S.; Wang, J.; Yu, Y.; Wang, F.; Dong, J.; Huang, R. D27E mutation of VTC1 impairs the interaction with CSN5B and enhances ascorbic acid biosynthesis and seedling growth in Arabidopsis. *Plant Mol. Biol.* **2016**, *92*, 473–482. [CrossRef]
43. Li, Y.; Chu, Z.; Luo, J.; Zhou, Y.; Cai, Y.; Lu, Y.; Xia, J.; Kuang, H.; Ye, Z.; Ouyang, B. The C2H2 zinc-finger protein SlZF3 regulates AsA synthesis and salt tolerance by interacting with CSN5B. *Plant Biotechnol. J.* **2018**, *16*, 1201–1213. [CrossRef] [PubMed]
44. Qin, H.; Wang, Y.; Wang, J.; Liu, H.; Zhao, H.; Deng, Z.; Zhang, Z.; Huang, R.; Zhang, Z. Knocking down the expression of GMPase gene OsVTC1-1 decreases salt tolerance of rice at seedling and reproductive stages. *PLoS ONE* **2016**, *11*, e0168650. [CrossRef]
45. Wang, J.; Li, D.; Peng, Y.; Cai, M.; Liang, Z.; Yuan, Z.; Du, X.; Wang, J.; Schnable, P.S.; Gu, R.; et al. The anthocyanin accumulation related ZmBZ1, facilitates seedling salinity stress tolerance via ROS scavenging. *Int. J. Mol. Sci.* **2022**, *23*, 16123. [CrossRef] [PubMed]
46. Sun, Y.; Zhou, Y.; Long, Q.; Xing, J.; Guo, P.; Liu, Y.; Zhang, C.; Zhang, Y.; Fernie, A.R.; Shi, Y.; et al. OsBCAT2, a gene responsible for the degradation of branched-chain amino acids, positively regulates salt tolerance by promoting the synthesis of vitamin B5. *New Phytol.* **2024**, *241*, 2558–2574. [CrossRef] [PubMed]
47. Dong, N.Q.; Sun, Y.; Guo, T.; Shi, C.L.; Zhang, Y.M.; Kan, Y.; Xiang, Y.H.; Zhang, H.; Yang, Y.B.; Li, Y.C.; et al. UDP-glucosyltransferase regulates grain size and abiotic stress tolerance associated with metabolic flux redirection in rice. *Nat. Commun.* **2020**, *11*, 2629. [CrossRef] [PubMed]
48. Xiao, G.; Wang, M.; Li, X.; Jiang, Z.; Zhang, H.; Gao, D.; Zhang, B.; Xia, G.; Wang, M. TaCHP encoding C1-domain protein stably enhances wheat yield in saline-alkaline fields. *J. Integr. Plant Biol.* **2024**, *66*, 169–171. [CrossRef]
49. Sun, W.; Zhang, H.; Yang, S.; Liu, L.; Xie, P.; Li, J.; Zhu, Y.; Ouyang, Y.; Xie, Q.; Zhang, H.; et al. Genetic modification of Ggamma subunit AT1 enhances salt-alkali tolerance in main graminaceous crops. *Natl. Sci. Rev.* **2023**, *10*, nwad075. [CrossRef]
50. Zhang, H.; Yu, F.; Xie, P.; Sun, S.; Qiao, X.; Tang, S.; Chen, C.; Yang, S.; Mei, C.; Yang, D.; et al. A Gγ protein regulates alkaline sensitivity in crops. *Science* **2023**, *379*, eade8416. [CrossRef]
51. Ellouzi, H.; Ben, H.K.; Hernandez, I.; Cela, J.; Muller, M.; Magne, C.; Abdelly, C.; Munne-Bosch, S. A comparative study of the early osmotic, ionic, redox and hormonal signaling response in leaves and roots of two halophytes and a glycophyte to salinity. *Planta* **2014**, *240*, 1299–1317. [CrossRef] [PubMed]
52. Tao, J.J.; Chen, H.W.; Ma, B.; Zhang, W.K.; Chen, S.Y.; Zhang, J.S. The role of ethylene in plants under salinity stress. *Front. Plant Sci.* **2015**, *6*, 1059. [CrossRef]
53. Qin, H.; Wang, J.; Chen, X.; Wang, F.; Peng, P.; Zhou, Y.; Miao, Y.; Zhang, Y.; Gao, Y.; Qi, Y.; et al. Rice OsDOF15 contributes to ethylene-inhibited primary root elongation under salt stress. *New Phytol.* **2019**, *223*, 798–813. [CrossRef]
54. Huang, Y.; Guo, Y.; Liu, Y.; Zhang, F.; Wang, Z.; Wang, H.; Wang, F.; Li, D.; Mao, D.; Luan, S.; et al. 9-cis-epoxycarotenoid dioxygenase 3 regulates plant growth and enhances multi-abiotic stress tolerance in rice. *Front. Plant Sci.* **2018**, *9*, 162. [CrossRef] [PubMed]
55. Huang, Y.; Jiao, Y.; Xie, N.; Guo, Y.; Zhang, F.; Xiang, Z.; Wang, R.; Wang, F.; Gao, Q.; Tian, L.; et al. OsNCED5, a 9-cis-epoxycarotenoid dioxygenase gene, regulates salt and water stress tolerance and leaf senescence in rice. *Plant Sci.* **2019**, *287*, 110188. [CrossRef] [PubMed]
56. Feng, Z.J.; Cui, X.Y.; Cui, X.Y.; Chen, M.; Yang, G.X.; Ma, Y.Z.; He, G.Y.; Xu, Z.S. The soybean GmDi19-5 interacts with GmLEA3.1 and increases sensitivity of transgenic plants to abiotic stresses. *Front. Plant Sci.* **2015**, *6*, 179. [CrossRef]

57. Yang, Y.; Ren, R.; Karthikeyan, A.; Yin, J.; Jin, T.; Fang, F.; Cai, H.; Liu, M.; Wang, D.; Zhi, H.; et al. The soybean GmPUB21-interacting protein GmDi19-5 responds to drought and salinity stresses via an ABA-dependent pathway. *Crop J.* **2023**, *11*, 1152–1162. [CrossRef]
58. Liu, Z.; Jiang, S.; Jiang, L.; Li, W.; Tang, Y.; He, W.; Wang, M.; Xing, J.; Cui, Y.; Lin, Q.; et al. Transcription factor OsSGL is a regulator of starch synthesis and grain quality in rice. *J. Exp. Bot.* **2022**, *73*, 3417–3430. [CrossRef] [PubMed]
59. Jiang, L.; Xiao, W.; Chen, H.; Qi, Y.; Kuang, X.; Shi, J.; Liu, Z.; Cao, J.; Lin, Q.; Yu, F.; et al. The OsGAPC1-OsSGL module negatively regulates salt tolerance by mediating abscisic acid biosynthesis in rice. *New Phytol.* **2024**, *244*, 825–839. [CrossRef]
60. Ma, H.; Song, L.; Shu, Y.; Wang, S.; Niu, J.; Wang, Z.; Yu, T.; Gu, W.; Ma, H. Comparative proteomic analysis of seedling leaves of different salt tolerant soybean genotypes. *J. Proteom.* **2012**, *75*, 1529–1546. [CrossRef] [PubMed]
61. Yang, C.; Ma, B.; He, S.J.; Xiong, Q.; Duan, K.X.; Yin, C.C.; Chen, H.; Lu, X.; Chen, S.Y.; Zhang, J.S. MAOHUZI6/ETHYLENE INSENSITIVE3-LIKE1 and ETHYLENE INSENSITIVE3-LIKE2 regulate ethylene response of roots and coleoptiles and negatively affect salt tolerance in rice. *Plant Physiol.* **2015**, *169*, 148–165. [CrossRef]
62. Li, C.H.; Wang, G.; Zhao, J.L.; Zhang, L.Q.; Ai, L.F.; Han, Y.F.; Sun, D.Y.; Zhang, S.W.; Sun, Y. The receptor-like kinase SIT1 mediates salt sensitivity by activating MAPK3/6 and regulating ethylene homeostasis in rice. *Plant Cell* **2014**, *26*, 2538–2553. [CrossRef] [PubMed]
63. Colebrook, E.H.; Thomas, S.G.; Phillips, A.L.; Hedden, P. The role of gibberellin signalling in plant responses to abiotic stress. *J. Exp. Biol.* **2014**, *217*, 67–75. [CrossRef] [PubMed]
64. Magome, H.; Yamaguchi, S.; Hanada, A.; Kamiya, Y.; Oda, K. The DDF1 transcriptional activator upregulates expression of a gibberellin-deactivating gene, GA2ox7, under high-salinity stress in Arabidopsis. *Plant J.* **2008**, *56*, 613–626. [CrossRef]
65. Shan, C.; Mei, Z.; Duan, J.; Chen, H.; Feng, H.; Cai, W. OsGA2ox5, a gibberellin metabolism enzyme, is involved in plant growth, the root gravity response and salt stress. *PLoS ONE* **2014**, *9*, e87110. [CrossRef] [PubMed]
66. Zhou, J.; Li, Z.; Xiao, G.; Zhai, M.; Pan, X.; Huang, R.; Zhang, H. CYP71D8L is a key regulator involved in growth and stress responses by mediating gibberellin homeostasis in rice. *J. Exp. Bot.* **2020**, *71*, 1160–1170.
67. Wang, J.; Qin, H.; Zhou, S.; Wei, P.; Zhang, H.; Zhou, Y.; Miao, Y.; Huang, R. The ubiquitin-binding protein OsDSK2a mediates seedling growth and salt responses by regulating gibberellin metabolism in rice. *Plant Cell* **2020**, *32*, 414–428. [CrossRef]
68. Toda, Y.; Tanaka, M.; Ogawa, D.; Kurata, K.; Kurotani, K.; Habu, Y.; Ando, T.; Sugimoto, K.; Mitsuda, N.; Katoh, E.; et al. RICE SALT SENSITIVE3 forms a ternary complex with JAZ and class-C bHLH factors and regulates jasmonate-induced gene expression and root cell elongation. *Plant Cell* **2013**, *25*, 1709–1725. [CrossRef]
69. Yamada, S.; Kano, A.; Tamaoki, D.; Miyamoto, A.; Shishido, H.; Miyoshi, S.; Taniguchi, S.; Akimitsu, K.; Gomi, K. Involvement of OsJAZ8 in jasmonate-induced resistance to bacterial blight in rice. *Plant Cell Physiol.* **2012**, *53*, 2060–2072. [CrossRef]
70. Peethambaran, P.K.; Glenz, R.; Honinger, S.; Shahinul, I.S.; Hummel, S.; Harter, K.; Kolukisaoglu, U.; Meynard, D.; Guiderdoni, E.; Nick, P.; et al. Salt-inducible expression of OsJAZ8 improves resilience against salt-stress. *BMC Plant Biol.* **2018**, *18*, 311. [CrossRef]
71. Li, X.; Li, C.; Shi, L.; Lv, G.; Li, X.; Liu, Y.; Jia, X.; Liu, J.; Chen, Y.; Zhu, L.; et al. Jasmonate signaling pathway confers salt tolerance through a NUCLEAR FACTOR-Y trimeric transcription factor complex in Arabidopsis. *Cell Rep.* **2024**, *43*, 113825. [CrossRef]
72. Li, X.; Wang, Z.; Sun, S.; Dai, Z.; Zhang, J.; Wang, W.; Peng, K.; Geng, W.; Xia, S.; Liu, Q.; et al. IbNIEL-mediated degradation of IbNAC087 regulates jasmonic acid-dependent salt and drought tolerance in sweet potato. *J. Integr. Plant Biol.* **2024**, *66*, 176–195. [CrossRef] [PubMed]
73. Deng, P.; Jing, W.; Cao, C.; Sun, M.; Chi, W.; Zhao, S.; Dai, J.; Shi, X.; Wu, Q.; Zhang, B.; et al. Transcriptional repressor RST1 controls salt tolerance and grain yield in rice by regulating gene expression of asparagine synthetase. *Proc. Natl. Acad. Sci. USA* **2022**, *119*, e2210338119. [CrossRef]
74. Lo, S.F.; Cheng, M.L.; Hsing, Y.C.; Chen, Y.S.; Lee, K.W.; Hong, Y.F.; Hsiao, Y.; Hsiao, A.S.; Chen, P.J.; Wong, L.I.; et al. Rice Big Grain 1 promotes cell division to enhance organ development, stress tolerance and grain yield. *Plant Biotechnol. J.* **2020**, *18*, 1969–1983. [CrossRef] [PubMed]
75. Mao, C.; He, J.; Liu, L.; Deng, Q.; Yao, X.; Liu, C.; Qiao, Y.; Li, P.; Ming, F. OsNAC2 integrates auxin and cytokinin pathways to modulate rice root development. *Plant Biotechnol. J.* **2020**, *18*, 429–442. [CrossRef] [PubMed]
76. Nishiyama, R.; Watanabe, Y.; Fujita, Y.; Le, D.T.; Kojima, M.; Werner, T.; Vankova, R.; Yamaguchi-Shinozaki, K.; Shinozaki, K.; Kakimoto, T.; et al. Analysis of cytokinin mutants and regulation of cytokinin metabolic genes reveals important regulatory roles of cytokinins in drought, salt and abscisic acid responses, and abscisic acid biosynthesis. *Plant Cell* **2011**, *23*, 2169–2183. [CrossRef]
77. Wu, X.; He, J.; Chen, J.; Yang, S.; Zha, D. Alleviation of exogenous 6-benzyladenine on two genotypes of eggplant (Solanum melongena Mill.) growth under salt stress. *Protoplasma* **2014**, *251*, 169–176. [CrossRef]
78. Yan, Z.; Wang, J.; Wang, F.; Xie, C.; Lv, B.; Yu, Z.; Dai, S.; Liu, X.; Xia, G.; Tian, H.; et al. MPK3/6-induced degradation of ARR1/10/12 promotes salt tolerance in Arabidopsis. *EMBO Rep.* **2021**, *22*, e52457. [CrossRef] [PubMed]
79. Yin, P.; Liang, X.; Zhao, H.; Xu, Z.; Chen, L.; Yang, X.; Qin, F.; Zhang, J.; Jiang, C. Cytokinin signaling promotes salt tolerance by modulating shoot chloride exclusion in maize. *Mol. Plant* **2023**, *16*, 1031–1047. [CrossRef] [PubMed]

80. Yang, R.; Yang, Z.; Xing, M.; Jing, Y.; Zhang, Y.; Zhang, K.; Zhou, Y.; Zhao, H.; Qiao, W.; Sun, J. TaBZR1 enhances wheat salt tolerance via promoting ABA biosynthesis and ROS scavenging. *J. Genet. Genom.* **2023**, *50*, 861–871. [CrossRef] [PubMed]
81. Yu, Y.; Wang, J.; Li, S.; Kakan, X.; Zhou, Y.; Miao, Y.; Wang, F.; Qin, H.; Huang, R. Ascorbic acid integrates the antagonistic modulation of ethylene and abscisic acid in the accumulation of reactive oxygen species. *Plant Physiol.* **2019**, *179*, 1861–1875. [CrossRef] [PubMed]
82. Kakan, X.; Yu, Y.; Li, S.; Li, X.; Huang, R.; Wang, J. Ascorbic acid modulation by ABI4 transcriptional repression of VTC2 in the salt tolerance of Arabidopsis. *BMC Plant Biol.* **2021**, *21*, 112. [CrossRef] [PubMed]
83. Peng, P.; Gao, Y.; Li, Z.; Yu, Y.; Qin, H.; Guo, Y.; Huang, R.; Wang, J. Proteomic analysis of a rice mutant sd58 possessing a novel d1 allele of heterotrimeric G protein alpha subunit (RGA1) in salt stress with a focus on ROS scavenging. *Int. J. Mol. Sci.* **2019**, *20*, 167. [CrossRef] [PubMed]
84. Nie, S.; Huang, W.; He, C.; Wu, B.; Duan, H.; Ruan, J.; Zhao, Q.; Fang, Z. Transcription factor OsMYB2 triggers amino acid transporter OsANT1 expression to regulate rice growth and salt tolerance. *Plant Physiol.* **2024**. [CrossRef]
85. Miao, Y.; Lv, D.; Wang, P.; Wang, X.C.; Chen, J.; Miao, C.; Song, C.P. An Arabidopsis glutathione peroxidase functions as both a redox transducer and a scavenger in abscisic acid and drought stress responses. *Plant Cell* **2006**, *18*, 2749–2766. [CrossRef]
86. Tian, Y.; Fan, M.; Qin, Z.; Lv, H.; Wang, M.; Zhang, Z.; Zhou, W.; Zhao, N.; Li, X.; Han, C.; et al. Hydrogen peroxide positively regulates brassinosteroid signaling through oxidation of the BRASSINAZOLE-RESISTANT1 transcription factor. *Nat. Commun.* **2018**, *9*, 1063. [CrossRef] [PubMed]
87. Liu, W.C.; Song, R.F.; Qiu, Y.M.; Zheng, S.Q.; Li, T.T.; Wu, Y.; Song, C.P.; Lu, Y.T.; Yuan, H.M. Sulfenylation of ENOLASE2 facilitates H(2)O(2)-conferred freezing tolerance in Arabidopsis. *Dev. Cell* **2022**, *57*, 1883–1898. [CrossRef]
88. Ren, Y.; Jiang, M.; Zhu, J.K.; Zhou, W.; Zhao, C. Simultaneous mutations in ITPK4 and MRP5 genes result in a low phytic acid level without compromising salt tolerance in Arabidopsis. *J. Integr. Plant Biol.* **2024**, *66*, 2109–2125. [CrossRef] [PubMed]
89. Li, S.; Wang, N.; Ji, D.; Zhang, W.; Wang, Y.; Yu, Y.; Zhao, S.; Lyu, M.; You, J.; Zhang, Y.; et al. A GmSIN1/GmNCED3s/GmRbohBs feed-forward loop acts as a signal amplifier that regulates root growth in soybean exposed to salt stress. *Plant Cell* **2019**, *31*, 2107–2130. [CrossRef] [PubMed]
90. Fichman, Y.; Mittler, R. Integration of electric, calcium, reactive oxygen species and hydraulic signals during rapid systemic signaling in plants. *Plant J.* **2021**, *107*, 7–20. [CrossRef]
91. Liang, X.; Li, J.; Yang, Y.; Jiang, C.; Guo, Y. Designing salt stress-resilient crops: Current progress and future challenges. *J. Integr. Plant Biol.* **2024**, *66*, 303–329. [CrossRef] [PubMed]
92. Mao, H.; Sun, S.; Yao, J.; Wang, C.; Yu, S.; Xu, C.; Li, X.; Zhang, Q. Linking differential domain functions of the GS3 protein to natural variation of grain size in rice. *Proc. Natl. Acad. Sci. USA* **2010**, *107*, 19579–19584. [CrossRef] [PubMed]
93. Wang, J.; Zhu, R.; Meng, Q.; Qin, H.; Quan, R.; Wei, P.; Li, X.; Jiang, L.; Huang, R. A natural variation in OsDSK2a modulates plant growth and salt tolerance through phosphorylation by SnRK1A in rice. *Plant Biotechnol. J.* **2024**, *22*, 1881–1896. [CrossRef]
94. Wang, Y.; Chen, W.; Xing, M.; Sun, J.; Wang, S.; Yang, Z.; Huang, J.; Nie, Y.; Zhao, M.; Li, Y.; et al. Wild rice GL12 synergistically improves grain length and salt tolerance in cultivated rice. *Nat. Commun.* **2024**, *15*, 9453. [CrossRef] [PubMed]

antioxidants

MDPI

Review

Higher Plant-Derived Biostimulants: Mechanisms of Action and Their Role in Mitigating Plant Abiotic Stress

Sara Esperanza Martínez-Lorente †, José Manuel Martí-Guillén †, María Ángeles Pedreño, Lorena Almagro and Ana Belén Sabater-Jara *

Department of Plant Biology, Faculty of Biology, Campus of Espinardo, University of Murcia, 30100 Murcia, Spain; sesperanza.martinez@um.es (S.E.M.-L.); josemanuel.martig@um.es (J.M.M.-G.); mpedreno@um.es (M.Á.P.); lorena.almagro@um.es (L.A.)

* Correspondence: anabelen.sabater@um.es; Tel.: +34-868884904

† These authors contributed equally to this work.

Abstract: Modern agriculture is being challenged by deteriorating edaphoclimatic conditions and increasing anthropogenic pressure. This necessitates the development of innovative crop production systems that can sustainably meet the demands of a growing world population while minimizing the environmental impact. The use of plant biostimulants is gaining ground as a safe and ecologically sound approach to improving crop yields. In this review, biostimulants obtained from different higher plant sources are presented under the term higher plant-derived biostimulants (hPDBs). Their mechanisms of action regulate physiological processes in plants from germination to fructification, conditioned by responses induced in plant mineral nutrition and primary metabolism, specialized metabolism, photosynthetic processes, oxidative metabolism, and signaling-related processes. The aim of this review is to collect and unify the abundant information dispersed in the literature on the effects of these biostimulants, focusing on crops subjected to abiotic stress conditions and the underlying mechanisms of action.

Keywords: agriculture; climate change; abiotic stress; reactive chemical species; plant biostimulants; higher plant-derived biostimulants; mechanisms of action

1. Introduction

One of the main challenges facing modern agriculture is finding a sustainable way of feeding the growing world population, which the United Nations estimates will increase by nearly 2 billion people over the next 30 years, considering the decrease in crop area [1]. The use of agrochemicals to boost food production is becoming increasingly unsustainable due to indiscriminate use, prompting stricter regulations. Moreover, there is growing consumer demand for ecological products, particularly since the COVID-19 pandemic. Products with "BIO" or "ECO" labels, indicating sustainable production systems free of agrochemicals, are regarded as healthier alternatives to conventionally produced foods [2]. The need to maintain product quality standards at the highest level is evidenced by the data provided by the Food and Agriculture Organization of the United Nations (FAO), which highlights that 90% of vitamin C and 60% of vitamin A consumed by the human population come from agricultural crops [3]. Therefore, it is important to guarantee both quality standards and food security of the population.

In the move toward sustainable agriculture, the Farm to Fork Strategy, included in the European Green Deal, has set a deadline of 2030 to reduce the use of chemical pesticides by 50% as well as soil nutrient losses by at least 50%, which in turn should cut the use of fertilizers by at least 20%, with the intention of allocating 25% of agricultural areas to organic farming [4,5].

These limitations and reductions in the use of agrochemicals have been the consequence of years of their abuse in agriculture, facts that have manifested themselves in the deterioration of soils, along with the potential damage to human health and the environment. Among them, the uncontrolled use of nitrogen-rich fertilizers has triggered eutrophication processes, with consequent severe environmental damage, as well as nitrate accumulation in plants, exceeding the regulation limits that allow their consumption. In fact, the nitrates present in these foods are metabolized into potentially carcinogenic compounds that are harmful to human health [6].

Another challenge facing modern agriculture is the diminishing availability of land for crop production due to climate change. Alterations in rainfall patterns and escalating global temperatures are leading to the aridification of arable lands, rendering them unprofitable for food production [7].

All these factors indicate there is an urgent need to develop more sustainable agricultural systems capable of providing for the growing population while minimizing the environmental impact. Overcoming the increasingly adverse anthropological and edaphoclimatic conditions that limit production performance is imperative. Moreover, the long processes of genetic improvement through breeding developed over the years in crops are reaching the limit of their potential. Given that improving crop tolerance to climate change by genetic modifications or in vitro selection takes years to accomplish, it is of paramount interest to search for alternative strategies with a more immediate impact [3].

2. Climate Change and Plant Abiotic Stress

Climate change, in essence, is a phenomenon that affects the entire planet at the global level. Its manifestation includes variations in weather conditions, especially the increase in global temperatures and the disruption of precipitation patterns, with serious consequences for global agricultural production. It results in a reduction in annual yields in terms of both quantity and quality, as well as the annual economic performance of the crops [8]. The adverse conditions caused by climate change result in abiotic stresses on plants. Among these stresses, extreme temperatures, prolonged water deficit, high soil salinity, and contamination by potentially hazardous elements (PHEs) stand out (Figure 1).

Stress due to extreme temperatures occurs when the temperature range for plant growth and development is exceeded above or below the basal maximum and minimum. Extreme cold and sudden drops in temperature, as well as extreme heat, which occurs more frequently, restrict and limit the physiological processes of temperature-sensitive plants, affecting their growth and development [9]. Stress due to water deficit or drought occurs when the availability of water in the soil decreases drastically, generating a water imbalance in the transpiration rate, which exceeds the intake of water absorbed by the root system, or causing the plants to close their stomata to avoid the previous, and thus preventing them from absorbing CO_2 for photosynthesis. As well as insufficient precipitation, water deficit or drought are caused by increased water evaporation under conditions of prolonged heat, leading to elevated soil salt concentrations that impede optimal plant growth [9,10]. Similar effects are caused by soil contamination with PHEs, such as boron (B), aluminum (Al), copper (Cu), zinc (Zn), lead (Pb), vanadium (V), lanthanum (La), nickel (Ni), arsenic (As), cadmium (Cd), and chromium (Cr), essentially arising from human activities near crop areas, such as industrial effluents and global pollution [11].

As sessile organisms, plants are exposed throughout their life cycle to different abiotic stresses, which can act individually, but generally occur in combination. This aggravates the damage and generates unique responses that cannot be directly extrapolated from the responses to the individually applied stresses, involving metabolic and physiological mechanisms that allow them to acclimate and, consequently, tolerate exposure to adverse conditions [12,13]. The immediate response of plants to stress is a massive increase in the production of reactive chemical species, including reactive oxygen (ROS), nitrogen (RNS), and sulfur (RSS) species. These molecules, present at low concentrations in plants under non-stress conditions, act as signaling molecules in cellular events, but under stress conditions, their overproduction disrupts redox homeostasis and damages essential biomolecules and cellular components. This oxidative, nitrosative, or nitro-oxidative stress can potentially lead to cell death. However, cells employ diverse mechanisms to ameliorate stress, achieving tolerance and acclimation, allowing plant survival. These mechanisms include the expression of genes related to antioxidant defense; the synthesis and activation of enzymes and antioxidant compounds (e.g., melatonin, glutathione, ascorbic acid, and flavonoids) that counteract the overproduction of reactive chemical species; and the transcription of genes encoding heat shock proteins, LEA proteins, osmoprotectant biosynthesizing enzymes, stress-related transcription factors, aquaporins, and ion channels, among others [14–17].

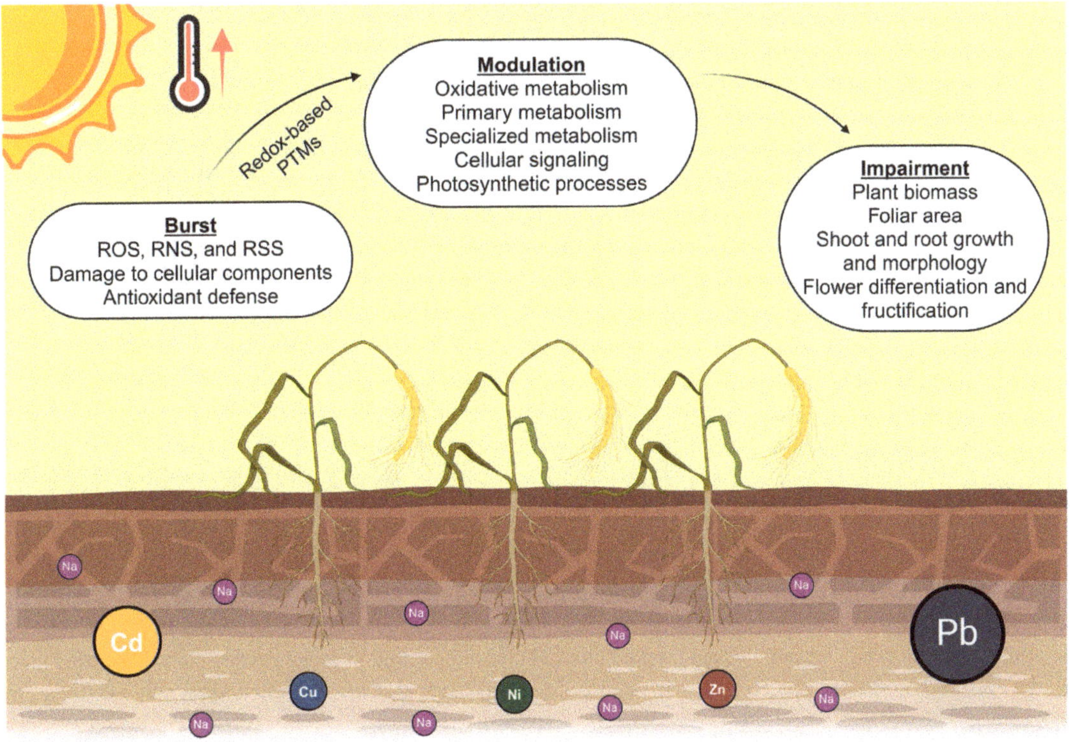

Figure 1. Schematic representation of the main abiotic stresses (drought, salinity, heat, and PHEs) mentioned in the text and the responses triggered in plants subjected to them. Created with BioRender.com, accessed on 1 March 2024.

In turn, ROS, RNS, and RSS have the capacity to generate post-translational modifications (PTMs) in target proteins, altering their physiological, structural, and functional properties, which affects their stability and affinity with other proteins, biomolecules, or related metabolites. The majority of PTMs are highly specific of each protein, affecting particular amino acid residues with physical–chemical properties that render them susceptible to these types of redox-based PTMs. The modifications induce conformational changes that can either promote or inhibit enzyme catalytic activity. After reactive chemical species burst, the proteome susceptible to these redox-based PTMs acquires new modifications that can compromise structural proteins and key enzymes, but also initiate the plant acclimation processes. It has been described that reactive species initiate plant acclimation processes by modifying key amino acid residues in proteins associated with oxidative metabolism, primary and specialized metabolisms, cellular signaling, photosynthetic processes, photorespiration, and Calvin–Benson cycle [18]. The modification of structural proteins and key enzymes of cellular metabolism determines the correct functioning of essential physiological processes such as cell division, photosynthesis, chlorophyll biosynthesis, photorespiration, balance between water absorption and transpiration, stomatal opening and closing, and translocation of essential nutrients [9,18].

Consequently, the imbalance of plant homeostasis induces morphological changes, such as reduced growth of aerial parts, including leaf expansion, thickness, and area; growth and deepening of roots in search of water; and disruption of the differentiation of male and female flowers. Ultimately, these changes negatively affect the total biomass, plant vigor, and fructification [19–21].

3. Plant Biostimulants: An Emerging Ecological Alternative

The use of biostimulants among sustainable agricultural practices is gaining ground as a promising, safe, and ecological alternative to improving crop production performance.

Prior to the term "biostimulant", the terms "biogenic stimulators" or "biogenic stimulants" were used to refer to substances synthesized in tissues under stressful, but not lethal, conditions, which stimulated the vital reactions of the organism [22]. The term "biostimulant" was used for the first time in a research article by Russo and Berlyn published in 1991 [23]. These authors went on to define biostimulants in 1992 [24] as "non-nutritional products that may reduce fertilizer use and increase yield and resistance to water and temperature stresses", also remarking that they "stimulate plant growth [when used] in relatively small amounts" [25].

Despite the recent exponential increase in research on biostimulants, controversy remained regarding their precise definition. Until official regulation by the European Union in 2019, the status of biostimulants differed between member states and they were marketed as biofertilizers in mixtures with nutritional elements [26,27]. Nevertheless, by 2019, a consensus on the key attributes of a biostimulant had been reached. These include the ability to act on plant homeostasis at low doses, improve plant growth, induce a more efficient use of nutrients and water, modulate abiotic stress response, and exert synergistic effects resulting from the combination of bioactive compounds [25]. Finally, Europe laid the foundations for the regulation of these products, with the publication of Regulation (EU) 2019/1009 of the European Parliament and of the Council of 5 June 2019, where biostimulants are defined as [28]:

"A plant biostimulant shall be an EU fertilising product the function of which is to stimulate plant nutrition processes independently of the product's nutrient content with the sole aim of improving one or more of the following characteristics of the plant or the plant rhizosphere:

(a) nutrient use efficiency,
(b) tolerance to abiotic stress,
(c) quality traits, or
(d) availability of confined nutrients in the soil or rhizosphere."

Accordingly, it was established that the main difference between a biofertilizer and a biostimulant is that a biofertilizer participates in the provision of nutrients to the plant, whereas biostimulants favor nutrient acquisition [29]. Furthermore, in contrast to plant defense elicitors, which provide a protection against biotic stress by inducing systemic acquired resistance, biostimulants act by providing tolerance to abiotic stress [30].

The lack of a clear definition due to the appearance of the concept of biostimulants in the scientific world has compromised the way of properly classifying them. The wide diversity of biostimulants, in both origin and composition, as well as the physiological functions triggered by their application, has complicated their classification, although the scientific community has widely accepted the proposal of Patrick du Jardin in 2015 [31], who divided biostimulants into seven categories. Humic substances, which include humic and fulvic acids, correspond to heterogeneous organic molecules resulting from the decomposition of organic remains in the soil. Protein hydrolysates (PHs) are mixtures of variable proportions of free amino acids, oligopeptides, and polypeptides, obtained through the chemical and enzymatic hydrolysis of proteins from various sources, both animal and plant, generally obtained from industrial by-products and waste. Seaweed and botanical extracts are variable and heterogeneous mixtures that, depending on the origin of the extract, include complex polysaccharides, phenolic compounds, and hormones, among others. Biopolymers, which include a wide diversity of molecules of highly variable size and characteristics, are obtained by extraction or industrial synthesis. Inorganic compounds correspond to beneficial elements, mainly Al, cobalt (Co), Na, selenium (Se), and silicon (Si), present as different inorganic salts. Beneficial fungi and bacteria constitute a very broad and varied group that include symbiotic microorganisms and plant growth promoters [31].

Despite their heterogeneity, biostimulants are generally applied in three different ways: through seed priming, application to soil (or to nutrient solution for hydroponic crops), and in foliar spray (Figure 2). In seed priming, the seeds are soaked in a solution containing the biostimulant to enable ingredient penetration. In the case of soil application, the product effectiveness depends on adequate soaking, while the application in hydroponic cultivation is more straightforward, with the desired concentration simply being added to the nutrient solution. In foliar spraying, the biostimulant is applied in solution when the seedling reaches a minimum age, typically after the generation of several true leaves. The frequency of application varies according to the biostimulant used [32]. The efficacy of a biostimulant hinges on the penetration of active ingredients into the seeds, roots, or leaves, and their assimilation is based on factors such as molecular structure, particle size, and solubility. Additives are commonly used to optimize solubility, absorption capacity, and penetration into the plant material [33].

Biostimulants from natural sources have attracted attention for their sustainability, ecological advantages, and biodegradability, ensuring minimum or null environmental impact. Among them, higher plant-derived biostimulants (hPDBs) stand out, above animal-derived biostimulants (ADBs), because their sustainability and profitability are superior. In fact, the production of ABDs generates more CO_2 emissions (+57%) and consumes more energy (+26%) and water [34]. Furthermore, animal-derived by-products, the source of most ADBs, represent a potential risk for the consumer through disease transmission [35]. Regarding microorganism-derived biostimulants, which have also attracted a lot of attention, there is controversy about their application, because some microbial products are not normally present in agricultural fields [22].

We define the term hPDBs as group of biostimulants that gather PHs, extracts from plant by-products, whole plants or specific organs, purified metabolites, and cell cultures derived from higher plants, included in the Tracheophyta phylum (Figure 2). Biostimulants, by definition, are effective in inducing physiological processes in plants that improve their growth and acclimation to stress, concluding in an increase in crop yield. Furthermore, the generation of abundant plant by-products after obtaining the marketable and consumable parts of the crop is very common, so they can be revalued and used to increase crop production and profitability while reducing the environmental impact of waste disposal [36]. hPDBs are of great interest due to the enormous variability of biomolecules that compose them, from peptides to specialized metabolites, which is the reason they can potentially induce a multitude of physiological processes that lead to an increased crop yield [32].

Therefore, this review brings together the knowledge, abundant but widely dispersed in the literature, concerning hPDBs and their mechanisms of action when they are applied to plants, focusing on crops subjected to abiotic stress conditions. Table 1 shows the main hPDBs used in different crops reviewed, as well as their application method.

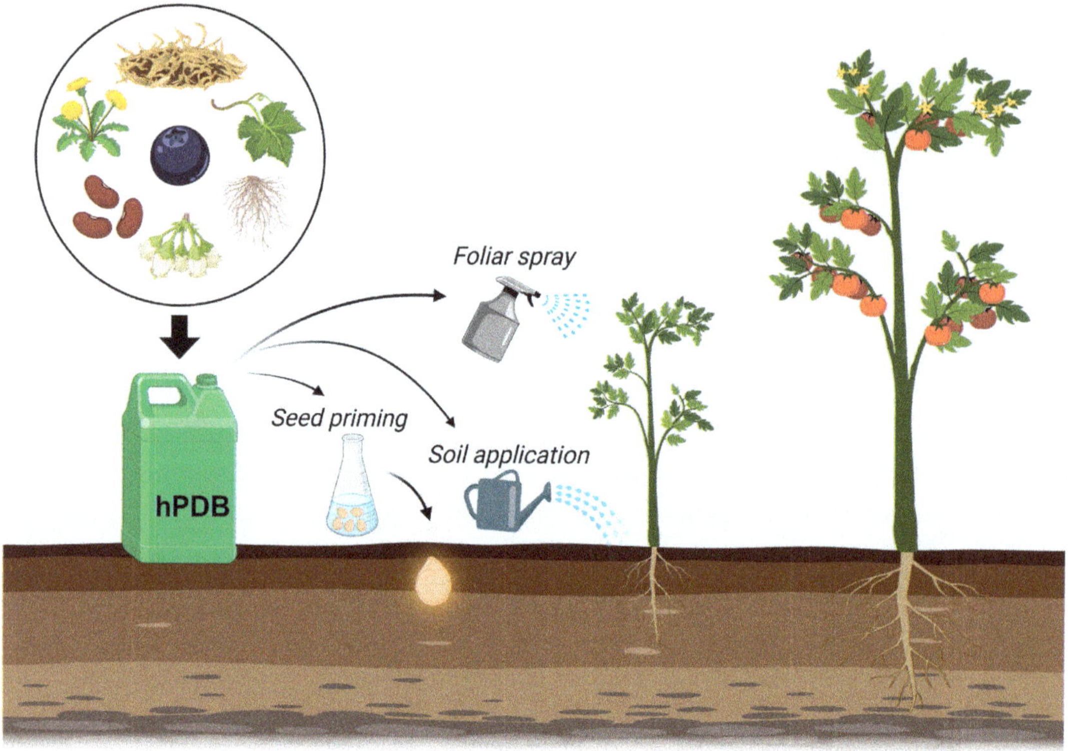

Figure 2. hPDB main sources and application method. Created with BioRender.com, accessed on 31 January 2024.

Table 1. Main hPDBs used in crops and their application method.

Crop	hPDB	Application Method	Ref.
Arabidopsis (*Arabidopsis thaliana*)	Moringa (*Moringa oleifera*) leaf extract, Trainer®, Vegamin®, and PHs from Fabaceae, Malvaceae, Brassicaceae, Solanaceae, and Graminaceae	Seed priming	[37,38]
Cabbage (*Brassica oleracea*)	Extracts from mugwort (*Artemisia vulgaris*), calendula (*Calendula officinalis*) flowers, purple coneflower (*Echinacea purpurea*) flowers and leaves, chamomile (*Matricaria chamomilla*) flowers, basil (*Ocimum basilicum*), giant goldenrod (*Solidago gigantea*) leaves, comfrey (*Symphytum officinale*) root, dandelion (*Taraxacum officinale*) flowers, leaves, and root, valerian (*Valeriana officinalis*) root, aloe vera (*Aloe vera*) leaves, chokeberry (*Aronia melanocarpa*) fruits, red beet (*Beta vulgaris*) root, horsetail (*Equisetum arvense*), common sea buckthorn (*Hippophae rhamnoides*) fruits, hypericum (*Hypericum perforatum*), red lentil (*Lens culinaris*) seeds, common bracken (*Pteridium aquilinum*) leaves, knotgrass (*Polygonum aviculare*), pea (*Pisum sativum*) seeds, broadleaf plantain (*Plantago major*), red clover (*Trifolium pratense*) flowers, and nettle (*Urtica dioica*) leaves and root	Foliar spray	[39,40]

Table 1. *Cont.*

Crop	hPDB	Application Method	Ref.
Camelina (*Camelina sativa*)	Shorgum (*Shorgum* sp.) water extract	Seed priming	[41]
Cannabis (*Cannabis sativa*)	Aloe vera (*Aloe vera*) extract and aloe vera, fish, and kelp complex	Nutrient solution	[42]
Cascading geranium (*Pelargonium peltatum*)	Trainer®	Foliar spray	[43]
Common bean (*Phaseolus vulgaris*)	Extracts from fish bean (*Tephrosia vogelii*) and tree marigold (*Tithonia diversifolia*), garlic (*Allium sativum*) cloves, licorice (*Glycirrhiza glabra*) root, moringa (*Moringa oleifera*) leaves, and tomato (*Solanum lycopersicum*) powder hydrolysate.	Foliar spray, seed priming, and substrate application	[44–52]
Cowpea (*Vigna unguiculata*)	Extracts from fennel (*Foeniculum vulgare*) and ammi (*Ammi visnaga*) seeds	Foliar spray	[53]
Eggplant (*Solanum melongena*)	Sugarbeet (*Beta vulgaris*) extract	Foliar spray	[54]
Genovese basil (*Ocimum basilicum*)	Trainer®	Foliar spray	[55]
Gladiolus (*Gladiolus grandiflorus*)	Extracts from moringa (*Moringa oleifera*) leaves, garlic (*Allium sativum*) extract, and licorice (*Glycyrrhiza glabra*) root	Foliar spray	[56,57]
Grapevine (*Vitis vinifera*)	Trainer®, Stimtide®, and soybean (*Glycine max*) hydrolysate and extracts from french oak (*Quercus sessiliflora*), vine-shoot (*Vitis vinifera*) waste, and carob germ (*Ceratonia silique*) and from mixture of maize (*Zea mays*) and sorghum (*Sorghum* sp.) distiller's dried grains.	Foliar spray and substrate application	[58–64]
Habanero pepper (*Capsicum chinensis*)	Red grape (*Vitis vinifera*) skin extract and alfafa (*Medicago sativa*) hydrolysate	Foliar spray	[65]
Kinnow mandarin (*Citrus nobilis* × *Citrus deliciosa*)	Moringa (*Moringa oleifera*) leaf extract	Foliar spray	[66]
Lamb's lettuce (*Valerianella locusta*)	Trainer®	Foliar spray	[67]
Lettuce (*Lactuca sativa*)	Auxym®, LISIVEG®, Trainer®, Vegamin®, and PHs from Fabaceae, Malvaceae, Brassicaceae, Solanaceae, and Graminaceae. Extracts from moringa (*Moringa oleifera*) leaves and leaves and flowers of borage (*Borago officinalis*).	Foliar spray, seed priming, root application, and substrate application before transplant.	[68–75]
Lilium 'Brindisi' (*Lilium longiflorum* × *Lilium x elegans*)	Alfalfa (*Medicago sativa*) hydrolysate	Foliar spray and substrate application	[76]
Maize (*Zea mays*)	Lignin nanoparticles from olive (*Olea europaea*) waste, Trainer®, hydrolysates from alfalfa (*Medicago sativa*), dry apple (*Malus domestica*), and extracts from carrot (*Daucus carota*) root, blueberry (*Vaccinium corymbosum*) fruits, duckweed (*Lemna minor*), hawthorn (*Crataegus monogina*) leaves, common grapevine (*Vitis vinifera*) grape skin, rosemary (*Rosmarinus officinalis*), white wormwood (*Artemisia herba-alba*), whortleberry (*Vaccinium arctostaphylos*) fruit, and willow tree (*Salix babylonica*) barks and leaves	Foliar spray, nutrient solution, seed priming, and substrate application	[72,77–87]
Mango (*Mangifera indica*)	Roselle (*Hibischus sabdariffa*), garlic (*Allium sativum*) clove, and algae extracts alone or combinated	Foliar spray	[88]

Table 1. *Cont.*

Crop	hPDB	Application Method	Ref.
Milk thistle (*Silybum marianum*)	Moringa (*Moringa oleifera*) leaf extract	Substrate application	[89]
Mung bean (*Vigna radiata*)	Cereal PHs and angosteen (*Garcinia mangostana*) pericarp extract	Foliar spray and seed priming	[90,91]
Okra (*Abelmoschus esculentus*)	Sugarbeet (*Beta vulgaris*) extract	Foliar spray	[92]
Olive (*Olea europaea*) tree	Duckweed (*Lemna minor*) plant extract	Foliar spray	[93]
Onion (*Allium cepa*)	Mimosa (*Acacia dealbata*) bark extract	Foliar spray	[94]
Pea (*Pisum sativum*)	Trainer® and extracts from licorice (*Glycyrrhiza glabra*) root, moringa (*Moringa oleifera*) leaves, and narrow-leaf cattail (*Typha angustifolia*) leaves	Foliar spray, seed priming, and shoot application	[79,95–97]
Perennial wall rocket (*Diplotaxis teniufolia*)	Auxym®, Trainer®, and their combination	Foliar spray	[98,99]
Potato (*Solanum tuberosum*)	Moringa (*Moringa oleifera*) seed extract	Substrate application	[100]
Pulasan (*Nephelium ramboutan-ake*)	ComCat®	Foliar spray	[101]
Radish (*Raphanus sativus*)	Extracts from leaves of mulberry (*Morus nigra*), brassica (*Brassica napus*), sorghum (*Sorghum bicolor*), and moringa (*Moringa oleifera*)	Foliar spray	[102]
Rice (*Oryza sativa*)	Cereal PHs	Foliar spray	[91]
Rocket (*Eruca vesicaria*)	Moringa (*Moringa oleifera*) leaf and twig extracts	Foliar spray	[103]
Rose-scented geranium (*Pelargonium graveolens*)	Moringa (*Moringa oleifera*) leaf extract	Foliar spray	[104]
Spinach (*Spinacia oleracea*)	Amalgerol® and Trainer®	Foliar spray	[67,105,106]
Squash (*Curcubita pepo*)	Moringa (*Moringa oleifera*) leaf extract	Foliar spray	[107]
Sweet cherry (*Prunus avium*)	Auxym®	Foliar spray	[108]
Sweet pepper (*Capsicum annuum*)	Radifarm®, Megafol®, Viva®, and Benefit® and extracts from moringa (*Moringa oleifera*) seeds and leaves and licorice (*Glycyrrhiza glabra*) root	Foliar spray, nutrient solution, and substrate application	[109–111]
Tomato (*Solanum lycopersicum*)	Auxym®, Trainer®, Vegamin®, PHs from alfalfa (*Medicago sativa*), Fabaceae, Malvaceae, Brassicaceae, Solanaceae, and Graminaceae and extracts from crushed maize (*Zea mays*) grain, garlic (*Allium sativum*) cloves, seagrass (*Zostera marina*), barley (*Hordeum vulgare*), and grain waste and processing residues from fennel (*Foeniculum vulgare*) and lemon (*Citrus limon*)	Foliar spray, seed priming, nutrient solution, cutting immersion, and substrate application	[68,79,112–123]
Wheat (*Triticum aestivum*)	Cereal PHs and shorgum (*Shorgum* sp.) water extract	Foliar spray and seed priming	[91,124]
White rocket (*Diplotaxis erucoides*)	Auxym® and Trainer®	Foliar spray	[6]
Wild mint (*Mentha arvensis*)	Calliterpenone from large-leaf beauty berry (*Callicarpa macrophylla*) extract	Sucker immersion	[125]
Wild mustard (*Brassica rapa*)	Soybean (*Glycine max*) waste hydrolysate	Root application	[126]
Wild tomato (*Solanum pimpinellifolium*)	Carob (*Ceratonia siliqua*) germ hydrolysate extract	Substrate application	[127]
Zucchini (*Curcubita pepo*)	Cypress (*Cupressus macrocarpa*) leaf extract	Seed priming	[128]

4. Higher Plant-Derived Biostimulants: Mechanisms of Action

Firstly, it is necessary to highlight the differences between mechanisms and modes of action. According to Yakhin et al. (2017) [22], the mode of action refers to the "specific effect on a discrete biochemical or regulatory process", while the mechanism of action involves the "impacts on general biochemical or molecular pathways or physiological processes". As biostimulants have not been characterized at the very specific level necessary to establish their modes of action, their functions in plants are mainly understood based on the underlying mechanisms of action. The research developed on the mechanism of action of biostimulants requires elucidating the composition and bioactive ingredients of the formulation, as well as the mechanism of action that induces modifications in plant metabolism. To perform these determinations, powerful omics tools (transcriptomics, metabolomics, proteomics, and phenomics) can generate abundant data on changes in mRNA transcripts, metabolites, proteins, and phenotypes. Thus, the application of these techniques is essential to be able to obtain this knowledge, crucial to customize formulations for specific crop applications [35,129,130]. Furthermore, determining the effect of biostimulants is not only recommended to obtain an optimal formulation, but members of the European Biostimulant Industry Council (EBIC) consider it an essential requirement for their commercialization [131].

The wide range of plant processes in plants modulated by biostimulants are presented here in two sections, beginning with the mechanisms of action at the molecular or cellular level and then in terms of the whole plant.

4.1. Cellular and Molecular Levels

4.1.1. Plant Mineral Nutrition and Primary Metabolism

Nutrient use efficiency and assimilation, as well as carbon (C) energy metabolism, are processes of plant primary metabolism, which constitutes a large part of the physiological reactions the plant carries out to perform its vital functions. The uptake and assimilation of nutrients from the soil play a decisive role in proper plant development, as nutrients are needed to produce essential metabolites and enzymes in addition to acting as cofactors. Various types of biostimulants are reported to enhance the uptake of phosphorus (P), sulfur (S), calcium (Ca), magnesium (Mg), potassium (K), iron (Fe), and Zn, increasing their concentration and intracellular bioavailability in spinach [105,106], perennial wall rocket [98,99], rocket [103], tomato [117,122], maize [83], olive tree [93], kinnow mandarin [66], cannabis [42], and rose-scented geranium [104]. In a study carried out in tomato plants treated with an alfalfa-based PH, an increase was observed in the transcription of transporters related to ion uptake, such as sulfate transporters *SULTR 2;1* and *SULTR 3;1*, Cu transporters, phosphate transporter *PT2*, Fe-phytosiderophore transporter protein yellow stripe 1 (*YS1*), K channels, and ABC transporters [123]. Thus, by improving fertilizer efficiency, biostimulants could reduce its application dosage. Additionally, hPDBs exhibit protective effects against the excessive accumulation of PHEs in plants grown in contaminated soils, modulating the levels of Cu [55], Cd [46], Pb, and Ni [111]. The application of biostimulants has also proven beneficial in mitigating the toxic effects of salt stress. An imbalance in the K/Na ratio causes ionic and osmotic stress in the cell, which also affects the correct assimilation of other ions. This imbalance was alleviated by biostimulants in maize [84,132], camelina [41], common bean [45,49], pea [97], lettuce [73], eggplant [54], and okra [92]. Moreover, this approach was also able to balance the levels of other essential ions such as Ca, S, P, K, Mg, and (Mn) [45,53,69,70].

Nitrogen (N) has special relevance among the nutrients assimilated by plants, as it is used to form amino acids after fixation as organic N from nitrate. The application of hPDBs has resulted in a higher N content in leaves of different crops, including cascading geranium [43], spinach [67], maize [72,83], dwarf pea [79], tomato [122], and rocket [103], an effect also observed under salt stress conditions [45,50,53,74]. This increase in N concentration allows for a reduction in fertilizer dosage, thus maintaining nitrate levels within the legal limits set by the European Commission [6,71]. Recent studies in tomato plants treated with different PHs found that greater N assimilation was not only related to the enhanced transcription of genes encoding transporters of nitrate (*NTR2*) and ammonium (*AMT1.1* and *AMT1.2*), increasing the translocation from soil to the root, but also to an upregulation of the genes involved in N metabolism and the synthesis of amino acids such as nitrate reductase (*NR*), nitrite reductase (*NiR*), aspartate aminotransferase (*AST*), glutamine-dependent asparagine synthetase (*ASN1*), ferredoxin-dependent glutamate synthase (*GLT*), NADH-dependent glutamate synthase (*GLS*), glutamate dehydrogenase (*GDH*), and glutamine synthetase (*GS1* and *GS2*) [115,123]. Moreover, an increase in the activity of NR, GS, and glutamate synthase (GOGAT) [44] was demonstrated in maize [85,86]. In fact, in maize plants subjected to saline conditions, treatment with an alfalfa-based PH also increased the activity of GS and GOGAT [87]. Therefore, an enhanced uptake of N and its consequent incorporation into amino acids also improves its bioavailability in plant tissues [52,105]. Furthermore, transcriptomic experiments have revealed that tomato plants treated with hPDBs have higher transcript levels of the amino acid transporter *AAT1* [115], organic cation/carnitine transporters, nodulin *MtN21*, glutathione-conjugate transporter *MRP4*, and other N-associated enzymes (aspartyl protease, glutamate dehydrogenase, and serine decarboxylase), as well as genes encoding proteins involved in protein synthesis and modification (aminoacyl-tRNA synthetases, elongation factors Tu and 1-alpha, translation initiation factors, and ubiquitin-conjugating enzymes) [123]. This surge in protein synthesis leads to an increase in total proteins, a parameter improvement described in maize [83], wheat, rice, mung bean, and common bean plants [44,91] following biostimulant application. Under stress conditions, higher protein [94,124] and amino acid [84] concentrations were reported in treated versus untreated plants, with a notable increase in the amount of proline, an amino acid involved in stress responses [50,74]. In grapevines grown under conditions of water scarcity, the application of commercial biostimulant formulations Trainer® and Stimtide® increased the accumulation of bifunctional aspartokinase/homoserine dehydrogenase 2, involved in protein metabolism [61].

Sugars and carbohydrates are essential for plant growth and development, serving as both a source of energy and essential components for the synthesis of structural molecules. The application of biostimulants can increase the levels of soluble sugars in plants, as has been shown in maize [81], kinnow mandarin [66], rocket [103], habanero pepper [65], and tomato [117], among others. Furthermore, in tomato plants, biostimulants enhanced the transcription of genes encoding enzymes involved in sugar metabolism (polygalacturonase, pectinesterase, starch synthase, sucrose synthase, cellulose synthase, and inositol oxygenase), as well as key enzymes of C metabolism (such as fumarate dehydrogenase, malate dehydrogenase, phosphoenolpyruvate carboxylase, and phosphoenolpyruvate carboxylase kinase 2) [123]. In monocotyledons, such as maize, the application of hPDBs resulted in increased transcription and activity of malate dehydrogenase, isocitrate dehydrogenase, and citrate synthase [85]. A striking response was observed in maize plants treated with cellulosolitic dry apple hydrolysate or a dry blueberry cool extract, which led to an increase in fructose concentration and a decrease in glucose and sucrose. By stimulating plant respiration, the treatment promoted the consumption of glucose, thus limiting its availability for sucrose synthesis and resulting in the accumulation of fructose [83].

Sugars and carbohydrates, in turn, are used for the synthesis of other primary metabolites, such as nucleosides, in whose chemical structure a ribose or deoxyribose pentose is attached, together with a nitrogenous base, constituting the different types of nucleosides. It has been shown that treatment with different hPDBs increases thymidine accumulation in habanero pepper fruits [65].

The application of hPDBs in crops subjected to stress conditions generally causes an increase in soluble sugar concentration, providing osmoprotective benefits in both dicotyledons [48,111] and monocotyledons [84]. In Arabidopsis plants exposed to salinity and treated with plant-derived PHs, maltose accumulation was observed. Maltose, a product of starch degradation [37], plays a protective role in response to stress by safeguarding the electron transport chain of the chloroplast, thereby preserving photosynthetic function [133].

The higher maltose levels could be related to the increased activity of α-amylase, which degrades starch to maltose units, as observed in camelina [41] and wheat [124] treated with sorghum water extract under salinity conditions. Furthermore, the application of narrow-leaf cattail extracts in pea plants subjected to salinity preserved metabolic activity, quantified as the amount of formazan accumulated in the viability test, and the respiration rate was higher compared to untreated plants [95].

With respect to lipid metabolism, an increase in the transcription of genes encoding lipases has been described in tomato plants treated with an alfalfa-based PH [123], whereas other studies in tomato reported an accumulation of structural lipids such as hexadecanediol and hydroxy stearate [112]. Under limited water availability, the pretreatment of tomato plants with Trainer® caused a modulation, both positive and negative, of membrane lipids and sterol accumulation [118].

4.1.2. Specialized Metabolism

Specialized metabolites, distinct from compounds essential for plant growth, development, and reproduction, are often restricted to a particular taxonomic group. They are produced from primary metabolites, normally at specific stages of development and in response to certain conditions. These compounds intervene in the dynamic interactions between plants and their environment, acting as structural or regulatory molecules or hormones. Specialized metabolites exhibit diverse ecological functions, serving as attractants, repellents, or even natural pesticides. A large proportion of specialized metabolites are pigments responsible for the colors of flowers and fruits. These pigments are essential for reproduction, attracting species crucial for pollination, and seed dispersal. Specialized metabolites also participate in protective functions under both abiotic and biotic stress conditions. Commonly, most of these compounds participate in responses to both types of stresses. Thus, the modulation in the content of these metabolites by hPDBs would not only determine their tolerance to abiotic stress, but also suggest a potential role as plant defense elicitors against biotic stress. Based on their chemical structure, specialized metabolites can be classified into three groups: terpenoids, phenolics, and nitrogen-containing compounds [134–137].

Terpenoids

Terpenoids are synthesized from acetyl-CoA, which is transformed via the mevalonic acid (MVA) and 2-C-methylerythritol 4-phosphate (MEP) pathways into the isoprene unit isopentenyl pyrophosphate (IPPP), the main terpene building block [136]. hPDBs exhibit a dual regulatory role in terpenoid accumulation. They are reported to negatively modulate terpenoid levels under various conditions, including those of control conditions [37,112], water scarcity [118], and saline stress [68], or increase their concentration under both control conditions [112,116] and Cu toxicity [82]. In tomato, the application of an alfalfa-based PH led to an upregulation of terpene biosynthetic genes, including hydroxy-3-methylglutaryl coenzyme A reductase, isoflavone reductase, and terpene synthase [123].

Carotenoids, a subset of terpenoids that act as antioxidants and pigments involved in photosynthesis, are positively influenced by various hPDBs. Biostimulant application enhanced carotenoid accumulation in leaves of different plant species, both under control [6,103,104] and stress conditions [95], and increased lycopene concentration in tomato fruits [113,114,117].

Terpenoid phytoalexins, which function as defense compounds, were found to accumulate after treatment with biostimulants [112,116]. The accumulation of phytoalexin precursors and terpenoid phytoalexins in lettuce was especially remarkable under abiotic stress conditions [69]. On the other hand, in Genovese basil, treatment with Trainer® alleviated terpenoid and phytoalexin accumulation after exposure to high Cu levels, which indicates the formulation exerted a protective effect against the oxidative stress induced by this metal [55].

Terpenoids are also among the volatile compounds responsible for the quality of certain crops. In different types of grapevine, the fruits had higher terpene and norisoprenoid concentrations after treatment with vine-shoot extracts, particularly of farnesol, citronellol, geraniol, and linalool [59,60]. Similarly, other plant extracts increased the concentration of menthol in wild mint [125] and geraniol, linalool, citronellol, and β-caryophyllene in rose-scented geranium [104].

Phenolic compounds

Phenolic compounds constitute one of the most important groups of plant metabolites, as they participate in a multitude of physiological processes. They can be synthesized by the shikimate/phenylpropanoid or polyketide acetate/malonate pathways [138]. In general, it has been observed that hPDBs can modulate phenolic concentrations in plant tissues. The application of higher plant-derived PHs and extracts has been found to increase phenolic compounds in different organs, such as seedlings [40], fruits [66,109,117,121], leaves [99,104,106], and roots [102,112], of multiple crops. Various types of hPDBs are reported to enhance the activity of phenylalanine ammonia-lyase (PAL), a key enzyme in the phenylpropanoid synthetic pathway, both in monocotyledons [81,87] and dicotyledons [75]. *PAL* transcription can also be promoted by hPDBs. In maize, the application of cellulosolitic dry apple hydrolysate or dry blueberry cool extract stimulated *PAL* activity and expression [83], while in tomato, an alfalfa-based PH induced the upregulation of *PAL* and *2-oxoglutarate-dependent dioxygenase* genes [123], resulting in higher levels of phenolic compounds. Furthermore, the induction of phenolic accumulation by hPDBs is also described in crops subjected to salinity [84,124], an effect which could alleviate the oxidative stress associated with this condition, given the antioxidant capacity of phenolic compounds. Conversely, hPDBs are also reported to reduce phenolic concentrations in spinach plants [105], tomato roots [112], cabbage seedlings [39], and Genovese basil subjected to Cu toxicity [55].

Flavonoids are low-molecular-weight phenolic compounds which participate in numerous physiological processes, mainly as antioxidant systems in response to stress [139], and several hPDBs are reported to increase flavonoid levels in plants [52,75,80]. For example, plants treated with hPDBs under conditions of salinity [95] or Cu toxicity [82] showed higher flavonoid levels compared to controls, indicating that the treatment reinforces plant antioxidant systems. Conversely, plants treated with hPDBs have also shown lower flavonoid levels than untreated controls [37] or those subjected to water scarcity [61]. A possible explanation is that the treated plants were healthier, with lower levels of ROS, given that the biostimulant acted to alleviate stress independently of stimulating flavonoid synthesis.

Lignin, the most abundant phenolic compound in nature, has an important structural function, so the correct regulation of its synthesis is essential for plant development. In tomato, the application of an alfalfa-based PH induced a positive regulation of *caffeoyl-CoA 3-O-methyltransferase*, a key gene in lignin biosynthesis [123].

The organoleptic quality of fruits is partly determined by their content of specialized metabolites. Treatment of habanero peppers with red grape skin extract resulted in increased levels of epicatechin, quercetin, dihydrocapsaicin, chlorogenic acid, *p*-hydroxybenzoic acid, *p*-coumaric acid, and capsaicin [65]. Likewise, a garlic extract enhanced the content of phenolic compounds in common bean pods, including protocatechuic acid, catechin, *p*-hydroxybenzoic acid, vanillic acid, and ferulic and cinnamic acid [51]. Vines treated with hPDBs produced grapes and wine with an enhanced concentration of anthocyanins in control conditions [63,64] or water scarcity [62]. In wine produced from Airén grapevines treated with vine-shoot extracts, an increase in the concentration of phenolic compounds was observed, including vanillin derivatives (vanillin and acetovanillone), volatile phenols (guaiacol and syringol), and phenolic acids (hydroxycinnamic acids) [59], whereas wine from Monastrell vines treated with oak extract showed a higher concentration of gallic acid, hydroxycinnamoyl tartaric acids, stilbenes, and the isomer piceid-*t*-resveratrol [63].

Nitrogen-containing compounds

N-containing compounds are mainly derived from amino acids such as phenylalanine, lysine, ornithine, tyrosine, and tryptophan [136]. Their accumulation in the plant can be modulated by the application of hPDBs, with the concentration either increasing [112,116] or decreasing [68,112]. It has been shown that under adverse conditions, the alkaloid content in plants tends to increase, as alkaloids help to alleviate stress by scavenging ROS [136]. Interestingly, although pretreatment with hPDBs is capable of inducing a higher alkaloid content in stressed plants [68,89,95], it may also result in a reduction, indicating that the biostimulant mitigates oxidative stress by an alternative pathway, obviating the need for enhanced alkaloid synthesis [55,68].

Glucosinolates are amino acid-derived specialized metabolites containing nitrogen and sulfur. Mainly found in plants of the *Brassica* genus, they play a relevant role in defense against both biotic and abiotic stresses [140,141]. Different higher plant PHs are reported to increase glucosinolates in unstressed tomato plants [112,116] and in lettuce grown under salt stress, where the upregulation of glucosinolate biosynthesis was evidenced by the accumulation of 5-(methylsulfanyl)-α-D-ribose, (E)-8-(methylsulfanyl)octanal oxime, 7-(methylsulfanyl)octanal oxime, and 7-(methylsulfanyl)heptyl-glucosinolate and the down-accumulation of their homomethionine precursors [74].

4.1.3. Photosynthetic Processes

Photosynthetic metabolism includes the chemical reactions within chloroplasts, involved in the reduction in CO_2 into carbohydrates by consuming ATP and reducing power in the form of NADPH, previously obtained through light-dependent reactions. Various biomolecules, including proteins from both photosystems and accessory pigments and ions acting as enzymatic cofactors, participate in these processes. As shown here, numerous studies have found that hPDBs can increase the components of photosynthesis, as well as improve its functioning, and more specifically, reduce or reverse the impact of abiotic stress on this vital process.

The commercial biostimulant Trainer® increased the total chlorophyll content in lettuce [71,72], pea [79], spinach [106], tomato [115], and lamb's lettuce [67], while other hPDBs have generated the same effect in maize [72,80,81], perennial wall rocket [98], cabbage [39,40], and common bean [52]. An increase in chlorophylls has also been observed after applying sorghum water extracts in camelina [41] and wheat [124] under salt stress; in the latter study, the extracts acted in synergy with benzyl aminopurine. Chlorophylls are part of light-capturing complexes, complemented by accessory pigments such as carotenoids (carotenes and xanthophylls) and anthocyanins, whose levels usually increase in parallel with those of chlorophylls after hPDB application. An increase in carotenoids was observed in treated lettuce [75], white rocket [6], rocket [103], sweet pepper [109], pea [96], maize [84], and rose-scented geranium [104]. The application of different hPDBs limited pigment loss in common bean [45–47,50], pea [95], maize [78], and milk thistle plants [89] under high-salinity stress. Similar results were obtained in sweet pepper plants under PHE stress [111], in squash in drought conditions [107], and in common bean plants stressed by high temperature [48].

The light absorbed by the pigments in photosystem II reaction centers is required for water photolysis, the electron transfer to the photosystems, and the electron transport chain in general. Electron transfer and photosynthetic efficiency can be quantified by measuring different parameters, many of which are improved by hPDB application. Thus, improvements were observed in photochemical efficiency (Fv/Fm ratio) in lettuce [74] and Genovese basil [55], photosystem II and photochemical efficiency in spinach [105], and net photosynthesis in olive trees [93]. In conditions of high salinity, hPDB application in Arabidopsis [37], maize [132], pea [97], okra [92], and lettuce [68] alleviated the stress impact, preserving their photosynthetic parameters, being comparable with those of control plants. In-depth studies in lettuce revealed a positive synergistic effect when hPDBs were applied with beneficial microorganisms for plant growth [117].

The reduction in carbon from CO_2 to carbohydrates takes place in the Calvin–Benson cycle, following the generation of reducing power (NADPH) and energy molecules (ATP) in preceding light-dependent stages. Carbohydrate production requires the capture of CO_2 through gas exchange phenomena such as transpiration and stomatal opening/closing, and specific hPDBs have been identified as enhancers of these processes. Trainer® application increased net CO_2 assimilation in tomato [117] and enhanced stomatal conductance, leaf transpiration, and net photosynthesis in cascading geranium [43]. Notably, hPDBs can either increase or maintain these parameters in plants growing under unfavorable conditions. Treatment with a cereal PH, for instance, reduced the water evaporation rate and improved drought tolerance in mung bean, rice, and wheat [91]. Under high-salinity stress, treated common bean plants maintained higher levels of relative water content, mitigating the impact of salinity [49], cowpea plants maintained high levels of net photosynthesis, transpiration rate, and relative water content [53], and eggplant plants preserved high levels of photosynthesis, transpiration rate, and stomatal conductance [54]. Maize treated with duckweed plant extract before the application of toxic concentrations of Cu preserved photosynthetic activity, evapotranspiration, stomatal conductance, and sub-stomatal CO_2 concentration to the levels of plants grown with normal amounts of Cu [82].

In non-stressed tomato plants treated with hPDBs, metabolomics and transcriptomics studies found higher levels of vitamin K1 and 4-hydroxycoumarin, as well as an increase in transcripts such as *phytochrome interacting factor 3-like 5* (*PIL5*), *light-harvesting complex protein LHCA5*, *ferredoxin-2*, *photosystem II 22kDa protein*, *chloroplastic ATP synthase chain precursor*, and *RuBisCO*. Additionally, higher levels of metabolites and transcripts participating in light-harvesting processes, electron transport chain, and CO_2 reduction were observed [68,123]. In zucchini plants treated with cypress leaf extract subjected to salinity stress, RuBisCO enzymatic activity was enhanced [128]. Similarly, in proteomic studies carried out in grapevine plants stressed by water deficit and treated with different PHs, an increase in proteins related to photosynthetic metabolism was observed, including RuBisCO large chain, chloroplastic ATP synthase subunit beta, photosystem II CP47 chlorophyll apoprotein, photosystem II D2 protein, chloroplastic RuBisCO activase, and chloroplastic glyceraldehyde-3-phosphate dehydrogenase A [61].

4.1.4. Oxidative Metabolism

Oxidative metabolism includes all the reactions that maintain cellular redox homeostasis, defending the plant from damage caused by reactive chemical species. These molecules are present in plant physiology, being part of signaling mechanisms or being generated because of cellular metabolism. However, under stress conditions, reactive chemical species are overproduced, potentially causing damage to cellular components. Molecular markers of this damage can be quantified by analyzing lipid peroxidation, membrane stability index, and electrolyte leakage. Plant antioxidant defense responses can be enzymatic, via the activity of catalase (CAT), superoxide dismutase (SOD), peroxidase (POX), ascorbate peroxidase (APX), glutathione peroxidase (GPX), and glutathione reductase (GR), among others. Non-enzymatic defense systems include enhanced levels of osmoprotectants, hydrophilic antioxidant molecules, such as ascorbic acid, glutathione, and polyphenols, and lipophilic antioxidant molecules, such as β-carotene, lutein, and α-tocopherol. The application of hPDBs can improve antioxidant defense systems in plants by increasing the concentration of these biomolecules in stress and non-stress conditions [14,17].

Antioxidant activity, both lipophilic and hydrophilic, determined as free-radical scavenging capacity, was reported to increase after treatment with hPDBs in white rocket [6], lettuce [71], tomato [117], sweet pepper [109], radish [102], cabbage [40], rose-scented geranium [104], and lamb's lettuce and spinach [67].

In general, hPDB mechanisms of action in crops modulate stress-related parameters, reducing electrolyte leakage and lipid peroxidation and increasing membrane stability index and the content of osmoprotectants (e.g., proline and soluble sugars) and antioxidants (e.g., ascorbic acid). Moreover, biostimulant application can enhance the transcription of genes encoding antioxidant enzymes (e.g., SOD, CAT, POX, APX, GPX, and GR) and their activities, resulting in a decrease in reactive chemical species. These results have been obtained in experiments conducted under non-stress conditions in perennial wall rocket [98], tomato [120,122,123], common bean [45], lettuce [70], and kinnow mandarin [66]. Furthermore, multiple studies have determined that the application of hPDBs increases salinity tolerance, improving the previously described parameters with respect to untreated plants and even reducing the total intracellular Na content. Such positive effects have been observed in cowpea [53], zucchini [128], milk thistle [89], pea [95,97], maize [78,84,132], camelina [41], lettuce [73], tomato [119], and common bean plants [47–50]. Similar results were obtained in common bean plants subjected to a combination of salinity and Cd contamination [46], or a single stress, such as a high temperature [48]. Regarding PHE contamination, a similar mechanism of action has been described for hPDBs applied to sweet pepper [111], potato [100], and maize plants exposed to Cu toxicity [82]; for tomato grown with limited water availability [118]; and for squash subjected to both water deficit and salinity [107].

A few studies indicate that hPDBs can also trigger a reduction in antioxidant enzyme activity compared to untreated plants, in the presence or absence of stress. This implies that the plant does not need to excessively activate the detoxifying enzymatic machinery to develop tolerance to stress, whose effects are mitigated by other detoxification mechanisms. Examples of this have been observed in treated rocket plants [103] and in maize under salt stress conditions [77,87].

4.1.5. Signaling-Related Processes

Correct plant growth and development, as well as reproduction, senescence, and stress tolerance, require careful coordination of several physiological and biochemical processes. This regulation is carried out to a large extent by phytohormones that function as signaling compounds, including abscisic acid (ABA), auxins, brassinosteroids, cytokinins, ethylenes, gibberellins, jasmonates, and salicylic acid (SA) [142].

The small peptides of certain biostimulants, both PH-based and those obtained from plant extracts, are reported to act as signaling molecules, eliciting auxin- or gibberellin-like activities [81,86,105], which stimulate different metabolisms and processes, including phenolic metabolism and plant growth. However, as not all biostimulants have peptides with hormone-like properties [126], it is also important to highlight the effect of hPDBs on phytohormone accumulation patterns. For instance, the application of garlic extract in common bean was found to stimulate an increase in growth-related phytohormones, namely auxins (indole-3-acetic acid (IAA) and indole-3-pyruvate), gibberellins (gibberellin A7), and cytokinins (*trans*-Zeatin and its riboside) [51]. Positive effects of hPDBs on the accumulation of growth hormones were also observed in rocket and tomato, which had higher levels of the IAA precursors tryptamine and 4-(indol-3-yl) butanoate; the gibberellin precursor ent-7α-hydroxykaur-16-en-19-oate; and the cytokinins lupinate and *trans*-zeatin-O-glucoside-7-N-glucoside [112]. In tomato plants, the increase in growth hormones was accompanied by an upregulation of development-related genes, such as expansins, growth-regulating factors 3 (*GRF3*) and 5 (*GRF5*), and *lob domain protein 1* [123]. However, certain biostimulant treatments may result in a lower concentration of these hormones and their precursors, as evidenced by a reduction in brassinosteroid levels in tomato [112,116].

Phytohormones have a role in mediating stress response mechanisms, even presenting a crosstalk in regulating both abiotic and biotic stress-induced signaling. ABA is a major phytohormone in mediating these processes, and its role as a master regulator stands out. In this sense, the effect of hPDBs on the phytohormone profile has also been investigated in plants growing under unfavorable conditions [143]. Pretreatment with moringa leaf extract of Arabidopsis plants grown under saline conditions enhanced the transcription of the *AtIAA*, *ABI5*, and *PR1* genes, related to the auxin, ABA, and SA synthetic pathways, respectively [38]. The application of different PHs in Arabidopsis induced the accumulation of the brassinosteroid precursors 6-deoxo-24-epicathasterone and campest-5-en-3-one and a reduction in derivative forms of plant hormones such as benzyladenine-7-glucoside, 16,17-dihydro-16α-17-dihydroxy gibberellin 12, and methylgibberellin 4; the IAA-derivate 4-(indol-3-yl)butanoyl-β-D-glucose; or the brassinosteroid castasterone compared to untreated plants [37]. The decrease in degraded or inactivated hormone products could signify a preservation of homeostasis regulated by the active hormone, whereas lower castasterone levels could indicate an amelioration of the effects of salt toxicity [37]. The preservation of auxin, cytokinin, gibberellin, brassinosteroid, and jasmonate levels under stress conditions by hPDBs has also been described in lettuce [69] and common bean subjected to salinity [49] and heat [48], maize exposed to Cu toxicity [82], and grapevines grown under water scarcity [61]. However, a negative modulation was noted in lettuce under salt stress [68] and tomato subjected to water scarcity [118].

The modulation of phytohormone levels can activate signaling cascades that regulate various processes, including stress response. A transcriptomic study in tomato plants proposed the induction of a signaling cascade by an alfalfa-based PH. Thus, this biostimulant originated an upregulation of genes involved in hormone synthesis, causing the activation of a phosphorylation cascade involving different protein kinases (e.g., CPK9, CPK28, CRCK3, LRR kinases, MAPKKK21, Pi kinase, PEPKR2, and WAK2), which regulate transcription factors (e.g., bHLH, APETALA2, homeobox-leucine zipper, PPR, WRKY, Myb, ERF2, and Rav). Together, these cascades would lead to the expression of genes encoding stress response proteins (ABC transporters, DC1 domain-containing proteins, alternative oxidase 1A, cytochrome 450, leucine-rich repeat proteins, heat shock proteins, L-threonine ammonia-lyase, aldo/keto reductase, threonine ammonia-lyase and chitinases, glutathione-S-transferases, and wound-induced proteins, among others) [123].

4.2. *Whole-Plant Level*

In addition to impacting cellular and molecular processes, biostimulants exert effects at the whole-plant level, influencing physiological processes such as germination, root and shoot growth and morphology, flowering, and fructification. From the agronomic perspective, these changes are highly important, as they are closely linked to crop yield (defined as the mass or volume of plant biomass per unit of cultivated land) and quality, thus determining economic profitability [144].

4.2.1. Germination

Germination is one of the most critical processes in agriculture, as it determines plant growth and development, as well as the subsequent crop yield and fruit quality. Therefore, it is essential to develop strategies that ensure a successful and homogeneous germination, even in plants commencing their life cycle under stress conditions [144]. The application of hPDBs has proven effective in promoting germination. For example, higher plant-derived PHs increased the germination percentage in mung bean and wheat [91]. Furthermore, studies performed in maize report that treatment with Trainer® increased the length of the coleoptile, a distinctive part of grass embryos, showing a similar effect to IAA. This result suggests that Trainer® could have an auxin-like function in the germination process [79].

Unfavorable conditions negatively affect the germination process, decreasing the total number of germinated seedlings and their size. hPDBs are also capable of counteracting the negative impact of abiotic stresses on germination. Several higher plant-derived PHs were found to maintain higher survival rates in primed Arabidopsis seeds subjected to high salinity, also ensuring a more homogeneous germination [37]. This beneficial effect, as well as better growth rates of seedlings, has also been observed under salinity in crops of agronomic interest such as zucchini [128], pea [95,97], camelina [41], wheat [124], and maize [78,132].

4.2.2. Root Growth and Morphology

Plant growth and development depend strongly on root absorption of nutrients and water, which is essential for achieving high biomass production and successful harvests [145]. Various hPDBs have shown the ability to promote rooting, resulting in increased root dry weight, length, area, and number in dicotyledons such as cabbage [39,40] and tomato [79,112,115,116,120,122] and monocotyledons such as maize [80,81,83]. This effect has also been verified in the soilless cultivation of lily plants, in which the application of alfalfa-derived biostimulants generated a longer root system compared to untreated plants [76].

The effective absorption of nutrients depends on the formation of adventitious roots and root hairs. In field mustard, bioactive peptides present in hydrolysates derived from soybean by-products not only increased root thickness, but also generated a higher number of adventitious roots and root hairs of greater length. In addition, an increase in trichoblasts and atrichoblasts was observed, although their location pattern did not change, which indicates that the biostimulant promoted root growth differently from ethylene [126]. In cannabis plants, treatment with a biostimulant formulation composed of aloe vera, fish, and kelp extracts increased the number of root tips and branch points, as well as the total surface area of the root system [42].

In plant species where the root is the edible part, promoting root growth improves crop profitability. The application of various leaf aqueous extracts in radishes increased root length, therefore resulting in an improved harvest yield. Moreover, the treated radishes had an improved content of bioactive compounds and greater antioxidant capacity [102].

The application of biostimulants may also protect roots from abiotic stresses, ensuring their optimal functionality. Tested in conditions of salinity, biostimulants promoted better root growth and morphology in maize [77,84], onion [94], camelina [41], common bean [47], and lettuce [74], as well as in maize subjected to Cu toxicity [82]. Under stress conditions associated with alkaline substrates, the application of Trainer® increased the effectiveness of microbial biostimulants in a synergistic process, improving lettuce root growth [70]. After treatment with moringa leaf extract, common bean plants subjected to high temperatures also showed a better response to stress than untreated plants, producing longer roots and more biomass [48].

4.2.3. Shoot Growth and Morphology

Often, shoot growth parameters are indicative of the future yield of the crop. As well as increasing plant height and stem diameter, the action of different biostimulants can positively modulate shoot dry weight and length. For example, Trainer® is reported to increase plant height and shoot dry biomass in maize [72], shoot dry weight and length in tomato and dwarf pea [79,115,121], and plant height of cascading geranium [43]. Biostimulants derived from plant by-products generated a similar response, exemplified by carob germ-based extracts, which increased stem diameter and plant height in tomato [122,127].

Regarding vascular tissues, cowpea plants treated with ammi seed extract showed larger leaf blades, more palisade and spongy tissue, and thicker phloem and xylem than the untreated controls, as well as a larger stem and vessel diameters and more vascular bundles [53].

The leaf area, a determining factor in photosynthetic capacity, can also be increased by biostimulants, as reported in cascading geranium [43], tomato [114,120,122], and maize [86]. In maize and olive trees, the application of hPDBs of different origins increased leaf biomass, another parameter of plant growth [80,81,83,93].

In the floriculture crop gladiolus, moringa leaf extracts increased plant height, corm biomass and diameter, and cormel production [56,57]. Similarly, in lily plants, the application of an alfalfa-based PH enhanced bulb size and resulted in greener and more expanded blades [76].

In leafy crops where the marketable/consumable part is mainly the shoot, such as cabbage, rocket, lettuce, or spinach, the positive effect of biostimulants on growth is of special interest as their action can increase crop yield. hPDBs are reported to increase the height and biomass of cabbage [39,40], rocket [103], perennial wall rocket [98,99], and white rocket shoots [6], increasing their total yield. In different types of lettuce and lamb's lettuce, the application of biostimulants also promoted growth and a higher fresh yield [67,71,75], an effect also observed in Genovese basil [55] and spinach, which also showed a greater leaf area [67,105,106]. In wild mint, large-leaf beauty berry extracts enhanced growth and leaf area and number [125], while in rose-scented geranium, moringa leaf extracts increased plant height, branch number, and leaf area, also enhancing overall yield and volatile oil content [104]. In addition, horticultural products had an improved nutritional quality, due to a higher concentration of minerals, sugars, and phenols and antioxidant capacity, as well as an increased concentration of chlorophylls, as mentioned above, which gives a better quality to the horticultural product [6,98,106].

Biostimulant treatments can also avoid the negative impact of abiotic stresses on plant size. Thus, under saline conditions, several hPDBs are reported to increase the height and biomass in maize [77,84], camelina [41], tomato [68], lettuce [68–70,73,74], mung bean [90], and milk thistle shoots [89]; the height and leaf biomass in onion plants [94]; the height, biomass, leaf blade size, phloem and xylem thickness, vessel diameter, and number of vascular bundles in cowpea [53]; and plant height and leaf elongation and expansion in wheat [124] and common bean, where the number of leaves and leaf area also increased [45,47,49,50]. hPDB effectiveness has also been shown under other unfavorable conditions. In tomato and squash plants, the application of Trainer® attenuated the negative effect of water scarcity, increasing the biomass of treated plants [107,118]. Similarly, duckweed extract promoted shoot growth in maize plants subjected to Cu toxicity [82]. Under PHE stress, moringa seed extract increased tuber yield and reduced PHE accumulation in potato [100]. Lastly, common bean plants subjected to heat stress produced longer shoots and more biomass after treatment with moringa extract [48].

4.2.4. Flowering

The effect of hPDBs on flowering is of great interest, not only because this process is closely linked to fruit production, but also because it is the main determinant of quality in floriculture crops and ornamental plants. Studies have found that biostimulants are capable of promoting flowering, increasing the number of flowers in various crops, from food crops such as common bean [51] and ornamental plants such as cascading geranium [43]. As well as increasing the number of flowers produced per plant [43,127] and the diameter of the flower buds, biostimulants can also shorten the crop cycle, thereby advancing the flowering process [76]. Treatment of gladiolus with moringa leaf extract not only resulted in earlier spike emergence and an increase in the number of spike florets per plant, but also extended its maximum vase life in sucrose [57]. Thus, hPDBs offer a broad spectrum of benefits for the cultivation of ornamental crops, ranging from accelerated flowering to improvements in flower quantity, size, and post-harvest characteristics.

4.2.5. Fructification and Fruit Quality

Among the myriad of actions that biostimulants perform at different stages of plant development, their role in boosting fructification is of particular economic importance. Their application can increase the weight, size, and number of fruits produced per plant [101,127] as well as improve overall fruit quality. In tomato plants, biostimulant treatments increased fruit production and enhanced the concentration of minerals (mainly K and Mg), bioactive compounds such as lycopene, phenols, ascorbic acid, organic acids (malate, oxalate, citrate, and isocitrate), and soluble solids. The treated tomatoes also had greater antioxidant activity and fruit brightness and redness [113,114,117,121].

Regarding improvements in fruit quality, the application of plant extracts resulted in a lower incidence of cracking in sweet cherries due to enhanced fruit firmness and Ca concentration. As well as a higher fruit production, a greater proportion of cherries were of the color preferred by consumers [108]. Similarly, treatment of kinnow mandarin with moringa leaf extracts led to a higher fruit yield, a reduction in fruit drop, and a higher content of fruit sugars, vitamin C, antioxidants, and phenols [66]. In mango fruits, a combination of roselle and garlic extracts generated a far greater response than the sum of individual effects when applied separately, improving fruit set, retention, yield, and quality, in addition to its nutritional content (mainly increased levels of N, P, and K) [88]. Likewise, the synergy of an alfalfa-based PH and red grape extracts applied to habanero pepper plants increased fruit weight and number and enhanced their bioactive properties due to a higher content of glucose and specialized metabolites [65]. Biostimulants also had a positive effect on sweet pepper plants, increasing total fruit yield, vitamin C, total phenolic content, and antioxidant activity [109,110].

In winemaking, the organoleptic properties and overall quality of wine are closely linked to the primary and specialized metabolite contents of grapes. In an innovative and sustainable approach involving the revalorization of a waste by-product, extracts from vine-shoot residues have been applied as biostimulants and found to modify the phenolic composition, mineral content, anthocyanin concentration, volatile and glycosylated compounds, and color characteristics of grapes [58–60,63,64].

An investigation into the effects of biostimulants on legumes is of considerable importance, given their widespread consumption in many regions [146]. The application of different higher plant extracts in common bean improved crop yield, increasing pod number, pod fresh weight, and seed yield [45,49,51,52]. Similar effects were also observed in pea plants treated with moringa extracts, with an increase in seed weight and contents of protein and minerals (especially K, Ca, Mg, Fe, and P) [96].

A primary consequence of stress is a negative impact on fruit production, leading to a reduction in both quantity and quality. Biostimulants have proven effective in mitigating these adverse effects, as described in numerous studies. Under conditions of water scarcity, biostimulants positively influenced vine crop yields and enhanced the anthocyanin content of grapes [62]. In legumes subjected to saline conditions, including common bean, mung bean, cowpea, and okra, different plant extracts have been successfully used to improve crop yield, increasing the number of pods and seeds per pod [46,47,50,53,90,92]. A protective effect of biostimulants against salinity has also been demonstrated in solanaceous plants such as eggplant and sweet pepper, with the treated plants producing more fruits of higher weight [54,111]. A similar positive impact was observed in sweet pepper and common bean subjected to PHE stress [46,111].

All the different metabolisms and processes affected by hPDB-induced mechanisms of action are summarized in Figure 3.

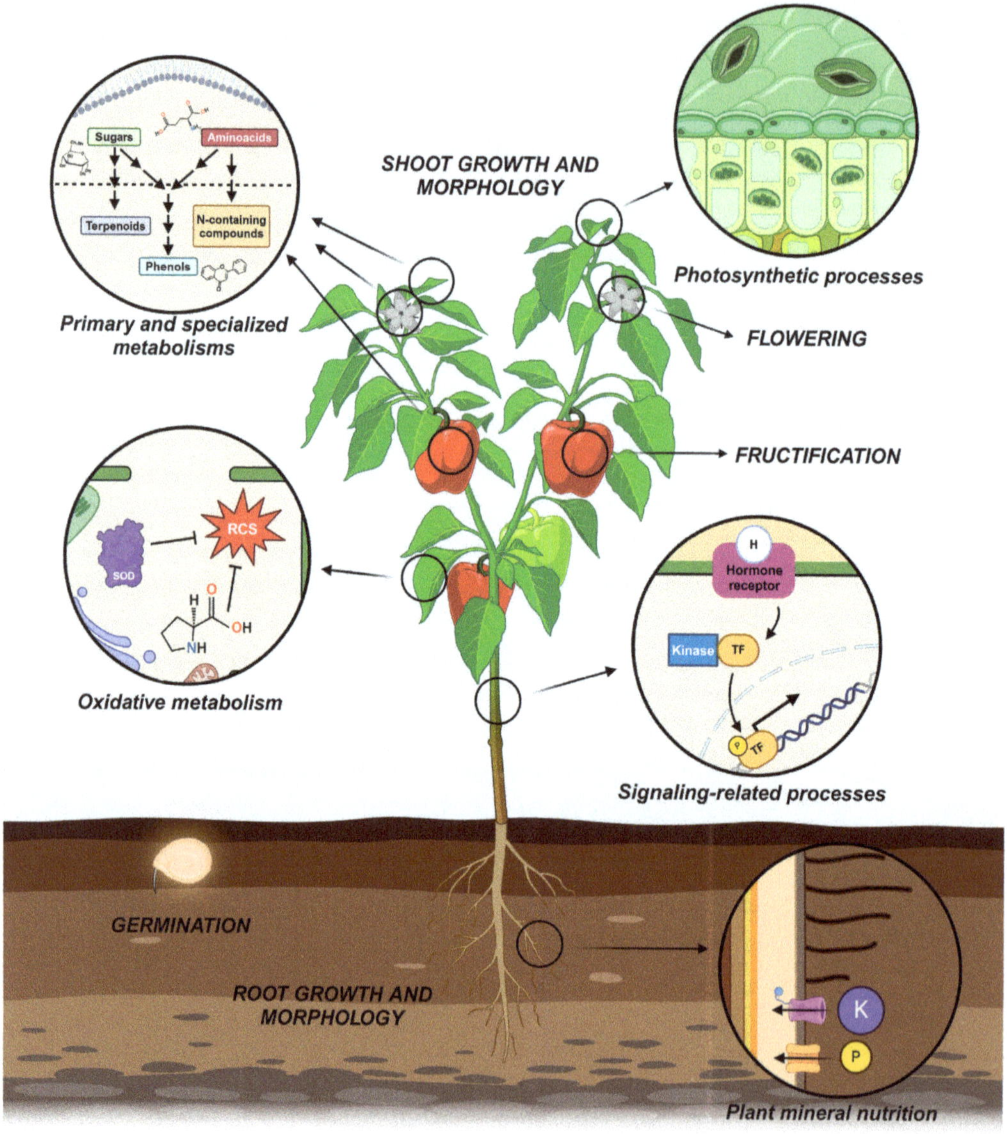

Figure 3. Schematic representation of the different metabolisms and processes affected by hPDBs in plants. Created with BioRender.com, accessed on 31 January 2024.

5. Conclusions and Future Perspectives

The application of biostimulants constitutes an alternative approach toward a more sustainable agriculture that is less reliant on fertilizers and agrochemicals. Those derived from natural sources are of particular interest, especially hPDBs, given their ecological advantages and broad spectrum of action on plant physiology. The ability of these biostimulants to modulate cellular reactions and physiological processes can alleviate the effects of abiotic stress on plants growing under unfavorable edaphoclimatic conditions. Understanding biostimulant mechanisms of action underlying these effects, and ultimately the modes of action, is vital for the design and customization of biostimulant formulations for each type of crop. With this aim, omics tools offer the possibility of elucidating the composition of the biomolecules in each formulation, as well as the modifications they induce in plant physiology. Simultaneously, it is also necessary to study different methods of biostimulant application, exploring key aspects such as the duration of the effect, optimal application frequency, the stage of the plant life cycle when application is most effective, and how biostimulants may interfere with the action of fertilizers and agrochemicals. Then, new sources of hPDBs should be uncovered to facilitate their large-scale industrial production.

Author Contributions: Conceptualization, S.E.M.-L., J.M.M.-G., M.Á.P., L.A. and A.B.S.-J.; figure software, J.M.M.-G.; writing—original draft preparation, S.E.M.-L. and J.M.M.-G.; writing—review and editing, S.E.M.-L., J.M.M.-G. and A.B.S.-J.; supervision, L.A. and A.B.S.-J.; funding acquisition, M.Á.P. All authors have read and agreed to the published version of the manuscript.

Funding: This study forms part of the Agroalnext Programme and was supported by MCIN with funding from European Union NextGenerationEU (PRTR-C17.I1) and by Comunidad Autónoma de la Región de Murcia–Fundación Séneca and by Fundación Séneca-Agencia de Ciencia y Tecnología from Murcia (project 22016/PI/22), via "Ayudas a proyectos para el desarrollo de investigación científica y técnica por grupos competitivos", included in "Programa Regional de Fomento de la Investigación Científica y Técnica (Plan de Actuación 2022)". S.E.M.-L. held grants by Ministerio de Ciencia, Innovación y Universidades of Spain FPU Grants (Grant No. FPU21/01593), and J.M.M.-G. held grants by Universidad de Murcia PhD contracts (Registry number 109144/2022).

Institutional Review Board Statement: Not applicable.

Informed Consent Statement: Not applicable.

Acknowledgments: All the figures displayed in this review, including graphical abstract, were created with BioRender.com, accessed on 31 January 2024.

Conflicts of Interest: The authors declare no conflicts of interest.

References

1. United Nations. *World Population Prospects 2019*; Department of Economic and Social Affairs: New York, NY, USA, 2019.
2. Gupta, S.; Bhattacharyya, P.; Kulkarni, M.G.; Doležal, K. Editorial: Growth Regulators and Biostimulants: Upcoming Opportunities. *Front. Plant Sci.* **2023**, *14*, 1209499. [CrossRef]
3. Bulgari, R.; Franzoni, G.; Ferrante, A. Biostimulants Application in Horticultural Crops under Abiotic Stress Conditions. *Agronomy* **2019**, *9*, 306. [CrossRef]
4. European Commission. A Farm to Fork Strategy for a Fair, Healthy and Environmentally-Friendly Food System. In *Communication from the Commission to the European Parliament, the Council, the European Economic and Social Committee and the Committee of the Regions*; European Commission: Brussels, Belgium, 2020.
5. European Commission. The European Green Deal. In *Communication from the Commission to the European Parliament, the European Council, the Council, the European Economic and Social Committee and the Committee of the Regions*; European Commission: Brussels, Belgium, 2019.
6. Di Mola, I.; Ottaiano, L.; Cozzolino, E.; Senatore, M.; Giordano, M.; El-Nakhel, C.; Sacco, A.; Rouphael, Y.; Colla, G.; Mori, M. Plant-Based Biostimulants Influence the Agronomical, Physiological, and Qualitative Responses of Baby Rocket Leaves under Diverse Nitrogen Conditions. *Plants* **2019**, *8*, 522. [CrossRef] [PubMed]
7. Ali, S.; Liu, Y.; Ishaq, M.; Shah, T.; Abdullah; Ilyas, A.; Din, I.U. Climate Change and Its Impact on the Yield of Major Food Crops: Evidence from Pakistan. *Foods* **2017**, *6*, 39. [CrossRef] [PubMed]
8. Del Buono, D. Can Biostimulants Be Used to Mitigate the Effect of Anthropogenic Climate Change on Agriculture? It Is Time to Respond. *Sci. Total Environ.* **2021**, *751*, 141763. [CrossRef] [PubMed]
9. Bhupenchandra, I.; Chongtham, S.K.; Devi, E.L.; R, R.; Choudhary, A.K.; Salam, M.D.; Sahoo, M.R.; Bhutia, T.L.; Devi, S.H.; Thounaojam, A.S.; et al. Role of Biostimulants in Mitigating the Effects of Climate Change on Crop Performance. *Front. Plant Sci.* **2022**, *13*, 967665. [CrossRef]
10. Akbari, M.; Najafi Alamdarlo, H.; Mosavi, S.H. The Effects of Climate Change and Groundwater Salinity on Farmers' Income Risk. *Ecol. Indic.* **2020**, *110*, 105893. [CrossRef]
11. Pardo-Hernández, M.; López-Delacalle, M.; Martí-Guillen, J.M.; Martínez-Lorente, S.E.; Rivero, R.M. ROS and NO Phytomelatonin-Induced Signaling Mechanisms under Metal Toxicity in Plants: A Review. *Antioxidants* **2021**, *10*, 775. [CrossRef]

12. Ahuja, I.; de Vos, R.C.H.; Bones, A.M.; Hall, R.D. Plant Molecular Stress Responses Face Climate Change. *Trends Plant Sci.* **2010**, *15*, 664–674. [CrossRef]
13. Mittler, R.; Blumwald, E. Genetic Engineering for Modern Agriculture: Challenges and Perspectives. *Annu. Rev. Plant Biol.* **2010**, *61*, 443–462. [CrossRef]
14. Martínez-Lorente, S.E.; Pardo-Hernández, M.; Martí-Guillén, J.M.; López-Delacalle, M.; Rivero, R.M. Interaction between Melatonin and NO: Action Mechanisms, Main Targets, and Putative Roles of the Emerging Molecule NOmela. *Int. J. Mol. Sci.* **2022**, *23*, 6646. [CrossRef]
15. Mittler, R. Oxidative Stress, Antioxidants and Stress Tolerance. *Trends Plant Sci.* **2002**, *7*, 405–410. [CrossRef]
16. Corpas, F.J.; Barroso, J.B. Nitro-Oxidative Stress vs Oxidative or Nitrosative Stress in Higher Plants. *New Phytol.* **2013**, *199*, 633–635. [CrossRef]
17. Zhou, X.; Joshi, S.; Patil, S.; Khare, T.; Kumar, V. Reactive Oxygen, Nitrogen, Carbonyl and Sulfur Species and Their Roles in Plant Abiotic Stress Responses and Tolerance. *J. Plant Growth Regul.* **2022**, *41*, 119–142. [CrossRef]
18. Martí-Guillén, J.M.; Pardo-Hernández, M.; Martínez-Lorente, S.E.; Almagro, L.; Rivero, R.M. Redox Post-Translational Modifications and Their Interplay in Plant Abiotic Stress Tolerance. *Front. Plant Sci.* **2022**, *13*, 1027730. [CrossRef]
19. Aslam, M.; Ahmad, K.; Akhtar, M.A.; Maqbool, M.A. Salinity Stress in Crop Plants: Effects of Stress, Tolerance Mechanisms and Breeding Strategies for Improvement. *J. Agric. Basic Sci.* **2017**, *2*, 70–85.
20. Sangiorgio, D.; Cellini, A.; Donati, I.; Pastore, C.; Onofrietti, C.; Spinelli, F. Facing Climate Change: Application of Microbial Biostimulants to Mitigate Stress in Horticultural Crops. *Agronomy* **2020**, *10*, 794. [CrossRef]
21. Challabathula, D.; Analin, B.; Mohanan, A.; Bakka, K. Differential Modulation of Photosynthesis, ROS and Antioxidant Enzyme Activities in Stress-Sensitive and -Tolerant Rice Cultivars during Salinity and Drought upon Restriction of COX and AOX Pathways of Mitochondrial Oxidative Electron Transport. *J. Plant Physiol.* **2022**, *268*, 153583. [CrossRef] [PubMed]
22. Yakhin, O.I.; Lubyanov, A.A.; Yakhin, I.A.; Brown, P.H. Biostimulants in Plant Science: A Global Perspective. *Front. Plant Sci.* **2017**, *7*, 2049. [CrossRef]
23. Russo, R.O.; Berlyn, G.P. The Use of Organic Biostimulants to Help Low Input Sustainable Agriculture. *J. Sustain. Agric.* **1991**, *1*, 19–42. [CrossRef]
24. Russo, R.O.; Berlyn, G.P. Vitamin-Humic-Algal Root Biostimulant Increases Yield of Green Bean. *HortScience* **1992**, *27*, 847. [CrossRef]
25. du Jardin, P. Plant Biostimulants: A New Paradigm for the Sustainable Intensification of Crops. In *Biostimulants for Sustainable Crop Production*; Burleigh Dodds Science Publishing: London, UK, 2020; ISBN 978-1-00-304786-5.
26. Colla, G.; Rouphael, Y. Biostimulants in Horticulture. *Sci. Hortic.* **2015**, *196*, 1–2. [CrossRef]
27. La Torre, A.; Battaglia, V.; Caradonia, F. An Overview of the Current Plant Biostimulant Legislations in Different European Member States. *J. Sci. Food Agric.* **2016**, *96*, 727–734. [CrossRef] [PubMed]
28. The European Parliament and the Council of the European Union. Regulation (EU) 2019/1009 of the European Parliament and of the Council of 5 June 2019 Laying down Rules on the Making Available on the Market of EU Fertilising Products and Amending Regulations (EC) No 1069/2009 and (EC) No 1107/2009 and Repealing Regulation (EC) No 2003/2003 (Text with EEA Relevance). *Off. J. Eur. Union* **2019**, *170*, 1–114.
29. Mącik, M.; Gryta, A.; Frąc, M. Chapter Two—Biofertilizers in Agriculture: An Overview on Concepts, Strategies and Effects on Soil Microorganisms. In *Advances in Agronomy*; Sparks, D.L., Ed.; Academic Press: Cambridge, MA, USA, 2020; Volume 162, pp. 31–87.
30. Qi, X.; Li, K.; Chen, L.; Zhang, Y.; Zhang, N.; Gao, W.; Li, Y.; Liu, X.; Fan, Z. Plant Defense Responses to a Novel Plant Elicitor Candidate LY5-24-2. *Int. J. Mol. Sci.* **2022**, *23*, 5348. [CrossRef]
31. du Jardin, P. Plant Biostimulants: Definition, Concept, Main Categories and Regulation. *Sci. Hortic.* **2015**, *196*, 3–14. [CrossRef]
32. Ahmad, A.; Blasco, B.; Martos, V. Combating Salinity Through Natural Plant Extracts Based Biostimulants: A Review. *Front. Plant Sci.* **2022**, *13*, 862034. [CrossRef]
33. Baltazar, M.; Correia, S.; Guinan, K.J.; Sujeeth, N.; Bragança, R.; Gonçalves, B. Recent Advances in the Molecular Effects of Biostimulants in Plants: An Overview. *Biomolecules* **2021**, *11*, 1096. [CrossRef]
34. Colantoni, A.; Recchia, L.; Bernabei, G.; Cardarelli, M.; Rouphael, Y.; Colla, G. Analyzing the Environmental Impact of Chemically-Produced Protein Hydrolysate from Leather Waste vs. Enzymatically-Produced Protein Hydrolysate from Legume Grains. *Agriculture* **2017**, *7*, 62. [CrossRef]
35. Xu, L.; Geelen, D. Developing Biostimulants From Agro-Food and Industrial By-Products. *Front. Plant Sci.* **2018**, *9*, 1567. [CrossRef] [PubMed]
36. Fritsch, C.; Staebler, A.; Happel, A.; Cubero Márquez, M.A.; Aguiló-Aguayo, I.; Abadias, M.; Gallur, M.; Cigognini, I.M.; Montanari, A.; López, M.J.; et al. Processing, Valorization and Application of Bio-Waste Derived Compounds from Potato, Tomato, Olive and Cereals: A Review. *Sustainability* **2017**, *9*, 1492. [CrossRef]
37. Sorrentino, M.; De Diego, N.; Ugena, L.; Spíchal, L.; Lucini, L.; Miras-Moreno, B.; Zhang, L.; Rouphael, Y.; Colla, G.; Panzarová, K. Seed Priming With Protein Hydrolysates Improves Arabidopsis Growth and Stress Tolerance to Abiotic Stresses. *Front. Plant Sci.* **2021**, *12*, 626301. [CrossRef]

38. Brazales-Cevallos, D.K.; Romero-Contreras, Y.J.; Vences-Guzmán, M.Á.; Torres, M.; Aviles-Baltazar, N.Y.; Sohlenkamp, C.; Serrano, M. Transcriptional Characterization of the Biostimulant Effect of Moringa Oleifera Leaf Extracts Using Arabidopsis Thaliana as a Model. *S. Afr. J. Bot.* **2022**, *144*, 250–256. [CrossRef]
39. Godlewska, K.; Biesiada, A.; Michalak, I.; Pacyga, P. The Effect of Plant-Derived Biostimulants on White Head Cabbage Seedlings Grown under Controlled Conditions. *Sustainability* **2019**, *11*, 5317. [CrossRef]
40. Godlewska, K.; Biesiada, A.; Michalak, I.; Pacyga, P. The Effect of Botanical Extracts Obtained through Ultrasound-Assisted Extraction on White Head Cabbage (*Brassica oleracea* L. Var. *capitata* L.) Seedlings Grown under Controlled Conditions. *Sustainability* **2020**, *12*, 1871. [CrossRef]
41. Huang, P.; He, L.; Abbas, A.; Hussain, S.; Hussain, S.; Du, D.; Hafeez, M.B.; Balooch, S.; Zahra, N.; Ren, X.; et al. Seed Priming with Sorghum Water Extract Improves the Performance of Camelina (*Camelina sativa* (L.) Crantz.) under Salt Stress. *Plants* **2021**, *10*, 749. [CrossRef]
42. Wise, K.; Selby-Pham, J.; Chai, X.; Simovich, T.; Gupta, S.; Gill, H. Fertiliser Supplementation with a Biostimulant Complex of Fish Hydrolysate, Aloe Vera Extract, and Kelp Alters Cannabis Root Architecture to Enhance Nutrient Uptake. *Sci. Hortic.* **2024**, *323*, 112483. [CrossRef]
43. Cirillo, C.; Rouphael, Y.; Pannico, A.; El-Nakhel, C.; Colla, G.; De Pascale, S. Application of Protein Hydrolysate-Based Biostimulant as New Approach to Improve Performance of Bedding Plants. In Proceedings of the ISHS Acta Horticulturae 1215: International Symposium on Greener Cities for More Efficient Ecosystem Services in a Climate Changing World, Bologna, Italy, 12–15 September 2017; pp. 443–448. [CrossRef]
44. Baglieri, A.; Cadili, V.; Mozzetti Monterumici, C.; Gennari, M.; Tabasso, S.; Montoneri, E.; Nardi, S.; Negre, M. Fertilization of Bean Plants with Tomato Plants Hydrolysates. Effect on Biomass Production, Chlorophyll Content and N Assimilation. *Sci. Hortic.* **2014**, *176*, 194–199. [CrossRef]
45. Rady, M.M.; Mohamed, G.F. Modulation of Salt Stress Effects on the Growth, Physio-Chemical Attributes and Yields of *Phaseolus vulgaris* L. Plants by the Combined Application of Salicylic Acid and Moringa Oleifera Leaf Extract. *Sci. Hortic.* **2015**, *193*, 105–113. [CrossRef]
46. Howladar, S.M. A Novel Moringa Oleifera Leaf Extract Can Mitigate the Stress Effects of Salinity and Cadmium in Bean (*Phaseolus vulgaris* L.) Plants. *Ecotoxicol. Environ. Saf.* **2014**, *100*, 69–75. [CrossRef] [PubMed]
47. Rady, M.M.; Varma, C.B.; Howladar, S.M. Common Bean (*Phaseolus vulgaris* L.) Seedlings Overcome NaCl Stress as a Result of Presoaking in Moringa Oleifera Leaf Extract. *Sci. Hortic.* **2013**, *162*, 63–70. [CrossRef]
48. Latif, H.H.; Mohamed, H.I. Exogenous Applications of Moringa Leaf Extract Effect on Retrotransposon, Ultrastructural and Biochemical Contents of Common Bean Plants under Environmental Stresses. *S. Afr. J. Bot.* **2016**, *106*, 221–231. [CrossRef]
49. Rady, M.M.; Talaat, N.B.; Abdelhamid, M.T.; Shawky, B.T.; Desoky, E.-S.M. Maize (*Zea mays* L.) Grains Extract Mitigates the Deleterious Effects of Salt Stress on Common Bean (*Phaseolus vulgaris* L.) Growth and Physiology. *J. Hortic. Sci. Biotechnol.* **2019**, *94*, 777–789. [CrossRef]
50. Rady, M.M.; Desoky, E.-S.M.; Elrys, A.S.; Boghdady, M.S. Can Licorice Root Extract Be Used as an Effective Natural Biostimulant for Salt-Stressed Common Bean Plants? *S. Afr. J. Bot.* **2019**, *121*, 294–305. [CrossRef]
51. Elzaawely, A.A.; Ahmed, M.E.; Maswada, H.F.; Al-Araby, A.A.; Xuan, T.D. Growth Traits, Physiological Parameters and Hormonal Status of Snap Bean (*Phaseolus vulgaris* L.) Sprayed with Garlic Cloves Extract. *Arch. Agron. Soil Sci.* **2018**, *64*, 1068–1082. [CrossRef]
52. Mkindi, A.G.; Tembo, Y.L.B.; Mbega, E.R.; Smith, A.K.; Farrell, I.W.; Ndakidemi, P.A.; Stevenson, P.C.; Belmain, S.R. Extracts of Common Pesticidal Plants Increase Plant Growth and Yield in Common Bean Plants. *Plants* **2020**, *9*, 149. [CrossRef] [PubMed]
53. Desoky, E.-S.M.; EL-Maghraby, L.M.M.; Awad, A.E.; Abdo, A.I.; Rady, M.M.; Semida, W.M. Fennel and Ammi Seed Extracts Modulate Antioxidant Defence System and Alleviate Salinity Stress in Cowpea (*Vigna unguiculata*). *Sci. Hortic.* **2020**, *272*, 109576. [CrossRef]
54. Abbas, W.; Ashraf, M.; Akram, N.A. Alleviation of Salt-Induced Adverse Effects in Eggplant (*Solanum melongena* L.) by Glycinebetaine and Sugarbeet Extracts. *Sci. Hortic.* **2010**, *125*, 188–195. [CrossRef]
55. Rouphael, Y.; Carillo, P.; Ciriello, M.; Formisano, L.; El-Nakhel, C.; Ganugi, P.; Fiorini, A.; Miras Moreno, B.; Zhang, L.; Cardarelli, M.; et al. Copper Boosts the Biostimulant Activity of a Vegetal-Derived Protein Hydrolysate in Basil: Morpho-Physiological and Metabolomics Insights. *Front. Plant Sci.* **2023**, *14*, 1235686. [CrossRef] [PubMed]
56. Abdou, M.A.H.; Badran, F.S.; Ahmed, E.T.; Taha, R.A.; Abdel-Mola, M.A.M. Effect of Compost and some Natural Stimulant Treatments on: Ii. Corms Production and Chemical Constituents of (*Gladiolus grandiflorus* cv. Peter Pears) Plants. *Sci. J. Flowers Ornam. Plants* **2018**, *5*, 115–126. [CrossRef]
57. Younis, A.; Akhtar, M.S.; Riaz, A.; Zulfiqar, F.; Qasim, M.; Farooq, A.; Tariq, U.; Ahsan, M.; Bhatti, Z.M. Improved Cut Flower and Corm Production by Exogenous Moringa Leaf Extract Application on Gladiolus Cultivars. *Acta Sci. Pol. Hortorum Cultus* **2018**, *17*, 25–38. [CrossRef]
58. Sánchez-Gómez, R.; Garde-Cerdán, T.; Zalacain, A.; Garcia, R.; Cabrita, M.J.; Salinas, M.R. Vine-Shoot Waste Aqueous Extract Applied as Foliar Fertilizer to Grapevines: Effect on Amino Acids and Fermentative Volatile Content. *Food Chem.* **2016**, *197*, 132–140. [CrossRef] [PubMed]
59. Sánchez-Gómez, R.; Zalacain, A.; Pardo, F.; Alonso, G.L.; Salinas, M.R. Moscatel Vine-Shoot Extracts as a Grapevine Biostimulant to Enhance Wine Quality. *Food Res. Int.* **2017**, *98*, 40–49. [CrossRef] [PubMed]

60. Sánchez-Gómez, R.; Torregrosa, L.; Zalacain, A.; Ojeda, H.; Bouckenooghe, V.; Schneider, R.; Alonso, G.L.; Salinas, M.R. The Microvine, a Plant Model to Study the Effect of Vine-Shoot Extract on the Accumulation of Glycosylated Aroma Precursors in Grapes. *J. Sci. Food Agric.* **2018**, *98*, 3031–3040. [CrossRef] [PubMed]
61. Bavaresco, L.; Lucini, L.; Squeri, C.; Zamboni, M.; Frioni, T. Protein Hydrolysates Modulate Leaf Proteome and Metabolome in Water-Stressed Grapevines. *Sci. Hortic.* **2020**, *270*, 109413. [CrossRef]
62. Boselli, M.; Bahouaoui, M.A.; Lachhab, N.; Sanzani, S.M.; Ferrara, G.; Ippolito, A. Protein Hydrolysates Effects on Grapevine (*Vitis vinifera* L., cv. Corvina) Performance and Water Stress Tolerance. *Sci. Hortic.* **2019**, *258*, 108784. [CrossRef]
63. Pardo-García, A.I.; Martínez-Gil, A.M.; Cadahía, E.; Pardo, F.; Alonso, G.L.; Salinas, M.R. Oak Extract Application to Grapevines as a Plant Biostimulant to Increase Wine Polyphenols. *Food Res. Int.* **2014**, *55*, 150–160. [CrossRef]
64. Parrado, J.; Escudero-Gilete, M.L.; Friaza, V.; García-Martínez, A.; González-Miret, M.L.; Bautista, J.D.; Heredia, F.J. Enzymatic Vegetable Extract with Bio- Active Components: Influence of Fertiliser on the Colour and Anthocyanins of Red Grapes. *J. Sci. Food Agric.* **2007**, *87*, 2310–2318. [CrossRef]
65. Ertani, A.; Pizzeghello, D.; Francioso, O.; Sambo, P.; Sanchez-Cortes, S.; Nardi, S. *Capsicum chinensis* L. Growth and Nutraceutical Properties Are Enhanced by Biostimulants in a Long-Term Period: Chemical and Metabolomic Approaches. *Front. Plant Sci.* **2014**, *5*, 375. [CrossRef]
66. Nasir, M.; Khan, A.S.; Basra, S.M.A.; Malik, A.U. Foliar Application of Moringa Leaf Extract, Potassium and Zinc Influence Yield and Fruit Quality of 'Kinnow' Mandarin. *Sci. Hortic.* **2016**, *210*, 227–235. [CrossRef]
67. Di Mola, I.; Cozzolino, E.; Ottaiano, L.; Nocerino, S.; Rouphael, Y.; Colla, G.; El-Nakhel, C.; Mori, M. Nitrogen Use and Uptake Efficiency and Crop Performance of Baby Spinach (*Spinacia oleracea* L.) and Lamb's Lettuce (*Valerianella locusta* L.) Grown under Variable Sub-Optimal N Regimes Combined with Plant-Based Biostimulant Application. *Agronomy* **2020**, *10*, 278. [CrossRef]
68. Sorrentino, M.; Panzarová, K.; Spyroglou, I.; Spíchal, L.; Buffagni, V.; Ganugi, P.; Rouphael, Y.; Colla, G.; Lucini, L.; De Diego, N. Integration of Phenomics and Metabolomics Datasets Reveals Different Mode of Action of Biostimulants Based on Protein Hydrolysates in *Lactuca sativa* L. and *Solanum lycopersicum* L. under Salinity. *Front. Plant Sci.* **2022**, *12*, 808711. [CrossRef] [PubMed]
69. Rouphael, Y.; Carillo, P.; Garcia-Perez, P.; Cardarelli, M.; Senizza, B.; Miras-Moreno, B.; Colla, G.; Lucini, L. Plant Biostimulants from Seaweeds or Vegetal Proteins Enhance the Salinity Tolerance in Greenhouse Lettuce by Modulating Plant Metabolism in a Distinctive Manner. *Sci. Hortic.* **2022**, *305*, 111368. [CrossRef]
70. Rouphael, Y.; Cardarelli, M.; Bonini, P.; Colla, G. Synergistic Action of a Microbial-Based Biostimulant and a Plant Derived-Protein Hydrolysate Enhances Lettuce Tolerance to Alkalinity and Salinity. *Front. Plant Sci.* **2017**, *8*, 131. [CrossRef] [PubMed]
71. Di Mola, I.; Cozzolino, E.; Ottaiano, L.; Giordano, M.; Rouphael, Y.; Colla, G.; Mori, M. Effect of Vegetal- and Seaweed Extract-Based Biostimulants on Agronomical and Leaf Quality Traits of Plastic Tunnel-Grown Baby Lettuce under Four Regimes of Nitrogen Fertilization. *Agronomy* **2019**, *9*, 571. [CrossRef]
72. Colla, G.; Rouphael, Y.; Cardarelli, M.; Švecová, E.; Reynaud, H.; Canaguier, R.; Planques, B. Effectiveness of a Plant-Derived Protein Hydrolysate to Improve Crop Performances under Different Growing Conditions. *Acta Hortic.* **2013**, *1009*, 175–179. [CrossRef]
73. Shalaby, O.A. Moringa Leaf Extract Increases Tolerance to Salt Stress, Promotes Growth, Increases Yield, and Reduces Nitrate Concentration in Lettuce Plants. *Sci. Hortic.* **2024**, *325*, 112654. [CrossRef]
74. Lucini, L.; Rouphael, Y.; Cardarelli, M.; Canaguier, R.; Kumar, P.; Colla, G. The Effect of a Plant-Derived Biostimulant on Metabolic Profiling and Crop Performance of Lettuce Grown under Saline Conditions. *Sci. Hortic.* **2015**, *182*, 124–133. [CrossRef]
75. Bulgari, R.; Morgutti, S.; Cocetta, G.; Negrini, N.; Farris, S.; Calcante, A.; Spinardi, A.; Ferrari, E.; Mignani, I.; Oberti, R.; et al. Evaluation of Borage Extracts As Potential Biostimulant Using a Phenomic, Agronomic, Physiological, and Biochemical Approach. *Front. Plant Sci.* **2017**, *8*, 935. [CrossRef]
76. De Lucia, B. Type of Bio-Stimulant and Application Method Effects on Stem Quality and Root System Growth in L.A. Lily. *Eur. J. Hortic. Sci.* **2012**, *77*, 10.
77. Mutlu-Durak, H.; Yildiz Kutman, B. Seed Treatment with Biostimulants Extracted from Weeping Willow (Salix Babylonica) Enhances Early Maize Growth. *Plants* **2021**, *10*, 1449. [CrossRef]
78. Pehlivan, N. Salt Stress Relief Potency of Whortleberry Extract Biopriming in Maize. *3 Biotech* **2018**, *8*, 89. [CrossRef]
79. Colla, G.; Rouphael, Y.; Canaguier, R.; Svecova, E.; Cardarelli, M. Biostimulant Action of a Plant-Derived Protein Hydrolysate Produced through Enzymatic Hydrolysis. *Front. Plant Sci.* **2014**, *5*, 448. [CrossRef]
80. Tolisano, C.; Luzi, F.; Regni, L.; Proietti, P.; Puglia, D.; Gigliotti, G.; Di Michele, A.; Priolo, D.; Del Buono, D. A Way to Valorize Pomace from Olive Oil Production: Lignin Nanoparticles to Biostimulate Maize Plants. *Environ. Technol. Innov.* **2023**, *31*, 103216. [CrossRef]
81. Ertani, A.; Pizzeghello, D.; Francioso, O.; Tinti, A.; Nardi, S. Biological Activity of Vegetal Extracts Containing Phenols on Plant Metabolism. *Molecules* **2016**, *21*, 205. [CrossRef] [PubMed]
82. Miras-Moreno, B.; Senizza, B.; Regni, L.; Tolisano, C.; Proietti, P.; Trevisan, M.; Lucini, L.; Rouphael, Y.; Del Buono, D. Biochemical Insights into the Ability of *Lemna minor* L. Extract to Counteract Copper Toxicity in Maize. *Plants* **2022**, *11*, 2613. [CrossRef] [PubMed]
83. Ertani, A.; Schiavon, M.; Altissimo, A.; Franceschi, C.; Nardi, S. Phenol-Containing Organic Substances Stimulate Phenylpropanoid Metabolism in *Zea mays*. *J. Plant Nutr. Soil Sci.* **2011**, *174*, 496–503. [CrossRef]

84. Abdel Latef, A.A.H.; Mostofa, M.G.; Rahman, M.d.M.; Abdel-Farid, I.B.; Tran, L.-S.P. Extracts from Yeast and Carrot Roots Enhance Maize Performance under Seawater-Induced Salt Stress by Altering Physio-Biochemical Characteristics of Stressed Plants. *J. Plant Growth Regul.* **2019**, *38*, 966–979. [CrossRef]
85. Schiavon, M.; Ertani, A.; Nardi, S. Effects of an Alfalfa Protein Hydrolysate on the Gene Expression and Activity of Enzymes of the Tricarboxylic Acid (TCA) Cycle and Nitrogen Metabolism in *Zea mays* L. *J. Agric. Food Chem.* **2008**, *56*, 11800–11808. [CrossRef]
86. Ertani, A.; Cavani, L.; Pizzeghello, D.; Brandellero, E.; Altissimo, A.; Ciavatta, C.; Nardi, S. Biostimulant Activity of Two Protein Hydrolyzates in the Growth and Nitrogen Metabolism of Maize Seedlings. *J. Plant Nutr. Soil Sci.* **2009**, *172*, 237–244. [CrossRef]
87. Ertani, A.; Schiavon, M.; Muscolo, A.; Nardi, S. Alfalfa Plant-Derived Biostimulant Stimulate Short-Term Growth of Salt Stressed *Zea mays* L. Plants. *Plant Soil* **2012**, *364*, 145–158. [CrossRef]
88. El-Sharony, T.F.; El-Gioushy, S.; Oa, A. Effect of Foliar Application with Algae and Plant Extracts on Growth, Yield and Fruit Quality of Fruitful Mango Trees Cv. Fagri Kalan. *J. Hortic.* **2015**, *2*, 162. [CrossRef]
89. Zahra, N.; Wahid, A.; Shaukat, K.; Hafeez, M.B.; Batool, A.; Hasanuzzaman, M. Oxidative Stress Tolerance Potential of Milk Thistle Ecotypes after Supplementation of Different Plant Growth-Promoting Agents under Salinity. *Plant Physiol. Biochem.* **2021**, *166*, 53–65. [CrossRef] [PubMed]
90. Suryaman, M.; Sunarya, Y.; Istarimila, I.; Fudholi, A. Effect of Salinity Stress on the Growth and Yield of Mungbean (*Vigna radiata* (L.) R. Wilczek) Treated with Mangosteen Pericarp Extract. *Biocatal. Agric. Biotechnol.* **2021**, *36*, 102132. [CrossRef]
91. Sharma, R.K.; Kothari, R.M. Recycled Cereal Proteins as Foliar Spray Enhances Quality and Production of Food Crops. *Resour. Conserv. Recycl.* **1993**, *9*, 213–221. [CrossRef]
92. Habib, N.; Ashraf, M.; Ali, Q.; Perveen, R. Response of Salt Stressed Okra (Abelmoschus Esculentus Moench) Plants to Foliar-Applied Glycine Betaine and Glycine Betaine Containing Sugarbeet Extract. *S. Afr. J. Bot.* **2012**, *83*, 151–158. [CrossRef]
93. Regni, L.; Del Buono, D.; Miras-Moreno, B.; Senizza, B.; Lucini, L.; Trevisan, M.; Morelli Venturi, D.; Costantino, F.; Proietti, P. Biostimulant Effects of an Aqueous Extract of Duckweed (*Lemna minor* L.) on Physiological and Biochemical Traits in the Olive Tree. *Agriculture* **2021**, *11*, 1299. [CrossRef]
94. Lorenzo, P.; Souza-Alonso, P.; Guisande-Collazo, A.; Freitas, H. Influence of Acacia Dealbata Link Bark Extracts on the Growth of *Allium cepa* L. Plants under High Salinity Conditions. *J. Sci. Food Agric.* **2019**, *99*, 4072–4081. [CrossRef]
95. Ghezal, N.; Rinez, I.; Sbai, H.; Saad, I.; Farooq, M.; Rinez, A.; Zribi, I.; Haouala, R. Improvement of Pisum Sativum Salt Stress Tolerance by Bio-Priming Their Seeds Using Typha Angustifolia Leaves Aqueous Extract. *S. Afr. J. Bot.* **2016**, *105*, 240–250. [CrossRef]
96. Merwad, A.-R.M.A. Using Moringa Oleifera Extract as Biostimulant Enhancing the Growth, Yield and Nutrients Accumulation of Pea Plants. *J. Plant Nutr.* **2018**, *41*, 425–431. [CrossRef]
97. Desoky, E.-S.M.; ElSayed, A.I.; Merwad, A.-R.M.A.; Rady, M.M. Stimulating Antioxidant Defenses, Antioxidant Gene Expression, and Salt Tolerance in Pisum Sativum Seedling by Pretreatment Using Licorice Root Extract (LRE) as an Organic Biostimulant. *Plant Physiol. Biochem.* **2019**, *142*, 292–302. [CrossRef]
98. Giordano, M.; El-Nakhel, C.; Caruso, G.; Cozzolino, E.; De Pascale, S.; Kyriacou, M.C.; Colla, G.; Rouphael, Y. Stand-Alone and Combinatorial Effects of Plant-Based Biostimulants on the Production and Leaf Quality of Perennial Wall Rocket. *Plants* **2020**, *9*, 922. [CrossRef]
99. Caruso, G.; De Pascale, S.; Cozzolino, E.; Giordano, M.; El-Nakhel, C.; Cuciniello, A.; Cenvinzo, V.; Colla, G.; Rouphael, Y. Protein Hydrolysate or Plant Extract-Based Biostimulants Enhanced Yield and Quality Performances of Greenhouse Perennial Wall Rocket Grown in Different Seasons. *Plants* **2019**, *8*, 208. [CrossRef]
100. Elrys, A.S.; Merwad, A.-R.M.A.; Abdo, A.I.E.; Abdel-Fatah, M.K.; Desoky, E.-S.M. Does the Application of Silicon and Moringa Seed Extract Reduce Heavy Metals Toxicity in Potato Tubers Treated with Phosphate Fertilizers? *Environ. Sci. Pollut. Res.* **2018**, *25*, 16776–16787. [CrossRef] [PubMed]
101. Olivera-Olivera, V.; Morales-Payan, J.P.; Robles, W.; Goenaga, R. Effects of Biostimulants on Fruits of Pulasan [Nephelium Ramboutan-Ake (Labillardière) Leenhouts]. *Acta Hortic.* **2014**, *1042*, 101–104. [CrossRef]
102. Ashraf, R.; Sultana, B.; Iqbal, M.; Mushtaq, M. Variation in Biochemical and Antioxidant Attributes of Raphanus Sativus in Response to Foliar Application of Plant Leaf Extracts as Plant Growth Regulator. *J. Genet. Eng. Biotechnol.* **2016**, *14*, 1–8. [CrossRef] [PubMed]
103. Abdalla, M. The Potential of Moringa Oleifera Extract as a Biostimulant in Enhancing the Growth, Biochemical and Hormonal Contents in Rocket (*Eruca vesicaria* subsp. *sativa*) Plants. *Int. J. Plant Physiol. Biochem.* **2013**, *5*, 42–49. [CrossRef]
104. Ali, E.F.; Hassan, F.A.S.; Elgimabi, M. Improving the Growth, Yield and Volatile Oil Content of *Pelargonium graveolens* L. Herit by Foliar Application with Moringa Leaf Extract through Motivating Physiological and Biochemical Parameters. *S. Afr. J. Bot.* **2018**, *119*, 383–389. [CrossRef]
105. Carillo, P.; Colla, G.; Fusco, G.M.; Dell'Aversana, E.; El-Nakhel, C.; Giordano, M.; Pannico, A.; Cozzolino, E.; Mori, M.; Reynaud, H.; et al. Morphological and Physiological Responses Induced by Protein Hydrolysate-Based Biostimulant and Nitrogen Rates in Greenhouse Spinach. *Agronomy* **2019**, *9*, 450. [CrossRef]
106. Rouphael, Y.; Giordano, M.; Cardarelli, M.; Cozzolino, E.; Mori, M.; Kyriacou, M.C.; Bonini, P.; Colla, G. Plant- and Seaweed-Based Extracts Increase Yield but Differentially Modulate Nutritional Quality of Greenhouse Spinach through Biostimulant Action. *Agronomy* **2018**, *8*, 126. [CrossRef]

107. Abd El-Mageed, T.A.; Semida, W.M.; Rady, M.M. Moringa Leaf Extract as Biostimulant Improves Water Use Efficiency, Physio-Biochemical Attributes of Squash Plants under Deficit Irrigation. *Agric. Water Manag.* **2017**, *193*, 46–54. [CrossRef]
108. Basile, B.; Brown, N.; Valdes, J.M.; Cardarelli, M.; Scognamiglio, P.; Mataffo, A.; Rouphael, Y.; Bonini, P.; Colla, G. Plant-Based Biostimulant as Sustainable Alternative to Synthetic Growth Regulators in Two Sweet Cherry Cultivars. *Plants* **2021**, *10*, 619. [CrossRef] [PubMed]
109. Parađiković, N.; Vinković, T.; Vinković Vrček, I.; Žuntar, I.; Bojić, M.; Medić-Šarić, M. Effect of Natural Biostimulants on Yield and Nutritional Quality: An Example of Sweet Yellow Pepper (*Capsicum annuum* L.) Plants. *J. Sci. Food Agric.* **2011**, *91*, 2146–2152. [CrossRef] [PubMed]
110. Aluko, M. ARPN Journal of Agricultural and Biological Science Moringa Leaf Extract on the Growth and Yield of Pepper (*Capsicum annuum* L.). *ARPN J. Agric. Biol. Sci.* **2016**, *11*, 107–109.
111. Desoky, E.-S.M.; Elrys, A.S.; Rady, M.M. Integrative Moringa and Licorice Extracts Application Improves *Capsicum annuum* Fruit Yield and Declines Its Contaminant Contents on a Heavy Metals-Contaminated Saline Soil. *Ecotoxicol. Environ. Saf.* **2019**, *169*, 50–60. [CrossRef]
112. Ceccarelli, A.V.; Miras-Moreno, B.; Buffagni, V.; Senizza, B.; Pii, Y.; Cardarelli, M.; Rouphael, Y.; Colla, G.; Lucini, L. Foliar Application of Different Vegetal-Derived Protein Hydrolysates Distinctively Modulates Tomato Root Development and Metabolism. *Plants* **2021**, *10*, 326. [CrossRef]
113. Colla, G.; Cardarelli, M.; Bonini, P.; Rouphael, Y. Foliar Applications of Protein Hydrolysate, Plant and Seaweed Extracts Increase Yield but Differentially Modulate Fruit Quality of Greenhouse Tomato. *HortScience* **2017**, *52*, 1214–1220. [CrossRef]
114. Caruso, G.; De Pascale, S.; Cozzolino, E.; Cuciniello, A.; Cenvinzo, V.; Bonini, P.; Colla, G.; Rouphael, Y. Yield and Nutritional Quality of Vesuvian Piennolo Tomato PDO as Affected by Farming System and Biostimulant Application. *Agronomy* **2019**, *9*, 505. [CrossRef]
115. Sestili, F.; Rouphael, Y.; Cardarelli, M.; Pucci, A.; Bonini, P.; Canaguier, R.; Colla, G. Protein Hydrolysate Stimulates Growth in Tomato Coupled With N-Dependent Gene Expression Involved in N Assimilation. *Front. Plant Sci.* **2018**, *9*, 1233. [CrossRef]
116. Lucini, L.; Miras-Moreno, B.; Rouphael, Y.; Cardarelli, M.; Colla, G. Combining Molecular Weight Fractionation and Metabolomics to Elucidate the Bioactivity of Vegetal Protein Hydrolysates in Tomato Plants. *Front. Plant Sci.* **2020**, *11*, 976. [CrossRef]
117. Rouphael, Y.; Colla, G.; Giordano, M.; El-Nakhel, C.; Kyriacou, M.C.; De Pascale, S. Foliar Applications of a Legume-Derived Protein Hydrolysate Elicit Dose-Dependent Increases of Growth, Leaf Mineral Composition, Yield and Fruit Quality in Two Greenhouse Tomato Cultivars. *Sci. Hortic.* **2017**, *226*, 353–360. [CrossRef]
118. Paul, K.; Sorrentino, M.; Lucini, L.; Rouphael, Y.; Cardarelli, M.; Bonini, P.; Miras Moreno, M.B.; Reynaud, H.; Canaguier, R.; Trtílek, M.; et al. A Combined Phenotypic and Metabolomic Approach for Elucidating the Biostimulant Action of a Plant-Derived Protein Hydrolysate on Tomato Grown Under Limited Water Availability. *Front. Plant Sci.* **2019**, *10*, 493. [CrossRef]
119. Sadasivam Vinoth, S.V.; Sundari, S.; Packiaraj Gurusaravanan, P.G.; Subiramani Sivakumar, S.S.; Govindarajan Siva, G.S.; Kumar, G.P.; Manju, M.; Velmurugan, K.; Lakshminarayana, V.; Jayabalan, N. Evaluation of Seagrass Liquid Extract on Salt Stress Alleviation in Tomato Plants. *Asian J. Plant Sci.* **2017**, *16*, 172–183. [CrossRef]
120. Hayat, S.; Ahmad, H.; Ali, M.; Ren, K.; Cheng, Z. Aqueous Garlic Extract Stimulates Growth and Antioxidant Enzymes Activity of Tomato (*Solanum lycopersicum*). *Sci. Hortic.* **2018**, *240*, 139–146. [CrossRef]
121. Abou Chehade, L.; Al Chami, Z.; De Pascali, S.A.; Cavoski, I.; Fanizzi, F.P. Biostimulants from Food Processing By-Products: Agronomic, Quality and Metabolic Impacts on Organic Tomato (*Solanum lycopersicum* L.). *J. Sci. Food Agric.* **2018**, *98*, 1426–1436. [CrossRef] [PubMed]
122. Rady, M.M.; ur Rehman, H. Supplementing Organic Biostimulants into Growing Media Enhances Growth and Nutrient Uptake of Tomato Transplants. *Sci. Hortic.* **2016**, *203*, 192–198. [CrossRef]
123. Ertani, A.; Schiavon, M.; Nardi, S. Transcriptome-Wide Identification of Differentially Expressed Genes in *Solanum lycopersicon* L. in Response to an Alfalfa-Protein Hydrolysate Using Microarrays. *Front. Plant Sci.* **2017**, *8*, 1159. [CrossRef]
124. Bajwa, A.A.; Farooq, M.; Nawaz, A. Seed Priming with Sorghum Extracts and Benzyl Aminopurine Improves the Tolerance against Salt Stress in Wheat (*Triticum aestivum* L.). *Physiol. Mol. Biol. Plants* **2018**, *24*, 239–249. [CrossRef]
125. Haider, F.; Bagchi, G.D.; Singh, A.K. Effect of Calliterpenone on Growth, Herb Yield and Oil Quality of Mentha Arvensis. *Int. J. Integr. Biol.* **2009**, *7*, 53–57.
126. Matsumiya, Y.; Kubo, M.; Matsumiya, Y.; Kubo, M. Soybean Peptide: Novel Plant Growth Promoting Peptide from Soybean. In *Soybean and Nutrition*; IntechOpen: Shanghai, China, 2011; ISBN 978-953-307-536-5.
127. Parrado, J.; Bautista, J.; Romero, E.J.; García-Martínez, A.M.; Friaza, V.; Tejada, M. Production of a Carob Enzymatic Extract: Potential Use as a Biofertilizer. *Bioresour. Technol.* **2008**, *99*, 2312–2318. [CrossRef]
128. ElSayed, A.I.; Rafudeen, M.S.; Ganie, S.A.; Hossain, M.S.; Gomaa, A.M. Seed Priming with Cypress Leaf Extract Enhances Photosynthesis and Antioxidative Defense in Zucchini Seedlings under Salt Stress. *Sci. Hortic.* **2022**, *293*, 110707. [CrossRef]
129. De Diego, N.; Spíchal, L. Presence and Future of Plant Phenotyping Approaches in Biostimulant Research and Development. *J. Exp. Bot.* **2022**, *73*, 5199–5212. [CrossRef]
130. Ertani, L.L. Begoña Miras-Moreno, Andrea Bioactive Compounds and Evaluation of Biostimulant Activity. In *Biostimulants for Sustainable Crop Production*; Burleigh Dodds Science Publishing: London, UK, 2020; ISBN 978-1-00-304786-5.
131. Ricci, M.; Tilbury, L.; Daridon, B.; Sukalac, K. General Principles to Justify Plant Biostimulant Claims. *Front. Plant Sci.* **2019**, *10*, 494. [CrossRef] [PubMed]

132. Panuccio, M.R.; Chaabani, S.; Roula, R.; Muscolo, A. Bio-Priming Mitigates Detrimental Effects of Salinity on Maize Improving Antioxidant Defense and Preserving Photosynthetic Efficiency. *Plant Physiol. Biochem.* **2018**, *132*, 465–474. [CrossRef] [PubMed]
133. Kaplan, F.; Guy, C.L. Beta-Amylase Induction and the Protective Role of Maltose during Temperature Shock. *Plant Physiol.* **2004**, *135*, 1674–1684. [CrossRef] [PubMed]
134. Wink, M. Plant Secondary Metabolites Modulate Insect Behavior-Steps Toward Addiction? *Front. Physiol.* **2018**, *9*, 364. [CrossRef]
135. Erb, M.; Kliebenstein, D.J. Plant Secondary Metabolites as Defenses, Regulators, and Primary Metabolites: The Blurred Functional Trichotomy. *Plant Physiol.* **2020**, *184*, 39–52. [CrossRef] [PubMed]
136. Khare, S.; Singh, N.B.; Singh, A.; Hussain, I.; Niharika, K.; Yadav, V.; Bano, C.; Yadav, R.K.; Amist, N. Plant Secondary Metabolites Synthesis and Their Regulations under Biotic and Abiotic Constraints. *J. Plant Biol.* **2020**, *63*, 203–216. [CrossRef]
137. Kochhar, S.L.; Gujral, S.K. Secondary Plant Metabolites. In *Plant Physiology: Theory and Applications*; Kochhar, S.L., Gujral, S.K., Eds.; Cambridge University Press: Cambridge, UK, 2020; pp. 590–610. ISBN 978-1-108-48639-2.
138. Naikoo, M.I.; Dar, M.I.; Raghib, F.; Jaleel, H.; Ahmad, B.; Raina, A.; Khan, F.A.; Naushin, F. Chapter 9—Role and Regulation of Plants Phenolics in Abiotic Stress Tolerance: An Overview. In *Plant Signaling Molecules*; Khan, M.I.R., Reddy, P.S., Ferrante, A., Khan, N.A., Eds.; Woodhead Publishing: Thorston, UK, 2019; pp. 157–168. ISBN 978-0-12-816451-8.
139. Shomali, A.; Das, S.; Arif, N.; Sarraf, M.; Zahra, N.; Yadav, V.; Aliniaeifard, S.; Chauhan, D.K.; Hasanuzzaman, M. Diverse Physiological Roles of Flavonoids in Plant Environmental Stress Responses and Tolerance. *Plants* **2022**, *11*, 3158. [CrossRef]
140. Variyar, P.S.; Banerjee, A.; Akkarakaran, J.J.; Suprasanna, P. Chapter 12—Role of Glucosinolates in Plant Stress Tolerance. In *Emerging Technologies and Management of Crop Stress Tolerance*; Ahmad, P., Rasool, S., Eds.; Academic Press: San Diego, CA, USA, 2014; pp. 271–291. ISBN 978-0-12-800876-8.
141. del Carmen Martínez-Ballesta, M.; Moreno, D.A.; Carvajal, M. The Physiological Importance of Glucosinolates on Plant Response to Abiotic Stress in Brassica. *Int. J. Mol. Sci.* **2013**, *14*, 11607–11625. [CrossRef]
142. EL Sabagh, A.; Islam, M.S.; Hossain, A.; Iqbal, M.A.; Mubeen, M.; Waleed, M.; Reginato, M.; Battaglia, M.; Ahmed, S.; Rehman, A.; et al. Phytohormones as Growth Regulators During Abiotic Stress Tolerance in Plants. *Front. Agron.* **2022**, *4*, 765068. [CrossRef]
143. Waadt, R.; Seller, C.A.; Hsu, P.-K.; Takahashi, Y.; Munemasa, S.; Schroeder, J.I. Plant Hormone Regulation of Abiotic Stress Responses. *Nat. Rev. Mol. Cell Biol.* **2022**, *23*, 680–694. [CrossRef] [PubMed]
144. Wozniak, E.; Blaszczak, A.; Wiatrak, P.; Canady, M. Biostimulant Mode of Action. In *The Chemical Biology of Plant Biostimulants*; John Wiley & Sons, Ltd.: Hoboken, NJ, USA, 2020; pp. 205–227. ISBN 978-1-119-35725-4.
145. Gilroy, S.; Jones, D.L. Through Form to Function: Root Hair Development and Nutrient Uptake. *Trends Plant Sci.* **2000**, *5*, 56–60. [CrossRef] [PubMed]
146. Amarowicz, R. Legume Seeds as an Important Component of Human Diet. *Foods* **2020**, *9*, 1812. [CrossRef] [PubMed]

MDPI AG
Grosspeteranlage 5
4052 Basel
Switzerland
Tel.: +41 61 683 77 34

Antioxidants Editorial Office
E-mail: antioxidants@mdpi.com
www.mdpi.com/journal/antioxidants

www.ingramcontent.com/pod-product-compliance
Lightning Source LLC
Chambersburg PA
CBHW061457210926
51850CB00008B/1746
9783725859832